Developments in Economic Geology, 8

PROSPECTING AND EXPLORATION OF MINERAL DEPOSITS

CZECHOSLOVAK ACADEMY OF SCIENCES

Developments in Economic Geology, 8

PROSPECTING AND EXPLORATION OF MINERAL DEPOSITS

MILOŠ KUŽVART

Associate Professor
Department of Geology at the Faculty of
Natural Sciences, Charles University, Prague

and

MILOSLAV BÖHMER

Professor
Department of Geology at the Faculty of
Natural Sciences, Komensky University,
Bratislava

ELSEVIER SCIENTIFIC PUBLISHING COMPANY
Amsterdam – Oxford – New York – 1978

Scientific Editor

Professor Ing. Mirko Vaněček, DrSc.

Published in co-edition with ACADEMIA, Publishing House of the Czechoslovak Academy of Sciences, Prague

Distribution of this book is being handled by the following publishers
       for the U.S.A. and Canada
Elsevier/North-Holland, Inc.
52 Vanderbilt Avenue
New York, N.Y. 10017, U.S.A.
       for the East European Countries, China, Northern Korea, Cuba, Vietnam and Mongolia
Academia, Publishing House of the Czechoslovak Academy of Sciences, Prague
       for all remaining areas
Elsevier Scientific Publishing Company
335 Jan van Galenstraat
P.O. Box 211, Amsterdam, The Netherlands

Standard Book Number 0-444-99876-4

Library of Congress Card Number 74-21861
With 204 Illustrations

Printed in Czechoslovakia

To Professor Dr. Jaromír Koutek,
the foremost teacher of Czechoslovak
economic geologists,
a man of profound human insight
and broad scientific interests,
whose work and life are an
example to us all,
on the occasion of his 75th birthday

CONTENTS

Spring came once more,
and at the end of all their wandering
they found, not the Lost Cabin,
but a shallow placer,
where gold showed like yellow butter
across the bottom of the washing pan.
They sought no farther ...

Jack London, The Call of the Wild

PREFACE

Prospecting and exploration for mineral resources can be effective remedial measures in the raw material and energy crisis which emerged in the autumn of 1973. Although the possibility of such a crisis had been discussed in the press for some time beforehand, it was not expected to occur until the distant future. The crisis contributed greatly to the inflation which has led to currency depreciation of the order of 6 to 30 per cent per year. The situation has been caused, among other things, by the fact that global mineral and energy sources cannot support an economic growth level of 5 to 10 % ad infinitum. This growth is, however, necessary for advance in civilization and the cultural level, particularly in developing countries, where any increase is almost cancelled by the rapid rise in the number of people. It is clear that this rapid economic development, which began at the beginning of the 19th century, cannot last forever, though it may be assumed that it will continue for other 25 to 100 years.

Dealing with the economic crisis is closely connected with control of the population explosion, which cannot be accomplished before the year 2045, when the global population will amount to 15—20 billion people. All long-term planning on an international level must reckon with these figures. The immense problems that are to be solved by such planning can lead a pessimist to ask "Shall we survive beyond the year 2000" or to proclaiming a new chiliasm, that is a doctrine of the end of civilization and mankind during the next decades. The problems of technical civilization, especially mineral resources, are therefore particularly important against this background.

Of the four billion inhabitants of the world, only few enjoy the wealth of the inhabitants of industrial countries with an annual income of over 1500 dollars *per capita*. The majority of people must content themselves with a tenth of this sum. The number of people in the poorer parts of the world, which includes three-quarters of the world's population, grows three times more rapidly than in the industrial countries, but their income increases at half the rate. The result is that the rich become richer and the poor poorer. This is unquestionably the fundamental disequilibrium in the present world. It is in the interest of the sated minority to satisfy the hungry and particularly to assist them to attain prosperity by their own efforts. The statement of Baruch Spinoza, the great humanitarian philosopher of the 17th century, that "there is no separate happiness" is especially valid at the present. The United Nations plan to support the developing countries so that their economic potential increases by five per cent annually. This aim can be achieved not only by control of the population explosion but primarily by increased consumption of raw and especially mineral materials.

In the year 2000 the Earth will be inhabited by six to seven billion people. If the average consumption of raw materials *per capita* remains at the present level, the exploitation of mineral resources must double within the next 30 years. If the consumption of the U.S.A. remains at the present level and that of the developing countries rises to it, the extraction of mineral materials will have to increase 30 times. The absurdity of this prospect is evident from the image of the existence of 30 times as many quarries and mines and the impact of such increased exploitation on the environment.

In the last 30 years the United States alone has consumed more minerals and fossil fuel than had the whole of mankind from the beginning of human history. In 1970 the consumption in the U.S. covered 40 % of global production of aluminum, 30 % of copper, 24 % of oil and 21 % of coal. Up to the end of the twentieth century, the consumption of mineral materials in the U.S. will rise by 3.4 to 5.5 per cent yearly.

The global consumption will increase at least at the same rate. Apart from gold, the consumption of all metals will double during the next 10 to 15 years. By the end of the century the consumption of Cu, Pb, Zn and coal will be three- to six-fold the present value. The value of mineral reserves extracted in one year will rise from 116.5 billion US dollars in 1968 to 645 billion dollars in the year 2000 (in 1968 dollars).

The total value of mineral resources consumed between 1968 and 2000 will reach almost 10,000 billion dollars. The largest annual increments are expected with metals (except iron) — 7 per cent, and power — 5 per cent.

The mineral resources so far discovered cannot cover this increased need. The export of mineral raw materials from the developing countries will not solve this problem, since their own industry will be developed by this time. Progress in the technological treatment of the present non-industrial materials will undoubtedly be helpful and more intensive exploration will surely find new mineral reserves. After all, no pessimistic forecast has yet come true. Some twenty years ago, for example, known oil reserves were estimated to suffice for 20 years' exploitation. Today they are predicted to be sufficient for more than 20 years' exploitation at the present increased extraction rate. However, any optimism based on this fact would be out of place, since the impact of even a small increase in the global economy on the demand for mineral materials cannot be estimated. A mere 6 % increase in consumption per year doubles the consumption every 12 years, provided the world's population does not increase.

A forecast of the increased demand must take into consideration not only the rate of economic development and the resulting increased consumption but also changes in the technology of mineral treatment and in the cost of raw materials. Therefore, estimates are liable to error. In 1926, for example, the consumption of copper in the year 2000 was estimated at 12 million tons; this amount will be probably reached in 1980.

Since 1900 the global demand for oil and natural gas has doubled every 10 years and this increase will undoubtedly be even higher in the future. It is questionable whether the building of giant oil tankers (ca. 200,000 gross tonnage, prospectively 800,000 g.t.) can keep pace with increased oil production (the present capacity of all tankers is ca. 150 million gross tonnage, ships of a further 70 million g.t. are under construction). The fossil fuel reserves known in 1970 will suffice for 170 years. This estimate does not take into consideration hydroenergetic and nuclear sources, which at present cover only 3 % of global power production. In this context it should be mentioned that the construction of atomic power plants has recently been limited drastically for fear of contamination of the environment. Oil and gas reserves will be exhausted in 45 years or in 170 years if the present-day reserves can be increased tenfold.

Tin reserves will suffice for 16 years (with the assumed maximum increase in consumption) or 26 years (if consumption increases at a minimal rate), the reserves of zinc for 16 (20) years, of copper for 21 (25) years, lead for 21 (28) years, tungsten for 24 (30) years, molybdenum for 30 (37) years, titanium for 31 (57) years, nickel for 54 (74) years, cobalt for 71 (124) years, vanadium for 91 (127) years and chromium for 108 (148) years. If, using intensive exploration, the present reserves are increased five times in the course of future decades (which is an optimistic assumption), these figures will not change appreciably. The greater reserves of tin will be exhausted in 55 years (in 121 years at a minimum increase in consumption). The corresponding figures are for Zn 51 (66), for Cu 56 (70), for Pb 66 (96), for W 73 (84), for Mo 78 (96), for Ti 81 (142) (large-scale production of supersonic air planes will cause an early lack of Ti), Ni 124 (171), Co 189 (394), V 212 (292) (the use of V as an absorbent of exhaust gases, will lower these figures appreciably) and Cr 250 (394) years (Govett, Govett, 1972).

The above survey shows that a fivefold increase in mineral reserves will, surprisingly, lengthen the interval between the present and the time of exhaustion of the ores mentioned only 2.5 times. Since the volume of mineral resources assessed in one year has decreased continuously under a steadily increasing exploration cost, the amount of exploratory, mainly geophysical surveys in the near future must be increased appreciably all over the world.

The reserves of metals in low-grade ores such as iron (especially in taconites), aluminum (in clay minerals), phosphates (in deposits on land and shelf) and potassium (in salt deposits) are more than sufficient even with the maximum increase in consumption. The cost of Fe and Al production, however, will increase since the treatment of taconites and clays will require more energy.

Let us digress from this serious contemplation of the prospects of human civilization and culture in the third millennium and move into a serene fairy world. Some five to seven thousand years ago, mankind, like the forever discontented fisherman's wife, was endowed by the golden fish with knowledge of metal (gold, copper, bronze) treatment resulting in the origin of craft and urban civilization. Later, the golden fish provided man with better knowledge of how to use fossil fuel, which led to the first industrial revolution. After that, the golden fish enabled us to achieve the second industrial (scientific-technical) revolution based on modern utilization of mineral resources, particularly of non-metallic and radioactive materials, for advanced cosmic, communication, computer, nuclear and submarine engineering.

If the discontented fisherman's wife (the perpetual human discontent, avidity for luxury, laziness and desire for power) persuades her husband (creator of technical civilization) to go to the lake (heavily strained mineral resources) for the fourth time, the golden fish (the lot of mankind) may refuse to deliver another motor car into the garage of a sated 100 million people living in luxury, and may degrade mankind to a primitive society with low material and cultural needs. The other possibility for mankind is to be contented with what nature affords and to endeavour to divide it more justly between the sated and the hungry. The appeals of statesmen to save on power consumption are the first promising indications of this policy.

In conclusion, a few words can be added on the actual conditions and prospects of survival for the human species, provided that people are willing to lead a reasonable, simpler life and direct their interest more to culture, science and art.

Until the fusion of hydrogen nuclei is achieved to increase the amount of power required for treatment of low-grade ores, it will be necessary to maintain a well-proportioned growth of the economy using known methods, especially geophysical, in the search for new deposits with normal metal contents. In non-socialist countries alone, 30 million dollars are expended annually on geophysical, generally aerial prospecting. Mining is experiencing a boom in Africa (gold, copper), in Latin America (Dominican Republic — a new nickel-bearing laterite has yielded 2.3 m. tons of ore annually since 1973), in Ocenia (porphyry copper ore, gold), in Asia (tin) and even in Europe. Prospecting figures for recent years in Europe and other continents are still imposing even if achieved at higher costs (World Mining Yearbooks 1972 – 1977). In Africa a large uranium deposit was discovered in Algeria, as was 50 m. tons of ore with 5.5 % Cu and 0.4 % Co in Zaire near Tenke Fungurume and a chrome deposit with $35 - 40 \% Cr_2O_3$ near Ingessana in Sudan. Discoveries in Asia include a deposit containing 185 m. tons of laterite with 1.49 % Ni and 0.11 % Co on Gag Island in Indonesia; 2 m. tons of 2 % copper ore at Basantgarh in Rajasthan (India); 12 m. tons of phosphate rock in the Alborr Mts. near Shemshak (Iran); 32 m. tons of 25 % copper ore (+Au, Ag) near Ertsberg open pit mine (West

Iran); three m. tons of ore with 3 % Cu, 2 % Zn and 1 % Pb in Japan (Matsuki) and 2 m. tons of ore containing 1 % Cu, 3 % Pb and 13.5 % Zn ibidem (Fukazawa); 307 m. tons of phosphates in Jordan (Ruseifa) and rediscovery of King Solomons gold mines in Biblical Ophir at Mahd adh Dhahab (Cradle of Gold) SE of Medina (Saudi Arabia). In North America 30 m. tons of ore containing 1 % Cu ($+$Au, Ag) were assessed in the Afton Mines in British Columbia, large coal reserves in Alberta and Saskatchewan, 12 m. tons of ore containing 7.6 % Zn and 1 % Cu ($+$Pb, Ag) near Matsabi in Ontario, 20 m. tons of ore with 20 % Pb $+$ Zn near Arvi, 4 m. tons of ore with 1.6 % WO_3 and 0.22 % Cu on Tungsten deposit in the Northwest Territories and 82 m. tons of ore containing 13 % Zn $+$ Pb near Bathurst in New Brunswick. There is a plan to produce mineral oils by heating bituminous shales in retorts in Colorado, Utah and Wyoming. In Latin America, 625 m. tons of ores containing 0.7 % Cu were assessed in Colombia, 3 billion tons of ore with $0.6-0.7$ % Cu near Cerro Colorado in Panama, an Fe-ore deposit in Para, a uranium deposit near Belo Horizonte in Minas Gerais, 1.47 billion tons of ore containing $10-30$ % TiO_2 near Tapira in Minas Gerais, 165 m. tons of bauxite near Trombetas in Para, 1.5 billion tons of ore containing $0.8-1.0$ % Cu near El Abra in Chile and similar deposits of porphyry-ore type in Argentina (Padron), Colombia (Pegadorcitos), Panama (Petaguilla) and Peru (Michiquillay). Australia contributed 32 m. tons of ore with 0.28 % metal content (Jubiluka), 35 m. tons of ore containing 2 % Ni (Selcast) and 9.5 m. tons of ore containing 17.6 % Zn, 6.5 % Pb and 3 ozs. Ag/t (Lady Loretta) to global reserves. Expensive exploration in the Arctic has assessed 50 m. tons of ore with 20 % Pb $+$ Zn on Little Cornwallis Island, 10 m. tons of ore containing 20 % Pb $+$ Zn on the western coast of Greenland, and the largest world fluorite ($+$Sn) deposit with 20 % CaF_2 in ore on the Seward Peninsula in Alaska. In Europe 50 m. tons of coal in seams $1-3$ thick were discoverd near Musselburgh (Scotland) as were diamonds in the ancient Palaeozoic Timen deposit west of the Urals (U.S.S.R.); 10 m. tons of phosphates with 15% P_2O_5 in Epirus (Greece) and 8 m. tons of Pb$-$Zn ore in the Besna Kobila Mts. (Jugoslavia). Prospecting for oil and gas carried out on the shelves of all continents gave positive results in the North Sea, Nigeria, Indonesia; the deposits discovered at the mouth of the Mackenzie River in Canada may in the course of time surpass large deposits found on the shelf north of Alaska.

Oceans and oceanic floors are also potential suppliers of metals and other minerals. Manganese nodules containing 37 % Mn, 1.6 % Cu, 1.6 % Ni and 0.3 % Co were already extracted from the sea floor near Hawaii at a smaller cost than on land. Exploitation of mineral resources from the sea floor at a depth of 5000 m is technically practicable but international legislation lags behind; a deposit should not be the property of the country whose flag flies on the mast of the mining ship and the resources in the ocean floors could not justifiably belong only to maritime states. The impact of submarine exploitation on the biological equilibrium in the sea should be studied as soon as possible. The utilization of the oceans, their mineral resources and sea water, which is essentially an immense compound liquid deposit, will require international co-operation possibly under the supervision of United Nations. In fact, utilization of oceanic resources raises fewer technical problems than launching a space craft on an orbit around the Earth and will be many times cheaper.

With a sufficient amount of cheap power, every common igneous rock becomes a potential compound ore containing 8 tons of Al, 5 tons of Fe, 540 kg of Ti, 32 kg of Cr, 15 kg of V, 3 kg of V, 3 kg of Cu and 1.8 kg of Pb per 100 tons of rock.

The solution of the power problem, so important in overcoming the present crisis, lies in mastering nuclear reactions and, to a minor degree, in installations for utilizing solar energy in the tropics, high- and low-tide energy in gulfs and geothermal energy in volcanic regions.

To tackle all the above-mentioned technical problems, economical co-operation within and between organizations such as COMECON and the ECM within the framework of United Nations and its specialized branches, such as UNIDO or the Economic Commission for Africa, will be necessary.

Possibilities for international scientific collaboration are provided by UNESCO and IUGS in the International Geological Correlation Programme, by International Federation of Economic Geologistss, Commission for the Metallogenetic Map of Europe, International Decade for Ocean Investigation, and other organizations.

Mankind's present civilization and culture can survive with the available sources of power and mineral raw materials for another thousand years if ingenuity is used and a sensible and simple mode of living is accepted. Minimum reserves of mineral materials, oceanic water, rocks of the earth's crust and the air will be available in sufficient amounts. In the near future, however, the talents of all people must be used by a peaceful world, in order to improve mining methods in underground workings, which still involve losses of 30 to 70 % of the useful material in safety pillars; to refine prospecting methods, particularly in underground ore geophysics, hydrochemistry and all aerial geophysical methods. Geologists should take an active part in the preparation of long-term plans for 20 or more years. If these tasks are fulfilled, the requirements for raw materials will be met successfully for the next 30 to 50 years.

During this period technologists must develop new, economically (especially in the use of energy) convenient methods for the treatment of low-grade ores (or rocks), for large-scale extraction of minerals from the sea and the ocean floors and for the exploitation of deposits at depths of about 4000 m. This will raise new problems related to economics, ecology and politics.

The problem discussed is of global significance and should be solved through international co-operation. So far this is utopian, since mankind lags behind development in technology not only ethically but also in the field of international relationships and law. At the most we can hope for international co-ordination of efforts to ensure the material basis of civilization. Governments should allot sufficient financial means and specialists to these investigations and should ensure legislative protection of natural resources. This implies, of course, permanent alleviation of international tension throughout the world.

Exploration of mineral resources for the use of people living in the third millenium of our era is not given the public attention that exploration of the cosmos receives. It lies outside current ideological and political controversies of today; it strives for a more important goal – to reduce the discrepancy between the affluent and the hungry.

This handbook, summarizing experiences in prospecting and exploration of mineral deposits in both socialist and Western countries, is intended as a small contribution to the solution of the problems facing human civilization in the coming decades.

PART I

INTRODUCTION

Prospecting and exploration of mineral deposits form the subject of a geological science which studies the genesis of mineral resources and the measures for estimation of economic deposits, i.e. those recoverable under present conditions. Together with economic geology, hydrogeology and engineering geology, it forms part of applied geology. The discipline uses the same principal method—field observation—as the geological sciences and geological considerations in the choice of methods come before economic and mining ones. It summarizes both theoretical and practical knowledge in geological and other fields (tectonics, stratigraphy, mineralogy, palaeontology, geomorphology, geophysics, geodesy, economics, mining) and utilizes this for a rational, scientifically substantiated search for mineral deposits and their qualitative and quantitative assessment. The task of exploration ends with the collection of those data and the development and exploitation of deposits lie in the fields of mining and mining geology.

Ore prospecting and exploration are undertaken in a logical time sequence. The target of prospecting is to find useful mineral deposits and appraise their suitability for further exploration. Economic deposits are then explored, the quality of the raw material is determined by sampling and the reserves are calculated by means of surface, underground and drilling works. Exploration should be distinguished from investigation or research of a wider nature (e.g. basic investigation) and laboratory work (e.g. mineralogical research). The discovery, study and exploration of a mineral deposit is a long-term procedure, which begins with the establishment of prospecting criteria and indications in the general and preliminary prospecting stages, and ends with the depletion of the deposit. This statement appears to be somewhat paradoxical until we realize that exploration only provides a "sample" of information in the mathematical sense. A deposit is a "total set" of information, which only becomes available after the deposit has been exploited. The amount of information needed to discover a deposit is substantially less than this "total set". The duration and cost of the individual prospecting and exploration stages can be seen from Tables 1 and 2.

One per cent of the cost in prospecting and exploratory work in Table 1 may range from US dollars 2,000 up to 100,000. Prospecting and exploration methods are of practical importance if the search is based on sound theoretical and empirical knowledge, and if the deposits are evaluated as completely as possible. In any case, all work should be directed to the utilization of the entire deposit. The economy

TABLE 1

Examples of time and cost of prospecting and exploration stages

Stage	Result	Cost as a percentage of total expenditure	Duration
prospecting (general, preliminary, detailed)	discovery of mineral concentration, differentiation of min. occurrences and deposits, calculation of hypothetical reserves (during detailed prospection)	3	1—20 years
prospecting — exploratory stage	differentiation of economic and subeconomic deposits, calculation of inferred (C_2) reserves	3	3 months to 3 years
preliminary exploration (often associated with the previous stage)	calculation of indicated (C_1) reserves	6	6 months to 3 years
detailed exploration (often associated with preparation for mining)	calculation of indicated (B) reserves	7	2—5 years
construction of mine and dressing plant during the preparation for exploitation	calculation of measured (A) reserves	80	2—5 years
beginning of operation	calculation of measured (A) reserves	1	1—6 months

of the whole procedure can be ensured by using a reliable theoretical basis for prospecting and exploration, and by performing the work in the correct order. Financial losses may occur in several ways: neglect of the principle of multistage exploration (if for example the prospecting-exploratory or preliminary exploration stage is omitted, or if a detailed map is only compiled when the drilling has been completed); an inaccurate location and orientation of exploratory works (location of a drill hole on the lying side of an inclined vein, or the driving of a trial trench parallel to the presumed strike of vein); performance of unnecessary, geologically unsubstantiated exploratory work, an inadequate sampling, storage or transport of drill cores (if, for example, samples are taken only from the parts containing macroscopic ore particles, or if insufficient protection from rain is given to the cores); inaccurate laboratory and technological determination of ores and wall rocks (especially if the wall rock is erroneously designated as economically unutilizable); drilling exploration of a deposit so near the working face that a previously worked block is thus assessed; or, finally, the unnecessarily detailed examination of a deposit

TABLE 2

Example of capital and time requirements for an industrial mineral processing plant project (ceramic factory, cement works)

Stage	Expenses (per cent of the total cost)	Time requirement (in months)	
a) preliminary feasibility study (I): marketing	0.5	1—2	3—4
b) preliminary prospecting and sampling of all known and newly found occurrences of mineral raw materials in the country	1	overlap 3	
c) preliminary examination of small samples taken during stage (b)	0.5	1—2	
d) contract for the project			
e) exploration, estimation of reserves and bulk sampling of prospects with the required quality	3	6	
f) pilot-plant experiments	1.5	3—5	
g) feasibility study (II): 1. quality, possible use and reserves of the deposits, transport, water and energy supply, labour; 2. technological study on the basis of pilot-plant experiments	3	3—4	
h) authorization to build	—	2	17
i) detailed mechanical and technological design, project for buildings	6	overlap 6	
j) construction of the buildings; delivery of the technical equipment	38	overlap 15	
k) cost and fitting of the technical equipment (cost of fitting = aprox. 12 % of the value of the machines)	46	12—16	
l) start up	0.5	2	
	Total 100 %	47—56 months	

"Overlap" indicates that several stages can run simultaneously.

in the exploration stage (when this could be undertaken at a smaller cost during the mining exploration). Many of these mistakes are discovered after exploration is finished, but only a few of them are reparable without further expense.

One can prevent investment in the construction of a dressing plant if, in assessing the final report on the deposit, it is found that the bulk of ore has been evaluated erroneously in some technical sense. Financial losses can be avoided altogether if all the existing data are thoroughly evaluated before the technical work begins, if the results of the latter are studied continuously and the ore samples are subjected to repeated technological, petrographic and mineralogical analyses. The amenability of ore to treatment should be examined on a laboratory scale during the calculation of C_2 (inferred) reserves (see below).

In conclusion, it should be mentioned that mineral deposits still are, and will be, discovered by good chance. Some deposits have been exploited for a long time without ever having been explored, while with others the trial exploration has been made simultaneously with the mining. A modern industrial society, however, cannot rely on chance or the legacy of the past. The results of its efforts in this century are imposing, and a list of the mineral deposits that have been discovered by systematic prospecting on a scientific basis includes the following: iron — Kursk, Kustanai (U.S.S.R.), Ungava (Labrador); titanium — Malawi; manganese — Zambia, Botswana; chromium — Greece (discovered using a geobotanical method), Rhodesia, Madagascar; nickel — Norilsk (U.S.S.R.), Finland (discovered by using a geobotanical method); copper — Chuquicamata (Chile), Zambia, Katanga (Copper Belt, discovered by means of an air photographic survey of the vegetation), Jerome (Arizona, U.S.A.), Lubin-Polkówice (Poland), Panama, Bougainville; lead and zinc — Nigeria; tin — the Far East and Far North-East, Siberia (U.S.S.R.), Wales and Devonshire (United Kingdom, discovered by using a geobotanical method); gold — western continuation of Witwatersrand (Rep. South Africa); tantalum and niobium — Nigeria; uranium — Great Bear Lake and the Athabasca area (Canada), Zambia, North Africa (U-bearing phosphates), Rum Jungle (Australia), Antarctica; diamonds — Yakutia (U.S.S.R.), Lesotho; oil — Vtoroe Baku (U.S.S.R.), Algeria, Nigeria, Libya, Queensland (Australia, L. J. Lawrence 1965); sulphur — Tarnobrzeg (Poland); bituminous coal — Lvov basin (U.S.S.R.), Lons le Saumier (France), Taimyr basin — Permo-Carboniferous, Central Asian basins — Jurassic (U.S.S.R.), Colomb Béchar and Abadla (Algeria).

It can be assumed that fifty years hence the world will not be suffering from a shortage of iron, titanium, manganese, chromium, nickel, cobalt and vanadium. The reserves of copper, lead, zinc, tungsten and tin are, however, relatively small. The crisis in the supply of energy and mineral materials, which first appeared in the autumn 1973, set the geologist-prospector a task of ensuring the necessary mineral resources for the world economy. Since 99 per cent of the deposits outcropping at the earth's surface in accessible regions have already been discovered, prospecting must be oriented towards concealed deposits. How many deposits similar to the Guanajuato lie buried by the rhyolite sheets of the Sierra Madre in Mexico? How many areas similar to that of Kirkland Lake in Canada await discovery beneath glacial sediments, or lie hidden beneath alluvium, such as the Wiluna deposit in West Australia? If erosion had stopped 30 m above the present surface, the Goldfield deposit would not have been found. How many similar deposits await discovery? Prospecting for concealed mineralization affords the greatest probability of finding large deposits, but the search will be expensive, long, and demand technological methods of a high standard and sound theoretical basis.

LITERATURE ON PROSPECTING AND EXPLORATION

The origin of economic geology as a separate branch of science can be dated from the publication of F. Pošepný's book "The genesis of ore deposits" (New York 1893). In it the author corrected erroneous ideas on the origin of endogenous ore deposits by lateral secretion from wall rocks. Although some correct theoretical conclusions reached in economic-geological and other geological studies had been applied to the search for mineral deposits even before this time, empirical experience, which in Antiquity and the Middle Ages had been the only guide, predominated. The experiences of prospectors were passed from one generation to the next, often within one family, and no attempts were made to provide an explanation of the success or failure of their efforts. An example of an adequate approach to prospecting is the work of Agricola "Twelve Books on Mining and Metallurgy" (1556), especially Book Two, where prospecting methods based on rock fragments and geobotany are described. In the works of M. V. Lomonosov, quite modern conclusions were drawn from penetrating observations of natural phenomena. As early as 1763, Lomonosov, in his book "O rudnykh mestakh i o priiske ikh" (On ore deposits and their location) described gossans and surface subsidence as indicators of mineralization.

The science of prospecting and exploration of mineral deposits, orientated to the economic evaluation of deposits, separated from genetic economic geology in the 1920's. This trend was most marked in the U.S.S.R., since a planned economy demanded a scientific approach to mineral prospecting. The first five years' experiences of prospecting and exploration in the U.S.S.R. were summarized in two publications "Materialy k metodologii poiskov i razvedok poleznykh iskopaemykh" (Methodology of prospecting and exploration of mineral deposits) (1931) and "Oprobovanie" (Sampling) (1932) by V. M. Kreiter and others. Kreiter also wrote the textbook "Poiski i razvedki poleznykh iskopaemykh" (Prospecting and exploration of mineral deposits) (1940), which was revised and reissued in 1961 (Vol. 1) and 1962 (Vol. 2) under the title "Poiski i razvedka mestorozhdenii poleznykh iskopaemykh" (Prospecting and exploration of mineral deposits), and as an abbreviated edition in 1964. A number of Soviet handbooks deal with similar problems. Instructions for calculating mineral reserves were presented in V. I. Smirnov's "Podschet zapasov mineralnogo syrya" (1950), an extended edition of which appeared in 1960. This was prepared in collaboration with several co-authors and entitled "Podschet zapasov mestorozhdenii poleznykh iskopaemykh" (Calculation of mineral reserves).

Compared with Soviet literature, Western literature on prospecting and exploration is less prolific. Every geologist employs methods that he has learned from his teacher of economic geology. The best known of the American handbooks are

"Mining Geology" by H. E. McKinstry (1948) and J. D. Forrester's "Principles of Field and Mining Geology" (1946). The "Handbook for Prospectors and Operators of Small Mines" written by W. W. von Bernewitz (1943) was intended for the laymen who want to be prospectors or have their own mine. A textbook on prospecting and exploration was published in Australia in 1965 (L. J. Lawrence — editor). Pearl's "Handbook for Prospectors" (1973) is the most recent publication in this area.*

In French literature, prospecting is treated in detail (about 200 pages) only in P. Routhier's "Les gisements métallifères" (vol. 2, 1963). In P. Despujols and H. Termier's "Introduction à l'étude de la métallogénie et à la prospection minière" (1946), the subject is covered in only 20 pages. Of the earlier handbooks, the "Recherche et étude économique des gites métallifères" (1943) of L. Thiébaut deserves to be mentioned.

In the German literature on economic geology, a comprehensive book "Prospektion und feldmäßige Beurteilung von Lagerstätten" was published by G. Zeschke in 1964 and another one — Geowissenschaftliche Methoden (Erster Teil des Zweiten Bandes des Lehrbuchs der angewandten Geologie) — was edited by A. Bentz and H. J. Martini in 1968. The handbook "Einführung in die Berechnung von Lagerstättenvorräten fester mineralischer Rohstoffe" by F. Stammberger (1956) deals only with the calculation of mineral reserves. Prospecting for pegmatites is treated briefly by H. Schneiderhöhn (1961) in "Die Erzlagerstätten der Erde", vol. 2, Die Pegmatite.

A number of Western authors have discussed special problems in prospecting and exploration. Among others the following are notable: R. Blanchard, P. F. Boswell, A. Locke — the assessment of outcrops; K. Fulton, T. S. Lovering, L. C. Huff — the prospecting for concealed deposits using geochemical methods; D. W. Brunton, T. A. Rickard, B. Prescott — sampling of ore; K. Raeburn and H. Milner — prospecting for and exploration of placers; M. H. Burnham, W. A. Jones, S. G. Lasky, C. O. Swanson — calculation of mineral reserves.

The following periodicals are devoted to problems of prospecting and exploration: Economic Geology, Razvedka i okhrana nedr, Geologiya rudnykh mestorozhdenii, Geologiya i geofyzika, Zeitschrift für angewandte Geologie etc. In 1956 a Symposium on Concealed Deposits was held in Moscow, and two years later a Symposium on Prospecting for and Exploration of Mineral Deposits was organized in Vancouver. The Association of Exploration Geochemists has organized seven Symposia: Ottawa in 1966, Denver in 1968, Toronto in 1970, London in 1972, Vancouver in 1974, Sydney 1976, Denver 1978. It also publishes the Journal of Geochemical Exploration.

---

* During the printing of this book (1977) a comprehensive overview of exploration and mining geology written by William C. Peters of the University of Arizona was published by John Wiley and Sons, Inc.

INDUSTRIAL TYPES OF MINERAL DEPOSITS

Of the many genetic types of mineral deposits only a part are of sufficient size and grade to be economic. The remaining types have too small or poor a concentration. Metamorphic iron deposits of the Krivoi Rog type, for example, which comprise over half of the relatively good-quality Fe-ore reserves, are the most productive type. On the other hand, the gossan-type deposits of hydrothermal pyrite and other sulphidic deposits contain small amounts of ore, and large-scale exploitation is uneconomic. Deposits that cover more than one per cent of the world's production are termed industrial. In general, sedimentary (including metamorphic-sedimentary) deposits are more economical than other types. Impregnated and metasomatic deposits usually have greater reserves than other hydrothermal deposits and are easier to follow than vein deposits. Residual deposits are also of great importance, because the useful mineral, which might be greatly dispersed in the primary deposit, becomes concentrated in them.

The designation of a deposit as industrially important depends on the shape, dimensions, quality and location of the deposit. Tables 3 to 33 (according to Kreiter 1960, 1964) include industrial deposit types of the Fe group (iron, titanium, manganese, chromium), alloy metals (nickel, cobalt, tungsten, molybdene, vanadium), non-ferrous metals (copper, lead, zinc, tin, mercury, antimony, bismuth), light metals (aluminum, beryllium, lithium, magnesium), precious metals (gold, silver, the platinum metals), radioactive elements (uranium, thorium), rare elements (tantalum and niobium, zirconium), rare earths, trace elements, mineral materials for the metallurgical industry (fluorite, graphite, magnesite, refractory clays and kaolins, foundry sands), materials for the chemical industry (phosphates, sulphur and pyrite, arsenic, boron, barite, witherite, salts) and other industries (asbestos, micas, piezo-electric and optic minerals, diamond, corundum and Al_2O_3-rich minerals, talc, feldspar, mineral fillers and filters, glass sands), building materials and caustobioliths.

From a national point of view the order of importance of industrial deposits can differ to such an extent from the global hierarchy that deposits classified as industrial in one country may be of no account in world production.

In countries with planned economies, requirements of the minimum reserves, the differentiation of deposits into size categories (small, medium, large), the thickness of the deposits and the content of economic minerals, and the largest admissible amount of impurities in crude ore (or in concentrate) are defined in standards and instructions on the quality of mineral materials, published by the Ministry of Mining and Ministry of Construction.

The quality of mineral materials requires changes with time and place in all countries. It is most reliably determined, although also most laboriously, by calculating the profitability of the future mine. For this, basic information obtainable on the discovery of the deposit is necessary. First, the approximate cost of mining and treating one ton of metal in the concentrate and its transport to the processing plant is calculated. This is then compared with the cost of the material in the country itself and on the international market. After considering possible fluctuations in world costs, it is then possible to decide whether or not it would be profitable to explore and develop the prospect. Taking all these parameters into account, a large deposit of high-grade ore, which is separated by inaccessible terrain from its destination, may

TABLE 3

World industrial types of Fe deposits

Type	Form	Economic minerals	Metal content in %	World reserves in thousands million tons	Percentage of world production	Examples
1. meta-morphogenic (with quartz)	beds	hematite, magnetite	50—60	40 (ores with 20—40 % Fe — 3000)	53	Krivoi Rog, Kursk (U.S.S.R.), Lake Superior (North America), Brazil
2. sedimentary (oolithic)	beds	hematite, siderite, chamosite	23—40	70	32	Kerch (USSR), Lotharingia (France)
3. skarn	beds and lenses	magnetite, hematite	30—40	12	10	Gora Magnitnaya (USSR); Kiruna (Sweden)
4. hydro-thermal-metasomatic	lenses, irregular bodies	siderite	30—40	6	4	Erzberg (Austria), Bilbao (Spain)
5. impregnated (basic rocks)	stocks and veins	titano-magnetite	30—60	30	1	Kachkanar (Urals), Canada, USA

Maximum content of impurities: S — 0.03 to 1 %, P — 0.007—1.8 %, As — 0,7 %, Sn — 0.08 %, Zn — 0.2 %, Pb — 0.1 %, Cr — 1 %

TABLE 4

Industrial titanium deposits

Type	Form	Economic minerals	Metal content	Examples
1. placers	beds and lenses in alluvium	ilmenite	10—100 kg/ton (content of mineral)	Travancore (India)
2. late magmatic	impregnated lenses and veins in basic and ultrabasic rocks	ilmenite, magnetite		Tanzania
3. metamorphogenic	impregnations in amphibolites and other metamorphic rocks	ilmenite, rutile		North Carolina (USA)

TABLE 5

Industrial manganese deposits

Type	Form	Economic minerals	Metal content in %	Examples
1. residual (70 % world production)	cover of Mn-bearing rocks or poor Mn deposits	psilomelane, pyrolusite, manganite	after treatment 40—50	India, Ghana, Brazil
2. sedimentary	beds and lenses	pyrolusite, manganite, rhodochrosite, manganocalcite, oligonite	15—40	Chiatury, Nikopol (USSR)

Maximum content of impurities in the raw material for the chemical industry: Fe 4 %, CaO — 2 to 3 %, Co, Ni, As-traces, CuO — 2 %.

TABLE 6

Industrial chromium deposits

Type	Form	Economic mineral	Metal content in %	Examples
magmatic segregation	impregnated beds, lenses, veins in ultrabasites and serpentinites	chromite	varying, mined at 30—60%	Bushveld, Great Dyke (Rep. S. Africa), Urals, Turkey

TABLE 7

Industrial nickel deposits

Type	Form	Economic minerals	Metal content in %	Percentage of world reserves	Percentage of world production	Examples
1. liquid-magmatic	impregnated beds, veins passing into mass ore in basic and ultrabasic rocks	pentlandite, chalcopyrite, pyrrhotite	1.3—4.6 (0.26 in impregnations)	70	80 (5% Cu)	Sudbury (Canada) Pechenga, Norilsk (USSR)
2. residual	cover of ultrabasites and serpentinites	garnierite, nepuite	1.3—4.0	30	20	Khalilovo (USSR), New Caledonia, Cuba, Brazil, Sklary (Poland)

Useful accessory elements: Cu, Co, Pt, Pd, Rh, Au, Se; impurities: Pb, Zn, As, occasionally Cu.

TABLE 8

Industrial cobalt deposits

Type	Character of occurrence	Economic minerals	Metal content in %	Examples
1a. hydrothermal Ni—Co—Bi—Ag; Ag—As; Pb—Zn formations	cobalt is usually a by-product from the mining of other metals	linnéite, cobaltite, smaltite	min. several tenths of %; 0.06—0.08 in complete utilization of ore	Khovakhsy (USSR), Cobalt (Canada), Burma, Jáchymov (Joachimstal) ČSSR (abandoned)
1b. Cu-bearing sandstones		heterogenite		Zaire
1c. liquid-magmatic Cu—Ni deposits				Sudbury (Canada)

TABLE 9

Industrial tungsten deposits

Type	Form	Economic mineral	Metal content in %	Percentage of world reserves	Percentage of world production	Examples
1. skarn	layers and veins at the granitoid-carbonate contact	scheelite	0.3—6.0 WO_3	60 (WO_3)	55 (WO_3)	Sang Dong (S. Korea), Ingichke (USSR), USA, Mexico, Brazil
2. hydrothermal (with quartz)	veins at the granitoid contacts	wolframite	0.4—4.0 WO_3	30 (WO_3)	25 (WO_3)	Burma, China, Dzhida (USSR), Cínovec (Zinnwald, ČSSR)
3. placers	eluvial-colluvial cover, alluvium	wolframite	min.0.01 WO_3	10 (WO_3)	20 (WO_3)	China, Dzhida (USSR)

Maximum content of impurities: P — 0.03 to 0.2 %, S — 0.3 to 3 %, As — 0.04 to 0.2 %, Sn — 0.08 to 1.5 %, Cu — 0.1 to 0.22 %.

TABLE 10

Industrial molybdenum deposits

Type	Form	Economic mineral	Metal content in %	Percentage of world reserves	Percentage of world production	Examples
1. hydro-thermal	stockworks in small intrusions (quartz, molybdenite and/or chalco-pyrite)	molybdenite	0.1—0.4; as accompanying metal 0.005—0.01	95	98	Climax (USA), Dzhida, Kounrad, Trans-baikalya (USSR)
2. skarn	platy bodies and veins at the granitoid/lime-stone contact	molybdenite	0.1—1.0	5 (?)	2	Azegour (Morocco), Tyrnyauz (USSR)

Maximum content of impurities in concentrate (47—50 % Mo): Cu — 0.5 to 2 %; P — 0.07 to 0.15 %, As — 0.07%, Sn — 0.07 %, quartz — 5 to 9 %.

TABLE 11

Industrial vanadium deposits

Type	Form	Economic minerals	Metal content in %	Examples
1. contact metamorphic	vein with asphaltite	patronite, Ca-vana-date	0.7 (V_2O_5)	Minas Ragra (Peru), is the only deposit
2a. polymetallic (hydrothermal)	V is usually a by-product of mining for Cu, Pb, Zn, Fe, Ti and from oil refining	vanadinite, descloizite (in oxid. zone of poly-metallic dep.)	1—3 (V_2O_5)	Broken Hill (Zambia), Tsumeb (Namibia)
2b. late magmatic		titanomagnetite	0.1—0.4 (V_2O_5)	Sweden, India
3. infiltrations	beds with dis-seminations	carnotite		Colorado Plateau (USA)

TABLE 12

Industrial copper deposits

Type	Form	Economic minerals	Metal content in %	Percentage of world reserves	Percentage of world production	Examples
1. disseminated copper	impregnations, small veins in intrusive rocks	chalcopyrite	0.8—2.2 (in USA min. 0.5 Cu)	40	42	Bingham (Utah, USA), Kounrad (USSR), El Teniente (Chile), Medet (Bulgaria)
2. Cu-bearing sandstones of red-beds type, Mansfeld shales, conglomerates	beds with impregnations and veins	bornite, chalcopyrite	3—5 (Zambia)	43	25	Zambia, Katanga, Dzhezkazgan, Udakan (USSR), Mansfeld (GDR), White Pine (Mich. USA), Lubin—Polkówice (Poland)
3. hydrothermal metasomatic	thick veins in metamorphosed effusives	chalcopyrite disseminated in pyrite	1.5—5 (in Spain min. 0.7)	9	19	Rio Tinto (Spain), Urals, Canada, Japan
4. hydrothermal	veins	Cu-sulphides	1—10	2	5	Butte (Montana, USA)
5. skarns	stockworks, veins, layers at the granitoid-limestone contact	chalcopyrite	2—8	1	1.5	Bisbee (Arizona, USA), Minusinsk area (USSR)

TABLE 13

Industrial lead and zinc deposits

Type	Form	Economic minerals	Metal content in %	Percentage of world reserves Pb	Zn	Percentage of world production Pb	Zn	Examples
1. hydro-thermal, regional-meta-morphosed	concordant beds and lenses in meta-morphics	galena, sphalerite (chalco-pyrite)	6—12 (Pb) 5—12 (Zn)	30	25	35	30	Sullivan (Canada), Broken Hill (Australia), Zambia
2. skarn	veins and stocks in limestones or at their contacts with volcanic rocks	galena, sphalerite (chalco-pyrite)	2.5—20 (Pb) 5—15 (Zn)	15	25	30	25	Tetyukhe, Altyn Topkan (USSR), Santa Eulalia (Mexico), Peru
3. telethermal impregnated	impregna-tions and lenses in limestones	galena, sphalerite	2—5 (Pb) 3—12 (Zn)	15	25	20	25	Pine Point (Canada), Mississippi Valley (USA), Mirgalimsai (USSR)
4. hydro-thermal	polymetallic veins	galena, sphalerite (chalco-pyrite)	5—20 (Pb) 12—25 (Zn)	20	15	10	20	Sadon (USSR), Freiberg (GDR), Sardinia, Příbram (ČSSR)
5. hydro-thermal	layers and lenses of sulphidic or quartz-carbonate Pb—Zn ores in volcanic rocks	galena, sphalerite (pyrite)	2—25 (Pb) 3—12 (Zn)	20	10	5	8	Altai, Salair (USSR), Bawdwin (Burma)

Accessory elements: Ag, Au, Cu, Bi, As, Sn, Cd, Zn, Ga, Ge, Mo, Co, Tl. Maximum content of impurities in Pb concentrate: Cu — 2 to 4 %, Zn — 8 to 12 %, Fe — above 25 %; in Zn concentrate Cu — 1 to 1.5 %, Fe — 8 to 9 %, Co — tenths of one per cent.

TABLE 14

Industrial tin deposits

Type	Form	Economic mineral	Metal content in %	Percentage of world reserves	Percentage of world production	Examples
1. placers	beds with cassiterite in eluvium, colluvium, alluvium and the littoral zone	cassiterite	0.05—0.8 Sn; min. 0.015 Sn in thick beds, 0.03 Sn in thin beds	60	70	Malaysia, Indonesia, Zaire, Nigeria, Far North-East (USSR)
2. hydro-thermal	veins with sulphides and chlorite or tourmaline; polymetallic veins	cassiterite	1—5	15	20	Bolivia
3. greisens	veins and stockworks in granitoids and at their contacts	cassiterite	1—4 Sn in veins, 0.3—1 m in stockworks (min. 0.13—0.2)	25	10	China, Cornwall (England), Altenberg(GDR), Congo, Indonesia, Cínovec (Zinnwald — ČSSR)

Maximum content of impurities in the concentrate: Pb;— 0.5 %, WO_3 — 5 %.

TABLE 15

Industrial mercury deposits

Type	Form	Economic minerals	Metal content in %	Percentage of world production	Examples
1. tele-thermal	beds, lenses and veins in sediments	cinnabar	0.3—8 (min. 0.15)	50	Almaden (Spain), Monte Amiata (Italy), Idrija (Jugoslavia)
2. hydro-thermal	stockworks, nests and mineralized breccia	cinnabar	0.2 (Cloverdale, dep. USA — 0.045)	50	New Almaden (California, USA)

TABLE 16

Industrial antimony deposits

Type	Form	Economic mineral	Metal content in %	Examples
1. telethermal	beds of mass and impregnated ores in limestones and sandstones	stibnite	5—7 (min. 2—3)	Si-kchuang-shan (Khu-nan, China), Kadamzhai (USSR)
2. hydrothermal	veins with quartz, esp. in carbonate rocks	stibnite	10—25	Bolivia, Mexico, Algeria, Antimony King (USA)

Maximum content of impurities: As — 0.25 %, Cu — 0.03 %, Pb — 0.08 %.

Table 17

Industrial bismuth deposits

Type	Form	Economic minerals	Metal content in %	Examples
hydrothermal Bi—Ni—Co—Ag—U formation	veins (Bi is obtained with other metals also from Cu, Pb—Zn, Sn and W ores)	bismuthinite	min. 0.3	Krušné hory (ČSSR), Torre Kimpos (Spain), San Gregorio (Peru)

TABLE 18

Industrial aluminum deposits

Type	Form	Economic minerals	Metal content in %	Examples
1. residual	beds and pockets of lateritic bauxite (in situ on platforms)	hydrargillite	min. 45 Al_2O_3, max. 12—15 SiO_2, Al_2O_3: SiO_2 = > 3	Jamaica, Guyana, Surinam Guinea, Ghana, Arkansas (USA), India
2. sedimentary	fillings of depressions in karstic limestone surface, in geosynclines (true bauxites)	boehmite, diaspore		France, Hungary, Romania, Jugoslavia
3. sedimentary	lenses and beds with grey sediments in effusives of unstable platforms	hydrargillite, boehmite		China, Tikhvin, Krasnaya Shapochka (USSR)

TABLE 19

Industrial beryllium and lithium deposits

Type	Form	Economic minerals	Metal content in %	Examples
1. granite pegmatites	crystals and crystal clusters in blocky and metasomatic pegmatites	beryl, spodumene		Brazil, Mozambique, Argentina
2. quartz veins, greisen zones	veins in acid granites	beryl, spodumene	0.1 Be (in complex ore 0.02 Be), 0.7—1.0 Li_2O	India
3. hybrid pegmatites	veins at the contacts of granite intrusions and in basites	beryl		

TABLE 20

Industrial magnesium deposits

Type	Form	Economic minerals	Examples
1. evaporites	beds in salt deposits	carnallite	GDR, Urals, Alsace (France)
2. hydrothermal-metasomatic	beds and lenses	magnesite, dolomite	Styria (Austria), Slovakia (ČSSR)
3. seawater			Freeport (USA)

TABLE 21

Industrial gold deposits

Type	Form	Economic minerals	Metal content	Percentage of world reserves	Percentage of world production	Examples
1. metamorphosed fossil placers	disseminated in the cement of Precambrian conglomerates	native gold	6—10 g/t	45	40	Witwatersrand (Rep. South Africa), Canada, Brazil
2. hydrothermal (old gold formation)	quartz veins, stockworks and vein zones	native gold	6—25 g/t (Juneau, Alaska — min. 1—2 g/t	35	25	Urals, Siberia (USSR), Mother Lode (USA), Zaire, Australia
3. subvolcanic (young gold formation)	veins	native gold, Au, Ag tellurides	6—50 g/t	10	5	Romania, Mexico, Indonesia, Cripple Creek (USA)
4. placers	gold-bearing beds in eluvium, colluvium, alluvium and the littoral zone	native gold	50 mg/m^3 to 5 several g/m^3	10 (20 from Cu and Pb—Zn ores)		Lena, Kolyma, Transbaikalya (USSR), Alaska (USA), Australia

Maximum content of impurities: Cu — 0.1 %, Zn — 0.05 %, As, Sb, coal substance.

TABLE 22

Industrial silver deposits

Type	Form	Economic minerals	Metal content in %	Percentage of world production	Examples
1. hydrothermal a) Pb—Zn ores b) Cu ores c) subvolcanic Au ores	veins and lenses	by-product of Pb—Zn, Cu, Au mining	0.04—0.05 (= 400—500 g/t)	a) 50 b) 15 c) 10	
2. hydrothermal	veins and vein zones	argentite, proustite, pyrargirite		25	Pachuca, Guanajuato, El Oro (Mexico)

TABLE 23

Industrial deposits of platinum-group metals and their alloys

Type	Form	Economic minerals	Metal content	Examples
1. liquid magmatic (70 % world production)	impregnations in ultrabasic and basic rocks	native metals and their alloys in sulphides	1.5 g/ton of sulphidic ore	Sudbury (Canada)
2. late magmatic	stocks, veins and nests in ultrabasites	native metals and their alloys in chrome-spinellides and hortonolite	5—15 g/ton (Bushveld)	Urals, Bushveld (Rep. South Africa)
3. placers	beds in alluvium	native metals and their alloys	hundreds of mg/m^3	Urals, Choco (Colombia), Goodnews Bay (Alaska)

TABLE 24

Industrial uranium and thorium deposits

Type	Form	Economic minerals	Metal content in %	Percent-age of world reserves	Percent-age of world produc-tion	Examples
1. meta-morphosed fossil placers	conglomerate bankets in quartz-ites	uraninite and others	$0.05—0.2$ U_3O_3 (Rand $0.005—0.1$)	55	50	Witwatersrand (Rep. South Africa)
2. infiltra-tion	beds and lenses in arkosic sand-stones and conglomerates	uran-vanadates, uraninite	$0.3\ U_3O_8$ (Colorado $0.1—1$ U_3O_8)	15	25	Katanga, Gabon, Monument Valley, Colorado Plateau (USA), GDR, ČSSR
3. hydro-thermal	quartz, quartz-carbonate, fluorite-barite veins and vein zones with Bi—Ni—Co—Ag—U, Pb—Zn—U, U—Cu, U—Mo formations	uraninite	$0.3\ U_3O_8$	20	20	Eldorado (Canada), Shinkolobwe (Zaire), Joachimstal (ČSSR)
4. metaso-matic pegmatites	nests at the peri-phery of a quartz core	uraninite	$0.3\ U_3O_8$	2	5	Canada, India, Argentina Argentina
5. placers	beds on marine beaches	monazite (with U, Th, Ce)	$0.2—0.4$ U_3O_8	8		Travancore (India), Brazil, Australia

Potential industrial types: phosphates (e.g. in the U.S.A. they contain 0.01 % U_3O_8, annual produc-tion—1 m. tons, immense reserves); coal shales and coal (e.g. Dakota, USA), alum shales (Sweden).

TABLE 25

Industrial deposits of tantalum and niobium

Type	Form	Economic minerals	Metal content	Examples
1. granite pegmatite	disseminations in pegmatite veins and lenses	tantalite, columbite		South America
2. carbonatites (largest Nb_2O_5 reserves)	irregular bodies in metam. and ultrabasic alk. rocks	pyrochlore	0.05—0.2 % Ta_2O_5 0.1—0.4 %	Norway, Canada, USSR
3. late magmatic	a) layers in stratified alk. intrusions b) albitized leuc. granites and albitites	loparite microlite	Nb_2O_5	Khibiny (USSR), China
4. placers	beds, mainly in alluvium	columbite, tantalite, pyrochlore	tens of grams to kg/m^3 Ta_2O_5	Jos Plateau (Nigeria) in the proximity of columbite granite; Congo

Maximum content of impurities in the concentrate (10 % Nb_2O_5): P — 0.05 %, Si — 1.5 %, Sn, Zr, Ti.

TABLE 26

Industrial deposits of zirconium and rare earths

Type	Form	Economic minerals	Metal content	Examples
1. placers	beds in littoral zone	zircon (baddeleyite)	0.5—10 % of mineral	India, Australia, USA, Brazil (baddeleyite)
2. residual	weathered surface of neph. syenite (with eudialite)	baddeleyite		Brazil only
Deposits of rare earths (Y, La, Ce, Pr, Nd, Sm, Eu, Gd, Tb, Dy, Ho, Er, Tm, Yb, Lu)				
1. placers	beds in littoral zone and alluvium	monazite, xenotime (in places with Sn, W, Au, Zr minerals)		

TABLE 27

Industrial deposits of trace elements

Element	Type	Source of element	Content	Examples
cesium	1. Li-pegmatite 2. evaporites	pollucite carnallite	1% Cs_2O in pegmatite	USSR, Namibia, USA, Sweden
germanium	1. hydrothermal 2. caustobioliths	sphalerite etc., Ge-bearing sulphides	largest in sphalerite $(0.1—0.3\%$ Ge); 5—7 g/t ore 5—7 g/t coal	Tsumeb (Namibia)
thallium	hydrothermal	chalcopyrite, sphalerite, galena, marcasite, Hg—Sb ores with Tl traces	$0.0n\%$ in suphides	
scandium	1. greisens 2. pegmatites	concentrates of Sn and W ores tortveitite	ca. 0.1% Sc_2O	Madagascar only
cadmium	hydrothermal	sphalerite	up to 5% Cd in sphal. (in USA — min. 0.002% in complex ore)	
selenium	1. hydrothermal 2. hydrothermal	traces in sulphides, blockite	$0.0n—0.00n\%$ Se tens of per cent of Se	Bolivia only
tellurium	hydrothermal	traces in Cu—PB—Zn—Ni sulphides	$0.00n—0.0n\%$ Te	
rubidium and cesium	1. pegmatites 2. evaporites	amazonite, lepidolite, pollucite, carnallite (sylvite)	up to 3.12% Rb_2O up to 1.73% Rb 30% Cs_2O $0.00n—0.04\%$ Rb in a deposit	Namibia, USSR, USA, Sweden
gallium	1. hydrothermal 2. residual	Cu, Pb—Zn ores bauxites	$0.00n—0.0n\%$ Ga	USA, FRG, Italy
indium	hydrothermal	in dressing poly- metallic ores, esp. sphalerite-bearing	0.002% In in suphidic ore	USA, USSR
hafnium	marine placers	zircon (Hf: Zr = 1 : 80 to 1 : 6)		USA
rhenium	hydrothermal	molybdenite concentrates	0.05—100 g/ton	Mansfeld (GDR)

TABLE 28

Industrial deposits of metallurgical minerals and rocks

Industrial mineral or rock	Type	Form	Usage, quality (impurities)	Other industrial usage (impurities)	Examples
fluorite	hydrothermal	veins and metasomatic lenses	flux (SiO_2, $BaSO_4$)	chemical ind. min. 98 % CaF (CaO max. 1 %; SiO_2—1 %; Pb, Ba, S—0 %); glass and ceram. ind. min. 95 % CaF_2 (SiO_2—3 %, CaO—1 % Fe_2O_3—1 %)	Harz (GDR), China, East Transbaikaliya (USSR), Mexico, France, USA, Newfoundland (Canada), Krušné hory (ČSSR)
graphite	1. regional-metamorphic	beds of crystalline and amorphous graphite in Precambrian rocks (5—10 % C)			Madagascar, India, USSR, Alabama, Texas (USA), ČSSR, Austria
	2. contact-metasomatic	veins and lenses of cryst. graphite in carbonate rocks at their contacts with intrusives (2—10 % C)	graphite crucibles; after treatment min. 85 % C (volatiles max. 3 %; CO_2—0.3 %)	lubricants, pigments and pencils; electrical industry	Botogolsk deposit (USSR), Canada
	3. contact-metamorphic	beds of amorphous graphite in coal-bearing sediments (70—85 % C)			Sonora (Mexico), Kureika (USSR), N. Korea

Table 28 (continued)

Industrial mineral or rock	Type	Form	Usage, quality (impurities)	Other industrial use (impurities)	Examples
magnesite	1. hydro-thermal-metasomatic (Veitsch type — 80 % of world production)	lenses	refractory bricks, min. 43 % MgO (CaO max 1.5.%)	source of Mg — min 87 % MgO after firing (CaO max. 1.8%; R_2O_3 — 2 %); cement manufacture, min. 75 % after firing	Styria (Austria), Slovakia, Satka (Urals, USSR), China
	2. hydro-thermal and residual (amorphous magnesite, 20 % of world production)	veins and nests			Khalilovo (Urals), Euboia (Greece)
clays and kaolins	1. sedimentary (clays)	beds in lacustrine sed. (overburden 3—10 times thicker than clay)	fireclay, refractoriness 1580—1750 °C, Al_2O_3 25—40 % (Fe_2O_3 max. 2—3 %, CaO — 1 %, MgO — 1 %, TiO_2 — 1 %)	building ind., ceramics, paper manufacture	Carboniferous claystones (ČSSR), United Kingdom
	2. residual (kaolins)	weathering crust of feldspathic rocks (overburden max. 3 times thicker than kaolin)			China, ČSSR, United Kingdom, FRG, France
foundry sands	sedimentary	beds in various sediments (overburden max. 50 % of the sand thickness)	moulds (max. 0.5 % $Na_2O + K_2O$, 1 % CaO + MgO)		

Metallurgical industry also uses flux limestones, bauxites, boron compounds and strontium minerals among other materials.

TABLE 29

Industrial minerals and rocks for the chemical industry

Industrial mineral or rock	Type	Form	Quality (impurities)	Other industrial uses	Examples
phosphates	1. bio-chemical sedi-mentary	concretions and their accumula-tions in beds; gravels; guano	8—40 % P_2O_5 (Fe_2O_3, Al_2O_3)	fertilizers, metallurgy	Ukraine, Karatau (USSR), Idaho, Florida (USA), Oceania
	2. late magmatic	beds of apatite with nepheline in alkali rocks	8—9 % P_2O_5 (CO_2 max 6 %)		Khibiny (USSR)
sulphur and and pyrite	1. bio-chemical sedi-mentary (sulphur)	impregnations and beds over sulphates with oil nearby		manufacture of paper and explosives; textile and chemical industries, fertilizers, metallurgy	Sicily, Louisiana, Texas (USA), Tarnobrzeg (Poland)
	2. hydro-thermal metaso-matic (type 3 of Cu dep.)	thick pyrite lenses in altered effusive rocks	5—6 % S in raw mat. for flotation (bitumens, As, Se, Te)		Rio Tinto (Spain), Urals (USSR)
arsenic	hydrothermal	secondary product from processing of Cu, Pb—Zn, Au, Ag ores		manufacture of pigments, medicines	
boron	1. sedi-mentary	beds with borax, kernite, etc. in tuffites		agriculture, medicine, manufacture of glass and ceramics, metallurgy, electrical ind., rocket fuel, reactors	Kramer (USA), Tibet (China), Turkey
	2. evaporites	disseminations and beds in cap rocks of salt dep. (ascharite, hydroboracite after kaliborite)			Stassfurt (GDR)
	3. potential type: datolite, ludwigite				
salts	evaporites (halite, sylvite)	beds and plugs in lagoonal sediments		food-stuff industry, agriculture	GDR, USA, USSR

Other materials used in chemical industry: Chile saltpetre, soda, strontianite and celestite, alum, barite, bauxite, dolomite, fluorite, glauconite, magnesite and rare elements.

TABLE 30

Industrial deposits of other industrial minerals

Industrial mineral	Type	Form	Quality	Industrial use	Examples
asbestos	1. hydrothermal (chrysotile) (90 % of world production)	transversely fibrous veins in serpentine; rarely in serpentinized dolomites	0.5—15 % asbestos; length of fibres— 0.7—18 mm	asbestos-cement sheets, pipes, fireproof fabrics, thermal and radiation insulations, filters for chemical ind.	Bazhenovo (USSR), Thatford, Black Lake (Canada), Barberton (Rep. South Africa)
	2. metamorphic (brittle amphibole asbestos)	transversely and longitud. fibrous veins in Fe-quartzite, sandstone and serpentinite			Pengo (Rep. South Africa), Rhodesia (crocidolite, amosite), Swaziland, USSR
micas	1. pegmatites (muscovite — 90 % of world production)	nests and zones in pegmatites	1—2 % musc. in deposit; min. 10—80 kg/m^3	electrical ind. filler in rubber, paper, paints	Mama (USSR), India, Brazil, Canada
	2. contact-metasomatic and metamorphic (phlogopite)	veins, zones, nests at the contacts of igneous rocks with dolomites	5—20 % phlogopite in deposit		Slyudyanka (USSR), Canada, Madagascar, India
	3. residual (hydro-thermal?) vermiculite	veins and lenses in ultrabasites		insulations (expanded vermiculite)	Buldym deposit (USSR), Libby, Montana (USA)
piezo-electric quartz	1. pegmatites; hydrothermal and metamor-phic quartz veins	quartz mono-crystals in druse cavities	monocrystals of 100—2000 g weight	communication engineering (radio and telephone instr.)	Minas Gerais (Brazil)
	2. placers	eluvium and colluvium			Brazil
iceland spar	1. subvolcanic	nests at the contact between pyraxene—zeolite rocks and trapps	min. size 25 × 12 × 12 mm	manufacture of nicols	Iceland, Siberia (USSR)
	2. metamorpho-genic (Alpine paragenesis)	nests in carbonate rocks			Tanatuva (USSR)

Table 30 (*continued*)

Industrial mineral	Type	Form	Quality	Industrial use	Examples
optical fluorite	1. hydrothermal metasomatic	druse cavities in veins in carbonate rocks	min. size $10 \times 10 \times 4$ mm	lenses (aplanatic, achromatic)	Illinois (USA), Tadzhikistan (USSR)
	2. pegmatites	druse cavities in drusy pegmatites			Kazakhstan (USSR)
diamond	1. placers	beds in alluvium	0.1—1 carat/m^3 ($= 0.02$ to to 0.2 g/m^3)	cutting, abrasion and drilling of hard materials; jewelry	carbonados: Brazil, Venezuela, Guyana; borts: Zaire, Ghana, Guinea, East Siberia (USSR)
	2. early magmatic	disseminated in kimberlites (a few to 800 m in diameter)	0.1—0.5 carat/ton ($= 0.02$ to 0.1 g/t)		Rep. South Africa, East Siberia (USSR)
corundum and other minerals with high Al_2O_3	1. contact-pneumatolytic (corundum)	nests in secondary quartzites	40—60% of mineral	abrasives	India, USA, Semiz Bugu (USSR)
	2. metamorphic (emery)	lenses and nests in marbles and at their contact with ultrabasites			Obukhov deposit (Salair, USSR), Greece, Turkey
	3. metamorphogenic (andalusite)	see type 1, margins of corundum deposits	min. 54% Al_2O_3 (Fe_2O_3 + FeO max. 1%; K_2O + Na_2O max. 1.5%; for the silumin production — min. 59% Al_2O_3)	fire-proof fabrics (silumin for airplanes, ships)	India (also sillimanite and kyanite deposits)

Table 30 (continued)

Industrial mineral	Type	Form	Quality	Industrial use	Examples
talc	1. hydrothermal (soapstone)	lenses and veins in serpentinites	min. 50 % of talc in soapstone	manufacture of paper, paints, rubber, textiles; chemical and metallurgical industry	Chernigov dep. (USSR), Virginia, Vermont (USA)
	2. hydrothermal (talc)	veins and lenses in crystal. magnesites and dolomites, and at their contact with crystalline rocks	minimum thickness of veins — 5 cm		Mautern (Austria), Modoc (Canada)
barite and witherite	1. hydrothermal	veins and lenses	min. 70 % BaSO$_4$ (for manufacture of paints — 95 %)	inert filler, manufacture of paints, chemical ind.	USSR, USA
	2. sedimentary	beds			Meggen (FRG)
feldspars	pegmatites	central parts of veins and lenses	min. 7 % of alkalis; max. 1.2 % of Fe$_2$O$_3$ + TiO$_2$	ceramic and glass industries; abrasives	USA, Karelia, Ukraine (USSR), Sweden, Norway

TABLE 31

Industrial deposits of building materials

Material	Type	Form	Quality
building stones	1. sedimentary	limestone, dolomite and sandstone beds	building stone: min. strength 39 MPa ($= 400$ kg/cm^2);
	2. magmatic	massifs of granite, diorite, gabbro, porphyry, basalt, tuffs	facing stone: min. strength 39 MPa ($=400$ kp/cm^2); blocks—min. size 0.1 m^3
	3. metamorphic	beds of marble, quartzite and roofing slates	
limestones	1. sedimentary 2. metamorphic	beds beds	lime production, min. 80 % CaCO$_3$, max. 10 % MgCO$_3$
brick loams and clays	1. marine, freshwater, eolian and slope deposits	beds (max. thickness of overburden = 20 % of loam thickness)	max. 30 % carbonates
cement raw materials	1. sedimentary (limestone, marl, clay)	beds (max. thickness of overburden = 10 % of limestone thickness)	$$\frac{CaO}{SiO_2 + Al_2O_3 + Fe_2O_3} = = 1.7 - 2.4$$ $$\frac{SiO_2}{Al_2O_3 + Fe_2O_3} = 1.7 - 2.7$$ max. 7 % MgO; 0.5 % alk.; 1 % S
	2. evaporites (gypsum)	beds and lenses	min. 60 % CaSO$_4$. 2H$_2$O (added to clinker)
gravel and sand	sedimentary (fluviatile)	beds	max. 5 % of clay

TABLE 32

Deposits of caustobioliths (compiled by V. Havlena)

	Raw material	Principal use	Type or kind	Main quality index
COAL	anthracite, high-rank bituminous coal	fuel	meta-anthracite, anthracite, anthracitic coal	heat value ash-content
	medium-rank bituminous coal	production of coke and lighting gas	coking coal, fat coal, gas coal	ash content, sulphur content, phosphorus content
	low rank bituminous coal, brown coal	fuel, chemical raw material, partly prod. of coke	parabituminous coal, all types of brown coal	ash content, content of hydrogen, water content
	peat	fuel, chemical raw material, agriculture	wood peat, herbaceous peat (or accord. to composition carex peat, sphagnum peat)	water content, amount of woody component
	bituminous shales	chemical raw material, fuel		content of bitumen (kerogen)
BITUMENS	solid	chemical raw material, road pavement	asphalt, mineral wax	content of inorganic admixture
	liquid (oil)	fuel, chemical raw material	paraffin-base, naphthenic, aromatic oil	sulphur content, content of some hydrocarbons
	gaseous	fuel, chemical raw material, production of helium	according to composition or oil admixture	CH_4 content

prove unprofitable to work whilst a small deposit of poor quality near a railway port may be exploited even if determination of its quality will be costly.

In areas of poor transport it is necessary to calculate "the radius of econo mining" at the very beginning of regional prospecting before examination of publis reports. If, for example, one ton of anthracite hauled in the mine costs 5 $ (incl. cost of mining) and in the nearest port 10 dollars, 5 dollars can be expended on tra port from the mine to the port. With a road transport charge of 5 cents per ton and the "radius of economic mining" is 100 km; a railway transport charge of 2 c increases the radius to 250 km. In mines at these distances the break-even poin reached. The closer the anthracite mine is to the port, the higher the profit. therefore reasonable to limit prospecting to an area within a radius of ab 90—220 km from the port.

Kreiter (1960) lists 120 industrial deposit types according to their shape, composition and location. Since many types have a very similar origin, it is poss to identify a relatively small number of geological conditions favourable to mineral tion. In addition to the factors that control the deposit type, the tectonics of ore field has a decisive influence. Kreiter, Gorzhevskii and Kozerenko (1 identified four main groups of favourable geological conditions: platforms, geo clines, block-fault zones (old geosynclines and platforms with a thin sedimen cover, which have been broken into horsts and grabens and penetrated by typ magmatites), and areas subjected to intensive weathering (Tables 34—37).

Table 32 (continued)

Type of deposit	Examples
folded formations of foredeeps and intermontane depressions	Pennsylvanian anthracite basin (USA), Donbas (USSR), West European coal zone (Ruhrland, Belgium, France, England)
folded formations of foredeeps, intermontane and intramontane depressions	Upper Silesian basin, Appalachian basin (USA), Kuzbas, West European coal zone, Saarland, Kaipin basin (China)
horizontal or slightly folded formations of intramontane depressions and platform covers	Illinois basin (USA), Sub-Moscow basin (USSR), Shan-si basin (China), basins in India and Rep. South Africa
horizontal layers in recent to subrecent formations	in all countries, especially in the temperate climatic zone
horizontal and folded formations of all structural elements of the earth's crust	Peri-Baltic basin (USSR), Fushun basin (China), Australia, Scotland
horizontal beds, impregnations or fissure fillings	Trinidad, West Ukraine, USA
a) lithologico-stratigraphic structures	Near East, Baku, USA, Venezuela, Indonesia
b) tectonic structures	West Ukraine, Romania

TABLE 33

Approximate world production in 1972 (according to Minerals Yearbook, 1974) and reserves of metals or minerals in mineral deposits (in metric tons unless stated otherwise)

	Output	Reserves
iron	768 mil. (ore)	428,000 mil. (Fe)
titanium	3.56 mil. (concentrate)	500—600 mil. (TiO_2)
manganese	21.45 mil. (ore)	4,508 mil. (Mn)
chromium	6 mil. (chromite)	800 mil.
nickel	639,000	90 mil.
cobalt	23,382	more than 1 mil.
tungsten	38,390	several mil.
molybdenum	79,737	28 mil.
vanadium	18,065	9.5 mil.
copper	6.631 mil.	312 mil.
lead	3.492 mil.	128 mil.
zinc	5.551 mil.	1,510 mil.
tin	243,847	20 mil.
mercury	9,514	836,000
antimony	68,000	more than 5 mil.
bismuth	4,078	large reserves relative to exploitation
bauxite	65.315 mil.	1,170 mil. (Al)
beryllium	4,139 (beryl)	several hundred thousand ore with more than 1 % BeO
gold	1,393	11,000
silver	9,293	170,000
platinum metals	132,6	1,000
uranium (non-socialist countries)	23,310	1.235 mil.

Table 33 (continued)

	Output	Reserves
cadmium	16,758	—
selenium	1,303	—
fluorspar	4.396 mil.	hundred of millions
graphite	358,000	90 mil.
magnesite	11.89 mil.	large
phosphates	93.612 mil.	19,776 mil.
sulphur	22.883 mil.	2,520 mil.
pyrite	20.403 mil.	900 mil.
arsenic	43,000	16 mil.
diatomite	1.52 mil.	—
potash	20.433 mil. (K_2O equivalent)	tens of milliards
gypsum	57.963 mil.	—
asbestos	3.737 mil.	200 mil.
mica	239,000	tens of mil.
diamond... gem industrial	12.192 mil. carats... 31.513 mil. carats	100 mil. carats
talc, soapstone; pyrophyllite	4.758 mil.	—
barite	3.959 mil.	173 mil. (Ba)
strontium min.	105,802	—
feldspars	2.912 mil.	1,016 mil.
building materials	approx. 10,000 mil.	huge
bituminous coal	2,062 mil.	
lignite	802 mil.	8600,000 mil.
anthracite	177 mil.	
peat	91 mil.	
oil	2,915 mil.	100,000 mil.

TABLE 34

Favourable geological conditions for mineralization on platforms

1. Deeply metamorphosed zones with acid intrusions	magnetite-hematite with phlogopite and tourmaline; phlogopite deposits; pegmatites with rare earths, auriferous quartz dep.; sulphidic Cu and Zn dep. with Au; monazite dep. in charnockites, granites and pegmatites; Fe-quartzites; Mn dep. in gondites
2. Shallow-metamorphosed zones	conglomerates of Witwatersrand type bearing gold, diamonds and uraninite
3. Zones affected by metamorphism to a variable extent and by Proterozoic folding, with acid intrusions	dep. of auriferous quartz and sulphide-bounded gold; granite pegmatites with cassiterite, beryl and tantal-niobates; quartz veins with cassiterite and wolframite; apatite-magnetite deposits
4. Zones affected by metamorphism to a variable extent and Proterozoic folding, with basic intrusions	Cu—Ni deposits; deposits of native copper in basic effusives
5. Zones affected by metamorphism to a variable extent and by Proterozoic folding, without intrusions	Fe-quartzites and jaspilites; reg. metamorphosed graphite dep.; Cu-sandstones with Co and V
6. Areas in anteclises and syneclises with slight magmatism	Al and Fe in the weathering crust in anteclises and their rewashed equivalents; eluvial placers of rutile and zircon; sedim. Mn deposits; dep. of phosphates, glauconite, coal, sulphur, oil, gas, salts, clay; Pb—Zn dep. of Mississippi Valley type
7. Areas of trapps on platforms with intensive magmatism	Cu—Ni dep.; (telethermal) Pb—Zn dep.; Iceland spar dep.; contact-metamorphosed graphite deposits.
8. Areas of ultrabasites on platforms with intensive magmatism	deposits of diamonds (together with pyropes)

TABLE 35

Favourable geological conditions for mineralization in geosynclines

9. Areas of volcanic rocks in eugeosynclines	sulphidic Cu (Rio Tinto) and Cu—Zn dep.; Au—Ag dep. with Se and Te in the Alpine eugeosyncline
10. Areas of ultrabasic and basic rocks in eugeosynclines	magmatic chromite dep. with Pt; residual Ni or Co dep.; Pt placers; hydrothermal asbestos and talc dep.; Ti-magnetite deposits
11. Areas of medium-acid, intermediate and subalkaline intrusives in geosynclines and geanticlines of volcanogenic type	calc-silicate skarns with magnetite deposits
12. Areas of acid intrusive rocks in eugeosynclines	W and W—Mo deposits of vein and stockwork types at endo- and exocontacts
13. Areas of biotite granite (deep facies) in flyschoid miogeosynclines	pegmatites with rare earths and rock crystals at endo- and exocontacts
14. Areas of alaskites in flyschoid miogeosynclines	apical greisens with Sn, Sn—W—Li deposits; greisens and albitized zones inside massifs with molybdenite, wolframite, cassiterite and beryl deposits
15. Contacts of granitoids with limestones and terrigenous rocks	skarns with Fe deposits, occasionally with B
16. Sedimentary rocks with small intrusions	Au deposits, often associated with dykes of intermediate rocks; cassiterite dep. with sulphides or silicates; Sb, Hg dep., in places telethermal Pb—Zn dep.; sedimentary bauxite and phosphate deposits.
17. Areas with small hypabyssal intrusions in geanticlines of volcanogenic type	deposits associated with secondary quartzites: Cu, Mo, polymetallic, Au, etc.
18. Areas with small hypabyssal intrusions in geanticlines of terrigenous carbonate type	some polymetallic and Au deposits; some sedimentary Fe-ore, bauxite, Mn-ore, phosphate and other deposits
19. Foredeeps and intermontane depressions of terrigenous type	Cu-sandstones; some bauxite, oil, gas, bituminous coal, salt and sulphur deposits
20. Foredeeps and intermontane depressions of volcanogenic type	deposits of young Au-formation; realgar-auripigment dep.; Hg dep.; deposits of sulphur and sassoline of volc. origin

TABLE 36

Favourable geological conditions for mineralization in block-fault zones

21. Areas of small ultrabasic and alkaline intrusions (originally subvolcanic bodies?)	carbonatites with Nb, Ce, etc.
22. Areas of alkaline rocks of the granitoid formation	alkaline granites and foyaites with Zr-bearing albitites . in shields; post-orogenic nepheline syenites with pyrochlore + zircon-bearing albitites; postorogenic alkaline granites with Nb—Ta-bearing albitites
23. Areas of granitoids emplaced along tectonic lines and accompanied by their hypabyssal and effusive equivalents	hydrothermal Mo, Mo—W, and Sn deposits

TABLE 37

Favourable geological conditions for mineralization in areas subject to strong weathering

24. Weathering crusts and eluvial placers	Fe-laterites; Mn-dep. of gossan type; residual bauxites silicate Ni and Co ores; kaolins; amorphous magnesites placers of tin, scheelite, wolframite and gold
25. Alluvial and marine placers (including fossil and metamorphosed types)	deposits of Au, Pt, cassiterite, diamonds, W, V, Ti and Zr minerals, and rare earths

PROSPECTING FOR MINERAL DEPOSITS

The term "prospecting" covers geological, geochemical and geophysical fieldwork plus the complementary laboratory studies directed to the discovery of workable mineral concentrations. The question of "what to look for" is discussed in the chapter on industrial deposit types, that of "where to look" is answered in the chapter on prospecting criteria, and the chapter dealing with prospecting methods gives instruction on "how to look". Depending on the character of local natural factors, the phase of work and the purpose of prospecting, different field and laboratory methods are used to discover direct indications of mineralization. These indications are followed by prospecting-exploratory workings and, in turn, by exploration of the deposit.

CRITERIA FOR ORE PROSPECTING

In the search for mineral deposits it is impossible to examine in detail every square km of the area or country by, for example, drilling. This would be too expensive, time-consuming and in most cases pointless. An area where the required mineral resources can be expected to occur is therefore delimited using prospecting criteria, that is, geological features which directly or indirectly suggest the presence of a given deposit. These criteria should be distinguished from indications of mineralization – prospects – which directly show the presence of ore (ore fragments, outcrops of mineral deposits). In some cases the criteria (e.g. hydrogeochemical) are equivalent to the indications.

Under prospecting criteria are summarized practical conclusions on stratigraphy, the petrography of sedimentary, igneous and metamorphic rocks, tectonics, structural geology and the facies concept, geochemistry, geomorphology, hydrogeology, geophysics, history and mining geology. During geological mapping it is possible to employ all these geological and auxiliary sciences plus prospecting experience in a rational, systematic and scientific search for mineral deposits. The criteria are subdivided according to the particular scientific branches.

1. STRATIGRAPHICAL CRITERIA

If a mineral material is known to occur in a certain palaeontologically or otherwise defined stratigraphic horizon, the first task of prospecting is to determine the surface occurrence and extent of this horizon by detailed mapping. This does not imply, however, that a deposit will always be found, because the horizon may be barren in places. The outcroping horizon must always be followed along the strike (to locate the beginning of the deposit) and only then after careful consideration of all conditions, down the dip to determine the position of concealed ore.

Stratigraphical criteria are important in the search for sedimentary deposits and hypogene deposits that are associated with beds of lithologically favourable sediments (see below).

Throughout the world, deposits of coal, sedimentary copper ore, uranium, lead and zinc, pyrite, sulphur, phosphates and bauxite, sedimentary iron and manganese ores, placers, clays, carbonates, vanadium and salts are restricted to several definite stratigraphical horizons (Fig. 1). In general, sedimentary deposits originate in periods of waning orogenic movements and at the beginning of transgressions. Geosynclinal coal deposits predominated in the Palaeozoic, transitional deposit types have come to the fore since the Mesozoic, and platform deposits were dominant in the Cenozoic. The Precambrian Fe-quartzites formed in oceans of lower salinity and a higher CO_2, that is with a lower pH, than modern oceans. As a result, trivalent iron migrated farther from the shore and was deposited with siliceous rocks. From the Precambrian to the Devonian, submarine exhalative iron ores were formed, which are unknown in younger formations. Bog ores have increased in importance since the Silurian and lateritic iron ores since the Jurassic. Industrial deposits of micas and of regional-metamorphic graphite occur only in Precambrian rocks. Typically tin, tungsten, mercury and antimony deposits are associated with Mesozoic and Cenozoic intrusions. In most cases, however, the age of mother intrusions (often uncertain) is not very important in prospecting. Prospecting for sedimentary deposits in a given area demands knowledge not only of world-wide stratigraphical criteria of the highest order, but also of local stratigraphical and mineralization peculiarities.

---

1. Survey of periods in which sedimentary deposits were formed (according to Strakhov) 1 — segments corresponding to 20 million years in the Mesozoic and Cenozoic, to 200 million years in the Palaeozoic and to 2,000 million years in the Precambrian; 2 — curve showing the fluctuation of coal formation with orogenic activity. A — Variscan orogeny in West Europe and the U.S.A.; B, C — Variscan orogeny in the U.S.S.R. and China; D — Cimmerian orogeny in the U.S.S.R. and China; E — Early Cretaceous orogeny in the north-eastern U.S.S.R.; F — Laramide orogeny in North America; 3 — segment corresponding to 1,000 milliard tons of coal; 4 — segment corresponding to 5 % of the world oil production in 1947 from reservoir rocks of a given formation (from Lalicker, 1949); 5 — segment corresponding to 10 milliard tons of iron ore reserves; 6 — Fe quartzites; 7 — marine hydrogoethite-chamosite-siderite ores; 8 — continental (residual and lacustrine Fe ores); 9 — siderites of paralic basins; 10 — submarine exhalative Fe ores; 11 — number of major Mn deposits; 12 — number of major bauxite deposits; 13 — sedimentary Cu ores; 14 — sedimentary Pb—Zn ores; 15 — relative intensity of halogenesis; 16 — proportion of world phosphorite reserves in individual formations.

59

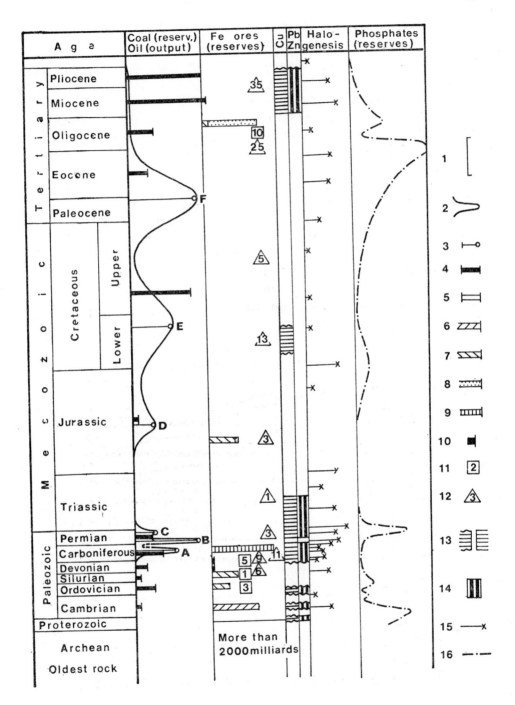

2. LITHOLOGICAL CRITERIA

The composition of sedimentary ores is always closely connected genetically with the lithology of the wall rocks. The potential occurrence of an ore can be inferred from the lithological character of the surrounding sediments. Figures 2 and 3 show the relationships between some deposit types and the rock facies. The study of the palaeoclimatic conditions of an area or formation is also important, since the association of deposits formed under an arid climate differs appreciable from that produced under humid conditions. Prospecting for manganese illustrates clearly the significance of palaeogeographical and facies conditions for the development of sedimentary deposits. Sedimentary manganese ores can originate from the destruction of rocks in the neighbourhood of a lagoon or shelf even if their Mn content is only one order higher than the clarke content. Sedimentary deposits are most commonly concentrated just above the transgression surface or just below the regression surface. Deposits of transitional types (which can be deduced from Figs. 2 and 3) are very rare (e.g. the transition bauxite—Fe ore—Mn ore is seldom found). When a deposit has been located, prospecting in its proximity should be carried out either in a higher or lower stratigraphical horizon.

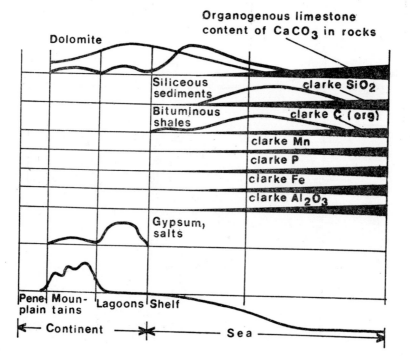

2. Facies profiles showing the formation of sediments in the arid zone (according to Strakhov, 1962).

Some sedimentary deposits are invariably over- or underlain by particular rocks. Thus, the oolitic manganese ores of the Chiaturi type are underlain by siliceous sediments such as marly sandstones, spongolites, or jaspers with radiolarians. In addition to positive criteria—based on a knowledge of the environment of their origin—there are also negative prospecting criteria, such as thick sandstone complexes for coal deposits. Evidence of changing denudation trends (e.g. polymictic sandstones) implies that a coal seam, if it ever existed, has been destroyed by erosion.

In prospecting for endogenic ore deposits, magmatogenic criteria (see below) and the lithology of rocks permeated by ore-bearing mother intrusions are of primary importance. Three properties of rocks make them favourable for mineralization: 1. permeability (sandstones, conglomerates, porous lavas, tectonically disturbed rocks); 2. chemical reactivity (on reaction with hydrothermal solutions, precipitation of ore minerals is induced; e.g. carbonate rocks); 3. brittleness (igneous rocks, quartzites and dolomites contrast with limestones and shales—Fig. 4). The presence of carbonate rocks close to acid and intermediate intrusions is extremely favourable for the genesis of polymetallic, copper, tin, tungsten, molybdenum, antimony and other deposits of contact-metasomatic (skarn) and hydrothermal-metasomatic types (e.g. crystalline magnesite). At the external contact of autometamorphosed granitoids

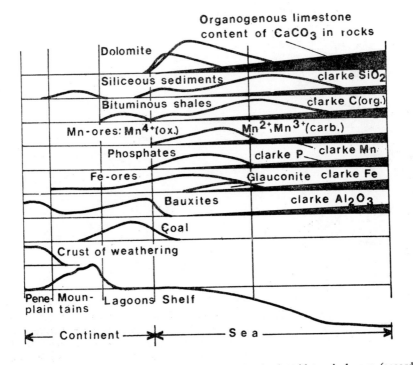

3. Facies profiles showing the formation of sediments in the humid tropical zone (according to Strakhov, 1962).

with layers of basic rocks, a maximum concentration of Sn (in addition to Cu and Zn) occurs. Lithological criteria are of special importance when certain beds of a carbonate complex are more richly mineralized because of their particular chemistry or permeability. Carbonate rocks contain not only ores but also non-metallic minerals, such as asbestos, talc, phlogopite, and emery. Pegmatites and Sn, Cu and Au deposits are rarely associated with pure limestones, but Sn deposits do occur in impure limestones, calc-silicate hornfels and skarns. Remarkably, a pure carbonate complex is not usually so intensely mineralized as a silicified rock, in which carbonate is almost in equilibrium with quartz. Sedimentary deposits of the "red beds" type are often associated with sandstones having a calcareous cement.

Graphite beds often form interlayers in paracrystalline rocks, which contain a graphite admixture around the deposit. This is probably due to the premetamorphic composition of the black shale facies sediments, which contain a variable proportion

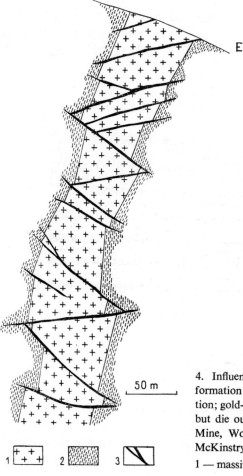

4. Influence of physical properties of rocks on the formation of fractures and localization of mineralization; gold-bearing veins fill fractures in brittle monzonite, but die out in the adjoining plastic slate. Morning Star Mine, Woods Point, Victoria, Australia (according to McKinstry, 1948).

1 — massive monzonite; 2 — slate; 3 — ore veins.

of organic matter in individual horizons. Deposits of crystalline magnesite of metasomatic origin are also common in graphitic slates and phyllites.

In conclusion it can be said that most mineral deposits are situated in rocks with alkaline feldspars (acid effusives and intrusives, arkoses, feldspar quartzites) or Mg and Ca-bearing carbonate rocks and that they are uncommon in clayey shales, phyllites and mica-schists (Fig. 5). Prospecting experience has shown that carbonate rocks do not contain major Sn, Cu and Au deposits (in contrast to Pb, Zn); that telethermal Cu deposits of the "red beds" type, insofar as they are not considered as sedimentary, are associated with sandstones, in contrast to telethermal Pb–Zn deposits which prefer carbonate rocks; that hydrothermal U^{4+} ores occur in rocks containing minerals with Fe^{2+} (diabase, chloritic and hornblende schists), and that Pb–Zn ores are common in quartzites and Au ores in chloritic rocks.

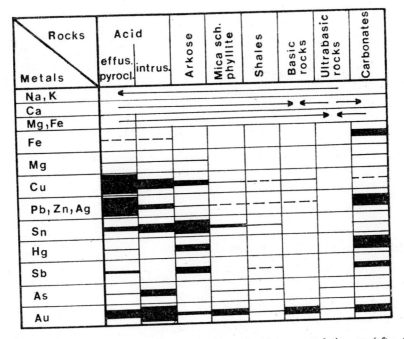

5. Rocks favourable for the deposition of hydrothermal and pneumatolytic ores (after Ozerov, 1949).

The lithology of rocks is of special importance for deposits of the placer type. In depressions of river beds, produced by the selective erosion of lithologically variable rocks, the natural washing of the alluvium results in the accumulation of heavy minerals and the formation of placers (Fig. 6). A clayey bed is impenetrable to heavy particles (for example of gold), which would otherwise move down to the valley floor, and so a *"false floor"* originates (Fig. 7). This can be gold-bearing as is the bedrock, or on its own if erosion reached a primary gold deposit in the later phase of aggradation. In this case the clay bed functions as a screen preventing the useful minerals from moving downwards. The opposite case occurs in the ore aureole of an intrusion, when the upward movement of hydrothermal solutions is obstructed by a bed of clayey shale or phyllite and their ore content is deposited beneath this barrier. Metasomatic deposits originate in this way, especially in carbonate rocks.

Similar impervious beds covering the reservoir rocks cause the formation of oil deposits in anticlinal elevations (see below).

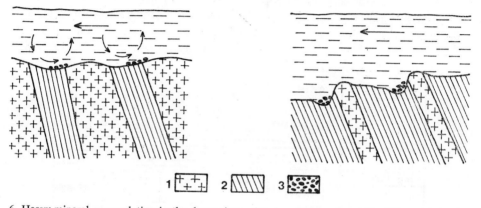

6. Heavy-mineral accumulation in the depressions of a river bed, produced by differential erosion. 1 — resistant rock; 2 — soft rock; 3 — heavy minerals.

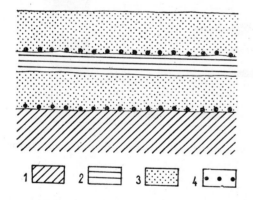

7. "False" and true bedrock floor with concentrations of heavy minerals. 1 — rock; 2 — clay interlayer; 3 — sand and gravel; 4 — heavy minerals.

3. STRUCTURAL CRITERIA

The structure of the earth's crust is often a controlling factor in the formation of ore deposits. Numerous types of metallic and non-metallic deposits of endogenic origin, for example, are confined to folded areas or, more precisely, to the magmatic bodies

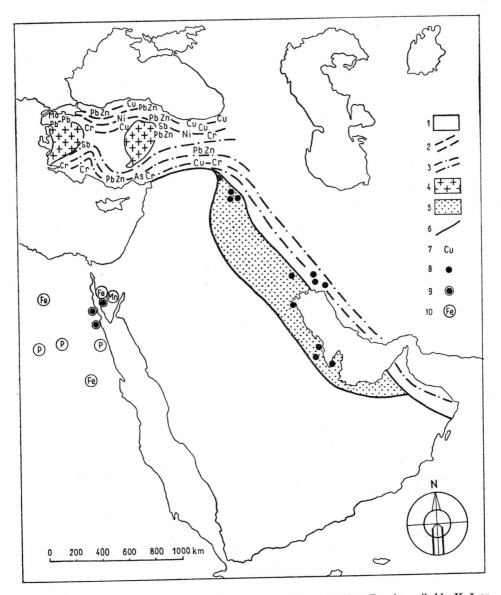

8. Relationship between ore deposits and the crustal structure in the Near East (compiled by Kužvar from published data). 1 — African shield (Arabian Peninsula, northern Africa); 2 — Pontide fold mountain range of Alpine age; 3 — Tauride fold mountain range of Alpine age; 4 — Variscan or older massifs; 5 — Alpine foredeep; 6 — approximate boundaries of major structural units; 7 — endogenic deposits related to intrusions in fold mountains; 8 — oil deposits associated with the transitional area (in part accompanying salt plugs); 9 — oil deposits occurring on platforms; 10 — sedimentary deposits on platform (Cretaceous—Tertiary; P — phosphates).

intruded into them. These deposits usually originate in the last orogenic cycle within an area. In contrast, coal, oil, carbonates, manganese, bauxite and some phosphate deposits are also found in transitional areas characterized by slight folding and cupolas (Fig. 8).

The ore deposits are concentrated into metallogenic zones, the course of which is consistent with that of the first order tectonic lines. The origin of metallogenic zones can among others be explained in terms of the new concept of global tectonics. According to this, there are six large lithospheric plates on the earth—the American, Eurasian (including pre-Alpine Europe), Pacific, Australian, East Atlantic (including Africa) and Antarctic—plus the small Nazca plate (Fig. 9). The metalogenic zones have developed along the margins of these plates, which may be accreting, transform (along horizontal displacements perpendicular to the margins where ultramafites are often intruded) or consuming. Other metallogenic provinces occur in the oceanic or continental segments of the plates, or along the trailing continental margins. The study of global tectonics is therefore of primary importance for determining the principal structural criteria that affect prospecting.

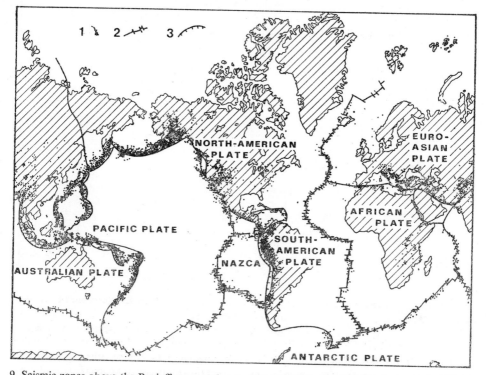

9. Seismic zones above the Benioff zone and principal lithospheric plates (according to Bullard in The Ocean, Scientific American 1969). 1 — epicentres of earthquakes; 2 — mid-oceanic rises, sites of oceanic crust accretion; 3 — deep-sea trenches in which the oceanic crust is subducted (subduction zones).

The area of the lithospheric plates remains constant, since the oceanic crust which accretes in the rifts or mid-ocean rises (e.g. the Mid-Atlantic Ridge) is consumed in submarine trenches (e.g. the Peru-Chilean Trench) by subduction of the oceanic crust beneath the continent with the formation of an Andean-type fold mountain range (Fig. 10). The oceanic crust can also be consumed in submarine trenches remote from any landmass (e.g. the Aleutian and Izu-Bonin Trenches). In this case an island arc is formed above the subduction zone and a marine basin of the Japan Sea type develops on the concave side of the trench, between the arc and the continent. A submarine trench does not develop where the oceanic crust is not thrust beneath a landmass (e.g. on both sides of the Atlantic, except for the Caribbean region). The four types of continental margins and the small disappearing oceanic basins between continents (the Tethydian zone resulting from the collision of the Eurasian and Gondwana plates) represent the five types of sedimentary or geosynclinal areas (Table 38). Sedimentation also occurs elsewhere apart from along the continental margins but only at a substantially lower rate; because of the greater depth of water organogenic calcareous ooze dissolves before it sinks to ocean floor. As a result of subduction, the pre-Mesozoic oceanic crust is probably not preserved anywhere on the ocean floor. Glikson (1974), however, considers the ultramafic and mafic rocks of the greenstone belts of West Australia, the Transvaal and Rhodesia to be the relics of a fossil oceanic crust more than two milliard years old.

When the ocean floor spreading ceases or the position of the subduction zone changes, the type of geosyncline is also altered (from the island arc and Japan Sea types into the Atlantic type). If in the Atlantic geosyncline the spreading or migrating lithosphere starts to become subducted, an island arc develops. The sediments of the continental rise will be intruded by calc-alkaline volcanics and a geosyncline of the island-arc type will develop, probably with a geosyncline of the Japan Sea type. If the spreading oceanic crust is thrust beneath the continental crust, an Atlantic type geosyncline will change into an Andean geosyncline. This model of the continental margins with five geosynclinal types has probably been valid since the Earth was differentiated into the core, mantle and crust, that is 2.5 to 3.5 milliard years ago. Only within formations of this period is it reasonable to seek, for example, the fossil analogues of modern island arcs or fold mountain ranges above the subduction zones and their metallic suites, especially porphyry copper ore deposits. The determination of metallogenic zones along the margins of shields would be of even greater significance, as the richest ore deposits are of Archean age. Their share in ore reserves led some scientists to believe that only the Archean epoch was ore productive and that all younger deposits are regenerations of, and chemically connected with, deposits from that epoch (Routhier et al., 1973). In seeking ancient shields and their metallogenic zones it is necessary to take into consideration, among other factors, the different properties of the marine sedimentary environment (pH, Eh, the lower salinity and the very small amount of organic matter) and the atmosphere, the different composition of terrestrial rocks, and the effect of submarine volcanics.

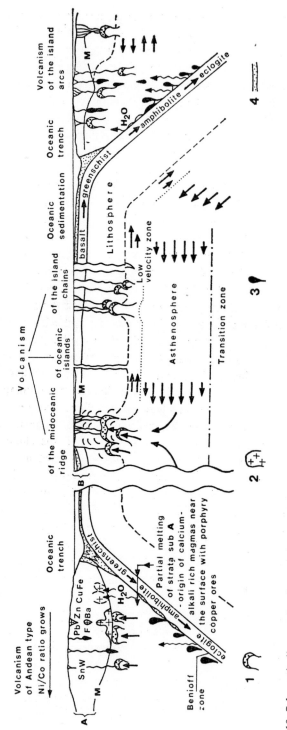

10. Schematic diagram showing petrochemical and metallogenetic processes in terms of plate tectonics (based on Green, 1972; Guild, 1972; Petraschek, 1973; Wright et al., 1973). 1 — segregation of basaltic magma above peridotite diapirs; 2 — rhyolite and dacite intrusions; 3 — dacite and andesite magmas; 4 — deep-sea sediments; M — Moho discontinuity; A — continental crust; B — oceanic crust: oceanic sediments with Mn concretions (±Cu, Ni, Co), basalt and gabbro.

TABLE 38

Characterization of geosynclines (according to Mitchell and Reading, 1969, and Mitchell and Garson, 1972)

Type of geosyncline	Atlantic		Andean	Island arc		Sea of Japan	Mediterranean	
	miogeosyncline: shelf and coastal plain	eugeosyncline: continental slope and rise, abyssal plain	arcuate fold mountain range on continental plain	chain of volcanic islands (eugeosyncline)	trench	continental margin with a limited basin	migeosyncline	eugeosyncline
recent examples	eastern coast of North and South America, eastern and western coasts of Africa, coast of India		Andes in Peru and Chile, Barisan Mts. (Sumatra), south-western part of the U.S.A.	New Hebrides, Solomon Is., Tongo, Aleutian Is., Lesser Antiles, Honshu, New Caledonia; inactive recent Benioff zones: N. Ireland, Bougainville (Pangura porphyry Cu-ore deposit), Philippines (Marcopper porphyry Cu-ore deposit)		Sea of Japan, Okhotsk Sea, Andaman Sea, Aleutian basin	sea between two continents or continental sea: Mediterranean Sea, Black Sea	
types of deposits	massive pyrites (with pillow-lavas); manganese concretions; oil from Tertiary sediments in depressions of the shelf; deep sea salt deposits (Gulf of Mexico, North Sea, area of hundreds thousand km², Pautot et al, 1970)		porphyry Cu-ores; veins with Cu, Pb, Zn and with Sn, U, Au farther from the trench (Bolivia, SE Asia)	volcanic-sedimentary Cu-pyrites	stratiform exhalative sulphides	submarine stratiform polymetallic dep. of Kuroko type, associated with tholeiites	few hypogenic deposits	
examples of fossil continental margins			NW part of Zagros (Iran) with porphyry Cu-ores (Sar Chesmeh); granite plutons of Hindu Kush and Karakorum; Krušné hory (ČSSR); Great Bear Lake — Ag, Ni, Co-arsenides (Badham, 1974); Norway (1.7 milliard years, Torske, 1974)	NW part of Borneo - Mamut porphyry Cu-ore dep.; Vancouver - Hardy porphyry Cu-ore dep.; basalt - Barberton (Rep. South Africa 3.2 to 3.5 milliard years)				

In addition, ancient metallogenic zones may be buried by later rocks, through which their regenerated ore aureole is barely perceptible.

The spreading of the ocean floors from the mid-ocean rises and their consumption in the trenches of the Peru-Chile type is very important in the genesis of ore deposits. The moving oceanic crust functions as a conveyor belt which continuously supplies the "melting pot" of the Benioff zone, beneath marginal fold mountains of the Andean type, with pelagic sediments enriched in metals from the upper mantle. These concentrate to form submarine-exhalative deposits close to mid-oceanic zones of volcanic activity (Wright et al., 1973). The metals contained in sediments are mobilized in the asthenosphere and ascend with the magmas of anatectic origin to form ore deposits. These are nearer the contact between the continent and the submarine trench, where the "conveyor belt" disappears, and are older where their mobilization demands a lower temperature and a smaller depth (Fig. 10). The centres of epizonal plutonism and volcanism move 0.6 − 1 mm per annum towards the centre of the continent, so that in the Andes the oldest ore deposits occur at the subduction zone (Lower Jurassic) and the youngest in the eastern part (Oligocene). This pattern is, unfortunately, greatly complicated by the later Mio-Pliocene mineralization phase (Clark et al., 1974). According to Corliss (1974), the melting temperature of sediments and the basalt in the ocean floor in the subduction zone [along the submarine trench] is decreased by alkalis from the sea water, which permeates not only the sediments but also the basalt to a depth of several kilometres. Corliss believes this explains why basic metals occur nearer the subduction zone than metals associated with acid postorogenic differentiates (Sn, W, M, Ba, Nb), which are formed along the contact of the fold mountains with the craton (Garson 1974). According to Sillitoe (1974), tin is liberated through partial melting of the oceanic lithosphere at a depth of 300 km, that is, at the extreme boundary where the subducted plate is consumed in the asthenosphere. If this is so, it would also account, for example, for the longitudinal zoning of the circum-Pacific metallogenic zone. The present Benioff zones have been in existence for about ten million years, and older deposits of porphyry copper ores are associated with the palaeo-Benioff zones (i.e. with the andesites containing 150 ppm Cu, which form above the Benioff zones).

A wide spectrum of ore deposits can develop where the basaltic magma, ascending from the mantle (above the Benioff zone) and the oceanic crust is enriched by anatectic magma from the continental crust (Petraschek in Tarling, Runcorn, 1973). In the latter the ore elements have been pre-concentrated by weathering, metamorphic, sedimentary, differentiation and lateral-segregation processes (e.g. in the Andean segment of the circum-Pacific metallogenic zone).

Sawkins (1974) associates stratiform polymetallic deposits (Fe, Cu, Zn, Ni, Co sulphides) of the Cyprus type with potassium-deficient basalts forming the upper parts of ophiolite complexes along oceanic rises; concentrations of chromite and platinum in peridotites result from magmatic segregation in the mantle (Thayer, 1974).

In Guild's opinion (1972), the post-Eocene deposits, for example, in Japan, Indonesia, New Guinea, Bougainville, Fiji and New Zealand, originated above the Benioff zone, which in these places lies at a depth of 100—200 km. He also believes that the scarcity of post-Eocene deposits in the Alpine-Himalayan fold mountain range is due to the fact that the range was formed by folding of the Tethys geosyncline, without subduction, during the collision of Eurasia with the fragments of Gondwana. Since for most of this mountain range's length the oceanic crust was unavailable, subduction did not occur (the continental crust cannot be consumed by the mantle because of its lower bulk density). As a result, the asthenosphere was not enriched by ore elements from the oceanic crust and ocean floor sediments, the lower continental crustal layer did not undergo anatexis and endogenic deposits were not formed. The only exception to this is the belt of endogenic deposits between the Balkans and Baluchistan, which includes the prominent province of porphyry copper ores in Iran. It can be postulated that the oceanic crust was present in this segment.

The examples mentioned above indicate that ore deposits often originate above subduction zones. In contrast, the accreting edges of plates are poor in ore deposits. An example to this are the hot ore-bearing brines of the Red Sea, which is thought to be an ocean *in statu nascendi*. According to this hypothesis, oceanic crust is forming along the whole length of the Red Sea in its deepest part. The landmass on either side of the sea, however, is so close that the brines can be enriched in ore elements and salts leached from the coastal rock complexes (Tooms, 1970); this opinion seems to be corroborated by the find of Cu deposits above the gypsum beds and brines with 23 ppm Cu in the north of the Ethiopian part of the East African Rift. The brines of the Salton Sea at the Gulf of California are analogous to the ore-bearing brines of the Red Sea (Fig. 9). The sediments of the East Pacific Rise are enriched in Fe, Mn, Cu, Cr, Ni and Pb, which originated at depth in the upper mantle and not on the continent.

The most important ore deposits associated with deep-seated assimilation and differentiation above the subduction zone in a deep-sea trench of the Peru-Chilean type, are those of porphyry copper. Deposits in northern Chile, southern Peru, Colombia, south-western U.S.A. and Panama are of this type. Other deposits such as those in Bougainville (Pangura deposit) or the Phillipines (Marcopper deposit) are related to trenches along the arc islands. Sillitoe (1972) anticipated the occurrence of similar deposits in Japan, New Zealand, the Aleutian Islands, Sumatra, Java, Banda, the Celebes, the New Hebrides, the Lesser Antilles, Kamchatka, Burma, Thailand, Turkey, Greece and Afghanistan. In many of these countries porphyry copper ore deposits have in fact been discovered in recent years.

The origin of diamond-bearing kimberlites is also connected with the subduction zones along the continent (Andean type). Sharp (1974) has suggested that a basaltic oceanic crust containing carbonate sediments submerged into the mantle by subduction. The kimberlite pipes extend parallel to the Andean type fold mountains far from the oceanic trench, since they originate at depths above 200 km, at a pressure of 30−56 kb and a temperature of 1150−1327 °C. The South African kimberlites, above the subduction zone beneath the Capides, are located far to the north of the Cape Mts. A similar situation probably exists in the Appalachian Mts. and Siberia.

The movements of lithospheric plates and the processes associated with them do not only induce the formation of new metallogenic zones but can also divide existing ones (Petraschek in Tarling, Runcorn, 1973). In following and predicting their course, the movements of continents must be taken into consideration. The Sn−W mineralization in Nigeria (Jos Plateau), for example, has its counterpart in the Brazilian tungsten deposits, and the auriferous conglomerates of the Ceara deposit in northern Brazil correspond to similar conglomerates at Tarkwa in Ghana. Tantalum-bearing pegmatites in Gabon and in Minas Gerais belong to the same province as do the Zr-bearing pegmatites in Travancore (India) and Greenbush (Australia) and placers with Zr minerals in Bushveld (Rep. South Africa). The gold-bearing Kolar mineralization in Mysore (India) is analogous to that in Kalgoorlie (Australia), the Fe deposits in Singbum (India) correspond to similar deposits in West Australia, and the graphite and lanthanide deposits in Ceylon are similar to those on the western coast of Australia. The carbonatites in India and Australia are associated with rift structures that were originally linked (Crawford, 1970). Sulphidic Pb, Zn, Fe and Cu ores in Ireland (Abbeytown, Trynagh, Gortdrum, Riofinex) have their counterparts on the opposite coast of the Atlantic in Nova Scotia (Russell, 1968). When the American continent began to move away from Europe in the Late Palaeozoic, Pb, Zn, Cu, Mo, Ag and Au ore deposits were formed in the Oslo area (Norway) at the intersection of deep faults of Caledonian or Skagerrak (NNE−SSW) and Fennoscandian (NW−SE) trends, which are evidence of the disintegration of Laurasia (Vokes, 1973).

The examples mentioned above show how important the study of the margins of lithospheric plates and of the accompanying geosynclines is for determining the course of first-order endogenous metallogenic zones. Recognition of continental movements is essential in tracing metallogenic zones that were originally continuous but subsequently have been torn apart.

The geosynclinal regime and its opposite−the platform regime−also play an important role in the formation of sedimentary bauxite, iron and manganese deposits. In earlier epochs the formation of these deposits was confined to geosynclines but the platforms gradually became the principal site of their formation. Geosynclinal deposits usually have larger ore reserves than platform deposits; Strakhov estimated geosynclinal Fe-ore reserves at 3000 milliard tons compared to 2 milliard tons for platform deposits.

A metallogenic zone generally embraces the whole fold mountain range (e.g. the Alpine-Carpathian metallogenic zone). A lower structural unit is a metallo-genic area, which is the highest monoparagenetic unit of platforms. This in turn is divided into metallogenic provinces characterized by mineralization of similar type and often of the same age in areas of similar geological structure. Metallogenic provinces are made up of metallogenic districts, which can be further subdivided into partial metallogenic districts. Many authors have developed their own classifications, but that introduced in this chapter is based on the system devised by Koutek (1964).

An ore belt associated with a fault may be a separate metallogenic province or part thereof (a district or partial district). The fault acted as a channel of ascent for hydrothermal solutions and the deposits themselves occur along the minor conjugate faults in the neighbourhood. This can be illustrated by the 200-km long fault of the main Caucasian ridge. Arsenopyrite, molybdenite, wolframite, antimonite and other deposits supplied by this feeding channel occur in parallel faults within 0.5 to 1.5 kilometre.

Metallogenic zones are longer and broader than the ore belts and pass gradually into the neighbouring rock formations. Ore belts are generally sharply delimited and their width ranges from 2.5 to 5 per cent of their length.

A special type of ore belts is associated with presumed deep faults, which often occur in the basement of fold mountains. This group includes, for example, the belt of silver deposits which extends for over 2000 km in Mexico and appears to continue south-eastwards into Peru and Bolivia. The belt cuts across all fold mountains and is unrelated to any known surface structures. The 500 km long belt of platinum deposits in Southeast Africa, and the Turkestan-Alaian belt of mercury and antimony deposits (U.S.S.R.) are similar.

Metallogenic provinces, districts and ore belts consist of ore fields, which usually comprise several ore deposits, but can consist of a single ore body. Their position in the higher metallogenic units is determined by tectonic factors. Of particular importance are places where the trends of anticlinal structures and tectonic zones change, axial zones of major anticlines, intersections of faults with anticlines or faults, periclinal closures of anticlines and anticlinoria, saddles of anticlines or the intersections of faults with favourable horizons. Ore fields can also be related to granitoid intrusions, small strongly differentiated intrusions, basic and ultrabasic rocks, narrow sunken zones, xenoliths of mantle rocks in intrusive bodies, and intersections of anticlinal zones with dykes.

In studying a new region, the prospector proceeds from the structures of fold zones (metallogenic zones) to the definition of metallogenic areas and the provinces they embrace. After establishing the ore districts, partial districts and ore fields he then searches for the ore deposits themselves. Finally the ore bodies are localized by studying the structures of the deposits and the ore shoots and nests are delimited.

Structural units of the lowest order of magnitude but of great importance for mineralization are faults, fissures (open) and joints (closed) and their sets, folds, cupolas and their combinations.

Structures are differentiated into pre-, syn- and post-mineralization types and are all the object of study during exploration. The first group of structures function as channels of supply and distribution of ore solutions and can be mineralized. The second group is relatively rare and the third disturb the deposits. The group of pre-mineralization geological structures includes regional faults along which ore belts develop (see above), folds, fissures, joints and zones of shattering. It is not out of place to discuss briefly the conditions of their origin.

Theoretical and experimental rock mechanics shows that under a high geostatic pressure even brittle rocks undergo plastic deformation without the formation of open fissures. Under these conditions even rocks such as marble and sandstone "flow", that is change their shape without rupture. The strength and plasticity of rocks show a linear increase with geostatic pressure. Under normal pressure, halite is brittle but it becomes plastic at a hydrostatic pressure of 2883.1 MPa ($=$ $= 29,400$ kp/cm^2) and a tension of 50 MPa ($= 510$ kp/cm^2). A cylindrical limestone sample stretches 1.76 times without rupture at a hydrostatic pressure of 2,745.8 MPa ($= 28,000$ kp/cm^2) and under tensile stress. Uniaxial pressure produces shear planes in marble oblique to the direction of pressure. The angle between these planes and the direction of pressure in steel is slightly less than 45°, and somewhat above 45° under tension. In a stressed homogeneous rock, open fissures are formed parallel to the direction of stress (Fig. 11) in addition to complementary closed shear joints. These are important for mineralization if ore solutions, particularly hydro-thermal ones, can replace the neighbouring rocks, or when mineralization is preceded by movements along their uneven surfaces, producing openings (Fig. 12). Fissures are often arranged en echelon and mineralized. Dislocation systems do not always develop with the same intensity. Their origin, position and orientation towards the axes of the strain ellipsoid are also affected by geostatic pressure, the temperature, rock anisotropy and time, among other factors. Since an increase in geostatic pressure is accompanied by an increase in plasticity, the rocks only undergo plastic deforma-tion, that is, they are disrupted by shear joints but not by open shear fissures. The increased temperature causes minerals to recrystallize and produces a change in their properties and those of the rocks. Mechanical anisotropy of the environment causes that angle α (see Fig. 11) changes as the joint passes through lithologically different rocks in a similar way to the index of refraction of a light beam in different mediums. This feature in rocks is related to their elasticity and brittleness which depends on stratification (the number of macroscopic planes of separation in 1 m^2 of rock), the character of fissures and the schistosity. From this point of view, rocks are divisible into brittle, plastic and highly plastic. Brittle rocks succumb readily to fracturing and their elastic limit is virtually equal to their ultimate strength (e.g. quartzite, effusives and some intrusive rocks). Plastic rocks are often bedded and liable to small-scale folding (e.g. argillaceous and sericitic shales) as are highly plastic rocks (e.g. wet argillaceous, sericitic and chloritic shales). The limit of elasticity and strength decreases under prolonged loading. No plastic deformation, however, occurs with a load below the critical load, even if it is operative for a very long time.

The tensile, bending and shear strengths of rocks are 7—50 times smaller than their compression strength. Sandstone, for example, with a compression strength of 147 MPa (= 1,500 kp/cm²) has a tensile strength of ca. 2.9 MPa (= 30 kp/cm²), bending strength equal to 9.7 MPa (= 100 kp/cm²) and a shear strength of 14.7 MPa (= = 150 kp/cm²). When these values are exceeded, the rock fractures or deforms plastically. Permanent plastic deformation caused by bending stress may produce, for example, a syncline (Fig. 13). The inner concave side of the bed undergoes compression resulting in intensive plastic folding up to crumpling, whilst the outer convex side is subject to tension which produces open fissures as a result of brittle deformation.

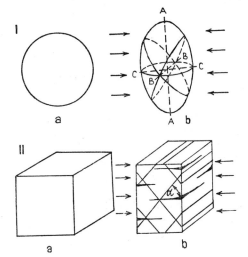

11. Scheme showing deformation of a spherical rock body by lateral pressure (from Smirnov, 1957.) Ia — original sphere; Ib — sphere deformed into a strain ellipsoid (A — axis of stretching, C — axis of compression; shear joints form along two circular sections which intersect on the B axis. Short open fractures are parallel to the direction of pressure), IIa — original cube; IIb — parallelepiped with long complementary shear joints and short fractures.

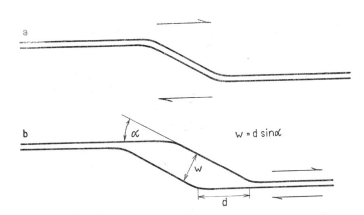

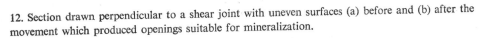

12. Section drawn perpendicular to a shear joint with uneven surfaces (a) before and (b) after the movement which produced openings suitable for mineralization.

During the folding of rocks whose beds are not smooth enough to glide along each other, cavities open in the anticlinal and synclinal bends (Fig. 14) and these may be later filled with ore. The ore solutions become concentrated in the anticlines and dispersed in the synclines. Statistical data show that most epigenetic deposits in folded areas occur in anticlines and cupolas. In the limbs of folds, especially if these are overturned, suitable openings for the deposition of ore minerals arise (Fig. 15a) as a result of crushing and small movements between rocks of different competency. Flexure folding also provides openings for mineralization (Fig. 15b).

V. I. Smirnov (1957) has divided mineralized structures into six major groups and twenty subgroups.

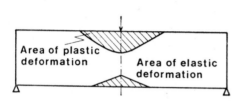

13. A prism supported at both ends bends under loading in the centre and plications (larger hatched area) and fractures (smaller hatched area) form simultaneously (after Nadai in Belevtsev, 1961).

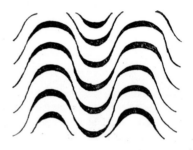

14. Formation of openings in anticlines and synclines during the folding of beds that do not glide easily over each other.

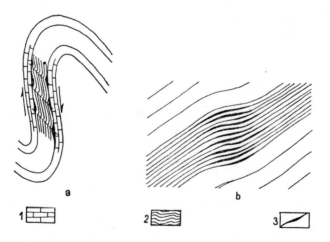

15. Origin of openings. a — during disharmonic folding of rocks showing different competency, b — at flexural bending: 1 — rocks yielding to folding; 2 — rocks yielding less to folding; 3 — openings.

I. Concordant structures of stratified rocks

1. *Saddle reefs of the Bendigo type* originate in anticlinal bends. On the Bendigo deposit itself, the post-folding movements followed the weakened bedding surfaces in the steep anticlinal limb; further movement produced moderately dipping faults. Mineralization took advantage of this tectonically deformed complex. The resulting form of saddle reefs is shown in Fig. 16. Among other places saddle veins are known in Krivoi Rog (Fe), the Altai Mts. (polymetallic veins), central Kazakhstan (polymetallic veins), Central Asia (Sb—Hg), Siberia (Au), Nova Scotia and Cariboo in Canada (Au).

0   30 m

16. Saddle reefs on Bendigo deposit, Australia (from Pabst in McKinstry, 1948).

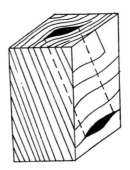

17. Elongated lens-shaped ore body in a flexure in steeply dipping beds (from Smirnov, 1957).

2. Lens-shaped bodies form in *flexures*, most often in shales and tuffs (Fig. 17). Deposits of this type include those at Hollinger in Canada (Au in quartz) and Arkansas in the U.S.A. (polymetallic veins, Fig. 18).

3. *In zones of interstratal gliding, plication* (Fig. 15a) and *crushing* (Fig. 19), mineralization is located in the fold limbs at the contact between two rocks of different competency. The polymetallic deposits in Transbaikalia, Kazakhstan (Uspenskoe — Cu, Dzhezkazgan — Cu), Central Asia and Jugoslavia are of this type.

4. *In favourable beds* (Fig. 20), mineral deposits form as a results of their chemical reactivity or permeability (see also Lithological criteria). Polymetallic deposits often occur in dolomites and Cu ores in sandstones (assuming these ores are epigenetic). Examples include the polymetallic deposits in Kazakhstan, Kirghizia and Abkhazia, the Mississippi valley (U.S.A.) and Mexico, the Cu-bearing sandstones in Dzhezkazgan (U.S.S.R.), Katanga and Zambia (Chambishi deposit), the impregnated arsenopyrite deposits of Central Asia, and the hematite deposits in Cumberland (U.K.).

5. Ore deposits can form *below impermeable beds* either as a result of lithological differences between them and underlying beds or because of a favourable structure (e.g. anticlinal).

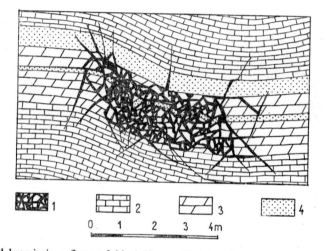

18. Mineralized breccia in a flexure fold, Arkansas, U.S.A. (after McKnight in Bateman, 1950). 1 — ore with dolomite fragments; 2 — limestone; 3 — dolomite; 4 — sandstone.

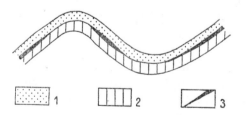

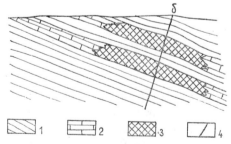

19. Lens-shaped ore bodies in the zones of interstratal gliding (accompanied by crushing). 1 — rock yielding to folding; 2 — rock yielding less to folding; 3 — ore (mineralized breccia).

20. Metasomatic ore bodies at the intersection of favourable rock with a fault (feeding channel). 1 — shale; 2 — limestone; 3 — ore; 4 — fault.

II. Ore bodies associated with regional faults (see also the paragraph on ore belts)

6. *Ore deposits in reverse faults* are, for example, the Mother Lode, California, U.S.A. and the mercury deposits in western Siberia, Central Asia, the U.S.A. and Spain. Deposits of this type are usually located in minor subsidiary faults and fractures rather than in regional faults themselves.

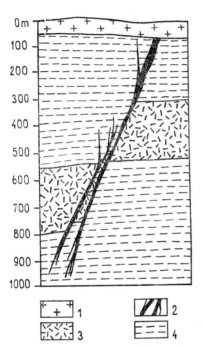

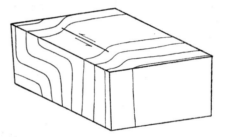

21. Vein of subvolcanic Au—Ag formation related to a normal fault, San Rafael, Mexico (from Lindgren, 1926). 1 — young andesite; 2 — ore vein; 3 — Miocene intrusive andesite; 4 — Jurassic shales and sandstones.

22. Formation of a fracture, later mineralized, in the extension of a flexure cutting folded rocks. Interstate Mine, Idaho, U.S.A. (from McKinstry—Svendsen in McKinstry, 1948).

7. *Ore bodies in normal and wrench faults* are relatively scarce, but where they occur they form veins or flat pipes. This group includes the gold deposits in Siberia and the Altai, the polymetallic veins of the Northern Caucasus, Trans-Caucasian region and the Harz Mts. (Germany), the auriferous veins of the subvolcanic formation in North America (Fig. 21), Morococha (Peru) and the Comstock Lode (Nevada, U.S.A.). Figure 22 shows a peculiar displacement in an intricately deformed flexure. Horizontal displacement produced a flexure in steeply dipping beds and a fault in moderately inclined or horizontal beds. The fault later became a locus of mineralization.

III. Ore bodies in tectonically stressed zones

8. *Ore bodies in open fractures* take the form of short irregular veins, usually joined into groups and are often branched. The bodies occur in zones under tensile stress, bends of the strata, transverse fractures of dykes (ladder veins) and in the cooling cracks of intrusions. Examples include the gold deposits in the Urals Mts. (Berezovsk), the chambered veins of polymetallic ore in the Mississippi valley, the Norseman deposit in West Australia (Fig. 23) and the slightly inclined cassiterite veins at Cínovec (Czechoslovakia, Fig. 24).

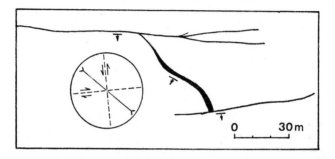

23. Mineralized fracture and one joint system (pressure acted in the direction of fracturing). Viking Mine, Norseman, West Australia (from McKinstry, 1948).

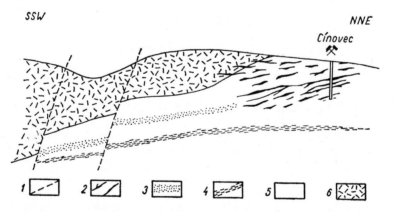

24. Contraction cracks mineralized by cassiterite and wolframite (from Tichý, Malásek in Čillík, Ogurčák, 1965). 1 — post-ore faults; 2 — veins with cassiterite and wolframite; 3 — greisen zone; 4 — zone with lithium micas; 5 — granite porphyry; 6 — quartz porphyry (Permian).

9. *Ore bodies in shear joints* of one system, along which only minor movement has occurred (Fig. 12). These often compose vein fields and are longer and more continuous than fracture veins. The auriferous quartz lodes in the Eagle deposit in Canada are of this type (Fig. 25).

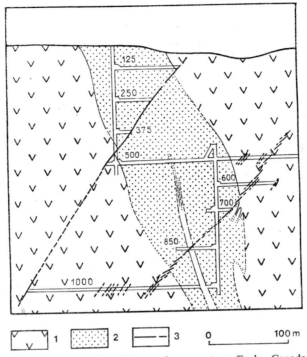

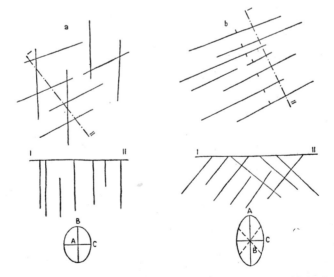

25. Veins of auriferous quartz in shear joints of one system. Eagle, Canada (from Norwood in Smirnov, 1957). 1 — granodiorite; 2 — greywacke; 3 — ore veins.

26. Ore veins in two systems of shear joints (from Smirnov, 1957). a — with horizontal A axis of dilatation (at depth), according to Sander—deformation (B_2); b — with vertical A axis of dilatation, according to Sander—deformation (B_1).

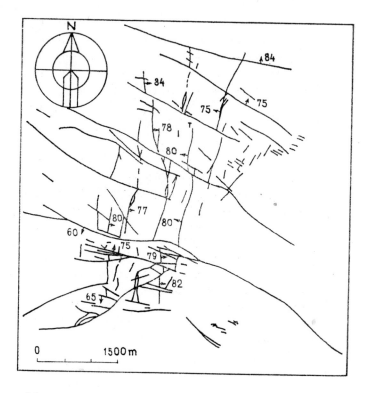

27. Veins located in two systems of shear joints (cf. with Fig. 26a). Real del Monte, Mexico (from Wisser in McKinstry, 1948).

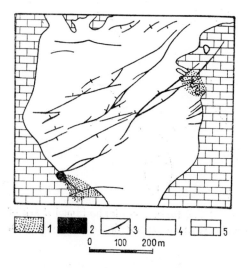

28. Copper veins located in two systems of shear joints (cf. with Fig. 26b). Morococha, Peru (after Smirnov, 1957). 1 : impregnated ore; 2 — mass ore; 3 — ore veins; 4 — quartz monzonite; 5 — limestone.

10. *Ore bodies in paired shear joints* which intersect approximately at right angles. If the joints have formed at depth, the pattern is revealed in a horizontal section (Fig. 26a) and if they have formed near the surface, it is visible in the vertical section. The Real del Monte in Mexico (Fig. 27) and the Cu-ore veins in the Morococha deposit in Peru (Fig. 28) are examples. If a stronger movement along a shear joint produces a fault, the thickness of the fault gouge is greater near the surface than at depth.

11. *Ore bodies in paired shear joints and fractures* are relatively rare. Some ore fields are associated with a larger number of joint and fracture sets, which have formed in several tectonic phases and been filled successively by ores of different types (e.g. the Freiberg Ag−Pb−Zn ore deposit in the German Dem. Rep.).

12. *Ore bodies in feather fractures,* which run diagonally to faults. Feather fractures can form if one of a paired set of joints develops into a fault (see the scheme in Fig. 29). The original fractures then become tension feather fractures with respect to it. The other set of paired shear joints can be suppressed from the very

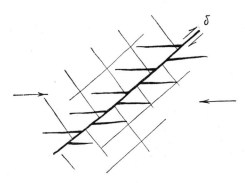

29. Formation of feather fractures along a fault (originally shear joints).

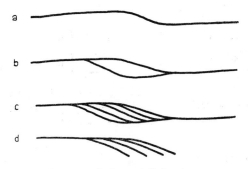

30. Development of a horse-tail structure.
a — cymoid curve; b — cymoid loop; c — multiple cymoid loop; d — horse tail structure (after McKinstry, 1948).

0   30 m

31. Veins arranged in a multiple cymoid loop (profile). Thin feather veins link the major veins. Pachuca, Mexico (after Thornburg in McKinstry, 1948).

84

beginning so that only the elements designated by heavy lines in Fig. 29 are observable. As a result of the inhomogeneity of the rock, feather fractures can develop only on one side of the fault. If feather fractures develop from the other set of paired shear joints, pressure feather fractures are formed. Examples of mineralized feather fractures: Au deposits in the Tien-Shan, and the horsetail structure (Fig. 30d) known, for example, from the Butte copper deposit in Montana (U.S.A.). Structure of this shape can be deduced from a composite "cymoid loop" (Fig. 30). Mineralizations of fracture systems related to a composite "cymoid loop" are also known (Fig. 31).

13. *In jointed zones,* where joints and fractures make up an irregular pattern, impregnation stocks of Mo, Sn, polymetallic and Cu ores occur (e.g. Chuquicamata Cu deposit in Chile).

IV. Ore deposits at igneous rock contacts

14. *The contacts of concordant intrusions* in carbonate rocks are the loci of tabular skarn deposits (e.g. Gora Magnitnaya in the Urals Mts.).

15. *At the contacts of discordant intrusions,* skarn bodies form lenses or nests. Moreover, the contact between the intrusion and mantle forms a plane of weakness along which an opening, suitable for the localization of later hydrothermal mineralization, can be produced by tectonic movements (Fig. 32).

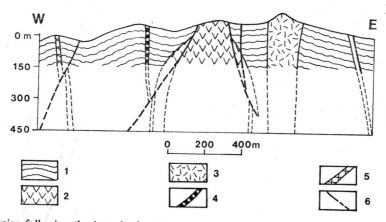

32. Ore veins following the intrusion/mantle contact. Tepezala, Aguas Calientes, Mexico (after Wanke and Moore in McKinstry, 1948). 1 — Mesozoic sediments; 2 — intrusive rhyolite; 3 — intrusive quartz porphyry; 4 — rhyolite dyke; 5 — porphyry dyke; 6 — ore veins.

V. Ore bodies in combined structures

16. *Ore bodies at the intersections of fractures and faults with a favourable rock* (examples include polymetallic Sierra Madre deposit in the U.S.A., Leadville in the U.S.A.—Fig. 33, Eschweiler in the Netherlands). The tectonic predisposi-

tion, combined with a favourable lithology, also controlled the localization of non-ferrous metal ores in the marginal parts of the Bohemian Massif (the Krušné hory Mts., Jeseníky, Slavkovský les Mts.).

17. *Ore bodies at the intersections and contacts of fractures, joints, faults with layers of favourable rock* take the form of stocks and ore shoots (Fig. 34). It has been shown statistically that 75 per cent of vein (originally fracture) intersections have been enriched, 13 per cent have been impoverished by leaching and 12 per cent have remained unchanged. Examples of enriched vein intersections: Au — Charters Towers veins, Australia, Ag–Pb in Guadelupe, Mexico, Cu — Rudnyi Altai, U.S.S.R.

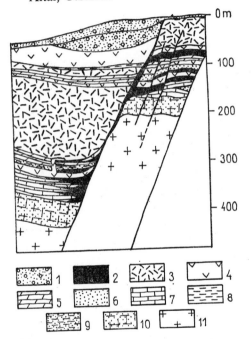

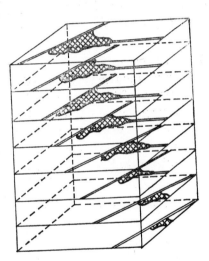

33. Ore bodies formed at the contact between favourable rock and a fault. Leadville, Colorado, U.S.A. (after Loughlin in Smirnov, 1957). 1 — alluvium; 2 — ore; 3 — white porphyry; 4 — grey porphyry; 5 — blue limestone; 6 — quartzite; 7 — white limestone; 8 — shale; 9 — limestone interbedded with quartzite; 10 — Cambrian quartzite; 11 — granite.

34. An ore stock develops at the contact of two faults or fractures.

VI. Ore bodies in intrusions

18. *In layered intrusions,* differentiation in situ gives rise to liquid-magmatic deposits of nickel (in norites), rare metals (in alkaline intrusives), nickel and copper (in basic and ultrabasic rocks), platinum (in basic and ultrabasic rocks, e.g. Merenski

reef in South Africa) and chrome-spinellides (in basic and ultrabasic rocks—Fig. 35). Examples: Bushveld, Rep. South Africa; Grand Dyke, Rhodesia; Stillwater Complex, Montana, U.S.A. The deposits form relatively thin but "stratigraphically" continuous concordant layers.

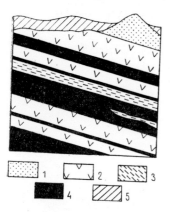

35. Layered basic intrusion with concordant chromite bodies (after Kupferbürger in Bateman 1950). 1 — anorthosite; 2 — diallage gabbro; 3 — bronzite-diallage bearing rock; 4 — chromite; 5 — overburden.

19. *Liquid-magmatic and late magmatic deposits* of the "offset vein" type form separate ore intrusions in the weakened zones around their basic and ultrabasic parent rocks. They are of tabular or pipe-like form. Examples: Cu — Ni ores of the Sudbury type, titanomagnetite and chromite ores, and chrome-spinellides with platinum.
20. *Ore bodies in basic and ultrabasic rocks with fluidal structures* have the character of protomagmatic ore streaks. They adapt themselves to linear and planar flows in the rock. Examples: chrome-spinellides and titanomagnetite ores.

Folds, faults and fractures can form simultaneously with mineralization but most endogenous deposits originate after the termination of orogeny. The opening of interstratal gaps in anticlinal bends and their simultaneous mineralization is extremely rare. Folding may have a greater influence on the formation of sedimentary deposits; a larger amount of useful mineral matter accumulates in deepening synclines than in updoming anticlines.

Synchronous faulting and mineralization is also very rare, but intra-mineralization movements along pre-mineralization faults, acting as supply channels, are more common.

Fractures formed during mineralization are filled with different mineral parageneses according to the time of their opening to hydrothermal solutions; the mineral content of these also changes with time. Consequently, ore veins of different composition can form within one ore field. Ore is precipitated initially in the earlier fracture system and other minerals are deposited in the later fracture system since by then the temperature of the ore solutions has fallen and the mineral content changed. This phenomenon makes the zoning of deposits fed by one parental magma more pronounced. The gradual extension of a fracture during mineralization is expressed by lateral changes in the composition of the vein material (e.g. the oldest part of the fracture is filled with high-temperature sphalerite, which in turn is succeeded by galena and finally by the lowest-temperature antimonite in the youngest part of the fracture). The gradual widening of the fracture during mineralization results in an increasing width of the developing vein. The ore shoots, described sub (17) in the survey of pre-ore structures, also form as a result of movement along an uneven joint plane (Fig. 12); this movement can be synchronous with mineralization.

Post-mineralization tectonic movements cause deformation, crushing or dragging out of the vein material. Sedimentary iron, manganese, bauxite and other deposits can be folded. Deposits of all types can be disrupted by post-mineralization faults. In contrast, pre-mineralization faults and fractures are mineralized or, at least, contain traces of metals. Post-mineralization structures are studied principally during exploration and mining. They affect the roof's strength and the permeability of beds, as well as other technical properties of the rocks.

Structural conditions also influence the formation of coal deposits, which often occur in tectonic depressions such as grabens or synclinal zones. In Bashkiria the occurrence of coal deposits is remarkable. They are located in depressions developed on the outcrops of a gypsum horizon in anticlinal cores.

4. MAGMATOGENIC CRITERIA

There is ample evidence of the intimate association of igneous rocks (even palingenetic) and ore deposits: the contemporary origin, their relationship to the same structures, a similar depth of formation, the same metamorphic grade in both metamorphosed rocks and ores, the association of mineralization with dykes and sills, the regular relationships in the chemistry of trace elements, accessory minerals and isotopic composition of elements, the regular arrangement of various deposit types around an intrusion, the same absolute age of ore deposits and parent intrusions, and the existence of regional granitoid zones of the same type and age with a certain metallization aureole.

Although the lateral-secretion theory of Sandberger, van Hise and other geologists of the late 19th century has recently found new supporters, it cannot be accepted as generally valid for explaining the genesis of most endogenous deposits.

Study of the chemical composition of the magma, its differentiation and crystallinity, alterations in the country rocks, grain size of the igneous rock, the size and fabric of the intrusion and the depth of magma congealing is important in prospecting for mineral resources.

a) The relationship between ore deposits and magma chemistry

Crystallization and differentiation processes in basaltic or granitic magma give rise to all types of igneous rocks and ore deposits (magmatic, pegmatitic, carbonatitic, contact-metasomatic, hydrothermal and sublimates).

Basaltic magma rich in Fe_2O_3, FeO, MgO and CaO can form by partial anatexis of the transitional zone between the upper and lower mantle beneath the oceanic crust (depth $200-300$ km, temperature about 1500 °C, pressure $50-125$ kb).

Granitic magma rich in SiO_2 and K_2O originates either by partial anatexis at the base of the continental crust (depth 45 km, temperature 600 °C, partial pressure of H_2O vapour 12 kb), or by granitization of older, even sedimentary, rocks. Both basaltic and granitic magmas can be contaminated by assimilation of the surrounding rocks.

Alkaline magma can originate in three ways: first by partial melting of sediments rich in alkalis; second by differentiation of basaltic magma poor in SiO_2; and third from granitic magma contaminated by country rocks containing alkalis or alkaline earths. In the last case, for example, Ca from the limestone reacts with Si from the magma to form pyroxene and the magma being thus impoverished in this element is relatively enriched in alkalis (Krauskopf, 1967).

Ore deposits associated with magmatic rocks form under two conditions: 1. If the magma contains a sufficient amount of ore elements (Sn, W, Mo, Pb, Zn Cu, Hg etc.), 2. if the magma contains enough gaseous components, that is F, B, Cl, S, As, to combine with ore components and give volatile compounds capable of concentrating ore elements in definite places.

Basic and ultrabasic magmas have a sufficient quantity of ore elements (some of them making the rock melanocratic) but are deficient in volatiles. Because of this they do not contain deposits formed by the concentration of ore elements through the co-action of volatiles, but rather deposits of Pt, Os, Ir, Cr, Ti, Ni and Cu ores. These elements do not form volatile coumpounds but accumulate in the basal part of the igneous body during magma differentiation and form segregation deposits. In addition to the metals mentioned above, some industrial minerals such as diamond, asbestos, corundum, talc and magnesite are associated with basic and ultrabasic rocks.

Acid magmas have a sufficiently high volatile content but only a small amount of ore elements. Greisen-type deposits (Sn, W, Li) are the only ore accumulations related to acid magmas. These are mainly associated with deposits of industrial minerals (mica, feldspar, quartz, beryl, monazite, etc.).

Intermediate magmas of granodioritic chemistry have the most favourable composition for the origin of ore deposits. About 95 per cent of endogenic deposits are related to intermediate and acid magmas. Intermediate rocks contain sufficient ore elements and volatiles and they bear Cu, Pb, Zn, Ag and other deposits, which are usually situated in the contact zone of the massif. Gold, lead-zinc and porphyry copper ($\pm$Mo) ore deposits often originate in small hypabyssal intrusions of intermediate and acid magma, which develop in the final phase of the geosynclinal evolution, usually at the geosyncline/platform boundary. Deposits of apatite, corundum, zircon and eudialyte are associated with intrusions of alkaline rocks. They occur inside the intrusions or close to them (Fig. 36) in the same way as deposits related to basic and ultrabasic rocks.

36. Relationship between ore deposits and intrusions of various composition (after Smirnov, 1957).

The spatial relationship between ore deposits and parental magma has stimulated the development of several classification schemes, the best known of which is that of Emmons (1940). Emmons, however, somewhat oversimplified the intricate relation-

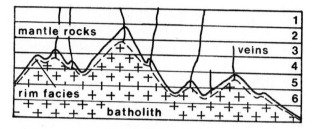

37. Six stages of batholith erosion (after Emmons, 1940). 1 — cryptobatholitic (telethermal deposits Pb—Zn and Cu in limestones, subvolcanic Hg, Sb, Au deposits, realgar, orpiment, barite and fluorite); 2 — acrobatholitic (many deposits derived from intermediate and acid magmas); 3 — epibatholitic (many types of ore deposits and industrial minerals); 4 — embatholitic (some deposits of Au, Cu, Zn; Sn—W); 5 — endobatholitic (Au deposits); 6 — hypobatholitic (only quartz and pegmatite veins).

ship between a batholith and ore deposits (Fig. 37). He erroneously considered "small" intrusions to be the apical parts of a batholith and disregarded the possibility of younger productive intrusions into the barren zone of the batholith; he also ignored other aspects such as the depth of cooling. In practice his zones 3 and 4 are often lacking. Despite these shortcomings, Emmons's scheme can be used to advantage, particularly for deposits related to the batholiths in anticlinoria.

In recent years S. S. Smirnov has developed a theory regarding the dependence of ore deposits on particular ("specialized") intrusions. A particular mineralization corresponds to a definite intrusion type or age. Tin deposits of the pegmatite and quartz-cassiterite types, for example, are related to acid magmas in which K predominates over Na. On the other hand, sulphidic tin deposits are associated with intermediate magmas of granodiorite or quartz-monzonite composition. Deposits of Sn, W, Be and Mo occur in more acidic intrusions of the adamellite-granite-alaskite series (containing $67-74\%$ SiO_2) than deposits of old gold formation, Mo, Cu, Pb and Zn, which are associated with rocks of the tonalite-granodiorite-granite series (containing 60 to 65 $\%$ SiO_2). The analysis of a granite sample reveals whether it is tin-bearing or not. The tin content in a granite containing tin deposits is $16-30$ g/ton or more, whereas the background content of a similar rock not associated with mineralization is only 5 g/ton. In the former case the main carrier of tin (mineral concentrator) is biotite, which contains from 80 to 200 g Sn per 1 ton. Lithium is another indicator of tin. The Cínovec massif (Czechoslovakia), for example, has a Li content ten times higher than the clarke value. Tourmaline, topaz, fluorite, albite and beryl also suggest the presence of concealed tin deposits. Tourmalinites in the crystalline complex above greisen bodies contain increased amounts of tenths of one in places even tin per cent. From the above it follows that prospection should be directed to areas containing intrusions of a definite chemical composition. Detailed mapping and prospecting should be carried out around such features.

b) The relationship between ore deposits and magma differentiation

High-temperature ore-bearing gases, vapours and solutions are the extreme differentiates of parental magma. They arise as a result of the complete differentiation of a magma, whereas diaschistic dyke rocks suggest half-completed differentiation. A varied suite of dyke rocks invariably indicates intensive differentiation, but only after a thorough study can it be decided if the process proceeded as far as the formation of ore solutions, because the chemistry of the magma is also an important factor. Consideration of the type of effusive rock can be a useful tool in prospecting. Gold deposits, for example, occur in areas of strong differentiation, where there are rhyolites, dacites and andesites side by side.

The relationship between dyke rocks and ore deposits is not always that of older and younger brothers but also of host to guest. The latter relationship results from the intrusion of dyke rocks into fault zones which, as planes of weakness are the sites of subsequent movement and thus accessible to younger ore solutions. An overestimation of the spatial connection between dyke rocks and ore veins gave rise to the "dyke theory", which presumed the occurrence of ore deposits only in association with dyke rocks. Dyke rocks are older than mineralization and only lamprophyres can be of similar or later age.

c) Rock alterations in the neighbourhood of deposits as prospecting guides

The action of hydrothermal solutions and high-temperature gases and vapours on rocks near developing ore deposits produces appreciable alterations in their petrographical composition. These changes can relate to structure and colour alone (e.g. recrystallization and leaching of bituminous limestones), without any change in mineral content, or they can result from the supply or removal of substances (e.g. the supply of sulphur to rocks with liberated iron gives rise to secondary pyrite). Garnet, hornblende, pyroxene, tourmaline and biotite form in this way near hypothermal deposits; sericite, chlorite and carbonate near mesothermal deposits; chlorite in excess of sericite, and carbonate, adular and alunite around epithermal deposits. These alterations cover a much greater area than the deposit itself and reveal the presence of a deposit before mineralization has been ascertained (Fig. 38); this is particularly important with concealed deposits (Fig. 39). In a horizontal section the rock alterations appear as a concentric target, generally of ellipsoid form. The target must not be too large or small with respect to the ringed deposit, if it is to be useful in prospecting (ideally 10—50 times larger than the deposit).

Alterations of rocks around deposits vary (Kurek, 1954, Boyle 1970) according to the activity of ore-forming processes and the rock characteristics. Alterations of basic and ultrabasic rocks include serpentinization, formation of reaction zones, listwänitization, carbonatization, chloritization and propyllitization. Alterations of intermediate and acid rocks include greisening, albitization, tourmalinization, beresitization, sericitization, formation of secondary quartzite, kaoliniza-

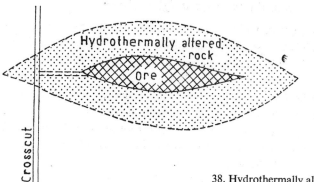

38. Hydrothermally altered rocks encountered by a crosscut may suggest the presence of an ore body.

39. Hydrothermal alteration around an outcropping barren fracture may indicate a concealed ore deposits. 1 — vein deposit; 2 — country rocks; 3 — hydrothermally altered country rocks; 4 — barren fracture in extension of the vein; 5 — ground surface.

tion, silicification, hematitization, alunitization and leaching. The alterations of carbonate rocks include the formation of skarns and reaction zones at the carbonate/silicate rock contact, scapolitization, silicification, baritization of limestones and dolomites, dolomitization and ophicalcitization.

Serpentinization often affects ultrabasic rocks. Regional serpentinization of autometamorphic derivation, manifested by the knitted texture of serpentinite, and dynamometamorphic serpentinization, which gives rise to antigorite, are of no importance in prospecting.

Hydrothermal serpentinization accompanied by chloritization and carbonatization is most useful in prospecting, since it is frequently associated with the formation of talc. Serpentinization which develops at the contact between ultrabasites and younger acid intrusions is a guide to deposits of chrysotile asbestos. Serpentinized harzburgites are more promising in this respect than lherzolites; pyroxenic rocks are of little importance. Prospecting for asbestos should also be concentrated at the margins between serpentinites and unaltered ultrabasites, and at the endocontact between serpentinites and acid dyke rocks.

Hydrothermal serpentinization often affects ultrabasic dyke rocks along the joints. Pyroxenite, microgabbro, diabase and plagioclasite are converted to garnet, chlorite-pyroxene-garnet, garnet-vesuvian and similar leucocratic rocks, which contrast with the neighbouring dark serpentinites and peridotites. Their light colour makes them easily distinguished in the field and indicates the circulation of solutions that could give rise to chrysotile asbestos in the surrounding serpentinites.

Reaction zones between an ultrabasic rock and the adjacent chemically different rocks consist of biotite (or one of the following: phlogopite, vermiculite and chlorite), actinolite, talc or serpentinitic rocks. They differ in colour from the ultrabasic rocks and often contain talc deposits.

Listwänitization is the hydrothermal alteration of Fe−Mg silicates in ultrabasic rocks, which produces Ca, Mg and Fe carbonates, quartz and talc. Outcrops of listwänite often form morphologically conspicuous ridges. Listwänitization accompanies copper vein deposits containing nickel and cobalt (e.g. in the Urals Mts.) and can provide the raw material for talc production.

Carbonatization usually affects basic and intermediate rocks and produces Fe−Mg carbonates. The process occurs in rocks adjacent to some gold, copper and polymetallic deposits. On weathering, the carbonates decompose into an ochre, which fills rhombic-shaped cavities. Carbonatization is less important in acid rocks and it is accompanied by the formation of sericite, chlorite and pyrite in feldspathic rocks.

Chloritization can serve as a guide only if it is produced by hydrothermal solutions. Regional, contact, retrograde and autometamorphic chloritization is unimportant. Hydrothermal chloritization, however, is only of small extent. Mineralization often appears in a chloritized rock as ore impregnations combined with sericitization, tourmalinization or silicification.

Individual types of chloritized rocks accompany definite types of ore deposits. Monomineral chloritic rocks originate in the neighbourhood of pyrite, sulphidic-cassiterite, lead-zinc and chromite deposits. Abundant quartz-chloritic rocks are associated with sulphidic-cassiterite, and, to a lesser extent, chalcopyrite and polymetallic deposits. Sericite-chloritic rocks (± quartz) occur close to pyrite deposits (as in the Urals and Altai Mts.). Tourmaline-chloritic rocks (± quartz) are characteristic of sulphide-cassiterite deposits, whilst biotite-chloritic rocks (± quartz) accompany the pyrite and porphyry copper ores.

The following zoning of altered rocks can serve as a prospecting guide for pyrite and polymetallic deposits: unaltered rock−slightly chloritized−quartz-chloritic−sericite-chloritic−quartz-sericitic rock−ore.

The mineralogical character of chlorite varies with individual deposits, according to the temperature and chemical conditions at its origin. The Fe-chlorites of the thuringite group are typical of sulphide-cassiterite ores. On sulphidic (especially Cu) deposits Mg – Fe chlorites of the sheridanite-ripidolite group are found. Mg-chlorites of the clinochlore-penninite-prochlorite series occur in chloritized rocks near Pb – Zn deposits. Gold deposits are accompanied by Fe – Mg chlorites, chromite deposits by Mg – Cr chlorite (kotschubeite) and skarns by Mg chlorites.

Propylitization chiefly affects andesites, dacites and basalts, less often rhyolites. It accompanies deposits of Au- and Ag-tellurides, arsenides and antimonides, but is often so extensive that it loses significance as a prospecting guide.

Greisening is an important aid in the search for Sn, W and Mo deposits, less often for As and Bi ores and rarely for Au ores. It is active along the outer and inner contacts of minor intrusions and apophyses of cupola-like shape, but does not affect large intrusive bodies. The petrography of greisenized rocks corresponds to acid granites and granodiorites, seldom to diorites. Greisening of more basic rocks has not been observed. Acid granitic intrusions, often accompanied by pegmatites, contain non-sulphidic cassiterite-wolframite deposits. Sulphidic Fe, Cu, As, Sn, Bi, Mo, Zn and Pb ores occur in greisenized granodiorites. According to its extent, greisening can be divided into two types. The first includes thin bodies located close to the supply channel and poor in ore minerals, and the second a more widespread alteration of the parent rock with the production of stocks and nests along the contraction fissures.

Individual greisen types, characterized by a non-metallic mineral, indicate the presence of a particular metal. Tourmaline and tourmaline-chloritic greisens indicate sulphidic tin ores, topaz-type greisens suggest the occurrence of non-sulphidic tin ores, the fluorite type is characteristic of tungsten mineralization and the fluorite-muscovite type of molybdenite veins.

Greisens containing sulphidic ores weather to limonite whilst those containing non-sulphidic ores disintegrate into sand bearing ore minerals, which becomes the source of cassiterite placers.

At present, the cupolas of greisenized granite massifs in the Krušné hory (Erzgebirge) Mts. and Slavkovský les Mts. (Czechoslovakia) are receiving particular attention (Janečka, 1964). The buried granite relief was traced by means of a gravimetric survey, but this failed to locate the Sn-bearing greisens, which can descend hundreds of metres below the granite surface. Drilling provides relatively reliable data on greisen deposits, but veins and narrow mineralized zones have to be explored by underground workings. Delimitation of the ore bodies is difficult since the mineralization is not visible. Underground works combined with detailed section sampling proved useful.

Albitization of acid intrusives indicates the occurrence of Sn, Mo and W-bearing greisens. It develops at the base of greisen cupolas and, with weak greisening, affects large parts of the intrusions along their contact with the roof. Albite forms from K-feldspars through the action of Na-rich solutions. In an environment saturated with SiO_2 and poor in alkalis, the K-feldspars are replaced by quartz and subsequently by muscovite. Na-rich solutions also cause the replacement of plagioclases by muscovite.

Tourmalinization is an indication of some Sn deposits, less often of Cu, Au, As, Pb–Zn, W, Mo and Co ores. This relatively rare type of alteration occurs, for example, in the proximity of Sn, Pb, Zn and Au deposits in the Pacific metallogenic zone (see Structural criteria). Tourmalinized rocks form lenses and stocks up to several square kilometres in area. Tourmalinization takes place along the margins of hypabyssal granite and granodiorite intrusions and in feldspathic effusives and sediments. Ore minerals are often absent.

Beresitization is the partial alteration of feldspars in plutonic and hypabyssal rocks to sericite and secondary quartz, occasionally with the simultaneous formation of carbonates and pyrite. Beresite is composed of quartz, mica and pyrite. The alteration is a guide to auriferous quartz veins and Au placers, less often to wolframite, polymetallic and copper veins.

Sericitization, often with silicification, accompanies almost all magmatogenic deposits. It affects alkali feldspars and acid plagioclases and to a lesser extent the coloured minerals of feldspathic rocks. The process is indicative of ore occurrence only when associated with other secondary minerals such as carbonates, chlorite (Fig. 40), epidote, tourmaline, fluorite, topaz, barite and sulphides. Regional sericitization of autometamorphic or retrograde-metamorphic origin is of no importance in prospecting.

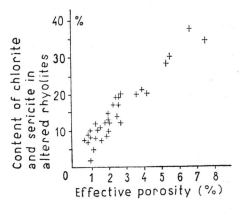

40. Hydrothermal mineralization may be associated with chloritized and sericitized rocks due to the increased effective porosity during pre-ore alteration (after Starostin, 1965).

Feldspathic rocks also become sericitized during subaerial weathering, when other minerals are also formed that are mistaken for sericite (illite, pyrophyllite, supergenic micas). It is thus necessary to distinguish between weathering products and those of hydrothermal alteration, since only the latter are important in prospecting. An exact determination demands precise laboratory methods. For field determination it is sufficient to know that in hydrothermal sericitization around ore deposits the feldspars are completely replaced by hydrothermal minerals, whereas weathering usually leaves the cores of feldspar crystals intact. Sericite formed in the neighbour-hood of hypothermal deposits corresponds chemically to muscovite. Near meso-thermal deposits (e.g. of Pb – Zn) phengite is formed and gold deposits are accompa-nied by V- or Cr-bearing sericite (fuchsite or mariposite).

Secondary quartzite formation is typical of hydrothermally altered effusives (lavas and pyroclastics). This alteration gives rise to quartz, pyrite, hematite, rutile, sericite, alunite, kaolinite, pyrophyllite, andalusite, diaspore, corundum, topaz and other minerals. Non-metallic deposits, especially of corundum and andalusite, are related to secondary quartzites derived from acid effusive rocks. Secondary quartzites with deposits of Cu (Kounrad), Pb, Zn and Au are derived from intermediate volcanic rocks. Quartzites of all types are characterized by a sharp dissected relief, a light colour with dark limonitized spots and white kaolinized patches. The interior of secondary quartzite masses shows the following zones: primary unaltered rock, propylite, sericite, pyrophyllite, kaolinite, alunite, diaspore, andalusite and corundum. The last zone, nearest the supply channel of hydrothermal solutions, formed at the highest temperature. Complete zoning is however rare. The presence of the sericite to alunite zones in rocks that were initially of intermediate chemistry suggests the occurrence of polymetallic, gold and copper ores.

Kaolinization affects feldspathic rocks in the neighbourhood of hydrothermal polymetallic, gold, barite and fluorite veins, or greisen tin deposits. Kaolinite accompanying hydrothermal and higher-temperature deposits seems to have originat-ed from the weathering of hydrothermal sericite. Kaolinization is also produced by the action of sulphuric acid liberated during the weathering of sulphides. Igneous and metamorphic rocks altered by supergene kaolinization point to the possible presence of an ancient weathering crust, which can contain deposits of Ni, Fe, Al, kaolin and refractory clays.

Silicification (without the formation of secondary minerals) is a frequent alteration associated with deposits of Cu, Pb – Zn, Au, Hg, piezoelectric quartz, fluorite and barite (witherite), and it occurs particularly around epithermal ore deposits in volcanic rocks. Evidence of *hematitization* is useful in prospecting for uranium deposits.

Alunitization occurs in the upper layers of hydrothermally altered intermediate and acid effusive rocks and is typical of polymetallic, gold and copper ores. Alunite is accompanied by quartz, sericite, and occasional kaolinite and pyrite.

The leaching of rocks is caused either by percolating sulphatic waters from sulphidic deposits or by descending kaolinization unassociated with mineralization.

The formation of skarns is often accompanied by mineralization. Skarns not only indicate certain contact-metamorphic processes but also themselves form mineral deposits. Ore-bearing skarns originate only where acid to intermediate intrusions (granite, granodiorite, quartz diorite, syenite) are in contact with limestone or dolomite interbeds of the mantle. They can develop at the contact itself or in the thermally metamorphosed mantle of the intrusion up to a distance of 200 – 400 m (exceptionally 2 km), or in carbonate xenoliths within the intrusion. Skarns generally occur in folded areas, less often along the platform margins, and are invariably related to faulting and jointing.

Skarns can be divided into simple and complex types. Simple skarns are composed mainly of garnet, pyroxene and ore minerals. The mineralogical character of the garnets and pyroxenes often indicates a certain type of mineralization. Skarns with dark garnets of andradite composition contain Fe, Pb – Zn and Co ores. In the Jumbo Basin deposit of Alaska, garnet in a magnetite-bearing skarn contains 96 – 100 % of the andradite component, whereas the andradite content of the garnet in a magnetite-free skarns ranges from 45 to 85 %; the remainder is the grossular component (McKinstry, 1948). The andradite component increases continuously towards the deposit. Skarns containing garnet of andradite-grossular chemistry are associated with Cu and W mineralization, whilst garnets with the predominance of the grossular component accompany most skarn W deposits. Isotropic garnets occur in skarns containing Fe deposits and anisotropic garnets occur with other metallic ores. Hedenbergite skarns are often related to rich Cu, Pb, Zn and W deposits and Mn-hedenbergite (more than 6 % MnO) indicates the proximity of Pb – Zn – Cu bodies. Axinite-bearing skarns are not accompanied by major mineralization.

Complex skarns develop through the gradual alteration of simple skarns and contain, among other minerals, epidote, actinolite and rhodonite. Some of them have undergone subsequent quartz-sulphidic mineralization with the formation of chalcopyrite, pyrrhotite, pyrite, sphalerite, galena, molybdenite, cobaltite, arseno-pyrite, scheelite, beryl, cassiterite or native gold.

Reaction zones at the contact between carbonate and silicate rocks indicate the presence of phlogopite. The following zone sequence occurs: unaltered carbonate rock (mainly dolomitized limestone) – calcite-forsterite zone (with a small portion of diopside and hornblende) – diopside zone (with phlogopite admixture) – phlogo-pite-diopside zone (with large crystals and aggregates of phlogopite) – quartz-scapolite zone (with feldspar and diopside) – alumosilicate rock (especially alaskite granite or granite pegmatite).

Scapolitization is also distinctive of phlogopite bodies; the scapolite zone follows immediately after the phlogopite one (see above).

Silicification of carbonate rocks, occasionally accompanied by fluoritization, is indicative of barite-witherite mineralization; silicification related to Hg and Sb deposits is characterized by the simultaneous formation of chalcedony.

Baritization of limestones and dolomites is an indication of Pb – Zn and, to a lesser extent, barite and witherite bodies. The large Pb – Zn ore deposit Mirgalim Sai in Kara Tau (southern Kazakhstan) was discovered by tracing a zone of baritized dolomites.

The *dolomitization* of limestones accompanies Pb – Zn deposits (e.g. Trepcha in Jugoslavia, Lower Silesia in Poland) as well as barite and witherite bodies. A pro-spector must distinguish between sedimentary dolomites and hydrothermal dolomites on the basis of their shape, texture and origin. Coarse-grained bodies of irregular

98

shape, related to fissures and faults, are particularly worthy of attention. These are almost invariably of hydrothermal origin. The intercalations of sedimentary dolomite in limestones should not be ignored since dolomite, being more brittle than limestone, is more intensely fractured and thus more readily accessible to ore solutions.

Ophicalcitization due to metasomatic processes at the contact between marbles and chrysotile-bearing serpentinites is accompanied by the formation of green, pink and yellow rocks which grade into marble on one side and serpentinite on the other. An intrusion of acid or, more commonly, basic magma provides the solution and heat required for ophicalcitization. Mg-bearing marbles involved in this process cause rapid serpentinization of the basic intrusion. Chrysotile is formed at a later stage. Pure marbles on the other hand are only partially silicified at the contact with the intrusion.

d) The relationship between ore deposits and the grain size of rocks

The grain size of igneous rocks is indicative of the rate of cooling. This depends on the depth of the intrusion, the distance from the contact with the country rock (the rock is finer-grained at the contact), the volume of the intrusion (the larger the volume, the slower the cooling), the heat conductivity of the country rocks and their ability to maintain the heat of the initial intrusion (a subsequent intrusion penetrating a heated environment is coarser-grained, all other conditions being equal). The rate

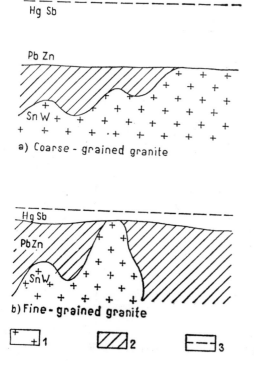

a) Coarse - grained granite

b) Fine - grained granite

41. Pb—Zn deposits related to coarse-grained granite (solidified at depth) were removed by denudation (Fig. 41a), while those associated with fine-grained granite (solidified near the surface) are preserved at depth (Fig. 41b). 1 — granite; 2 — mantle of granite; 3 — ground surface at the time of granite intrusion.

of magma congealing controls the moment of segregation of the ore solutions. The temperature at the time of segregation in turn governs the spacing of isotherms and thus the spacing of individual deposit types and their distance from the magma. The farther apart the individual isotherms, the greater the distance will be, for example, between perimagmatic and apomagmatic deposits. From the grain size of the rock, the prospector can infer the depth of the intrusion and from this the distance of tin and tungsten deposits from Pb and Zn or Hg and Sb deposits, and the distance of deposits of all types from the magma (Fig. 41). These relationships should be used in prospecting, because they make it possible to estimate whether an ore formation, which may be presumed to derive from magma of a certain chemistry, was removed by denudation or lies concealed at depth. Several correlations of practical significance have been established in Hungary, where the grain size/mineralization relationship has been studied (Szadecky—Kardoss, 1941). Gold-silver deposits were preserved at depth when the grain-size of the parental magmatic rocks was less than 5 mm, whereas they were denuded at a higher crystallinity. If the grain-size of the magmatic rock equals 5.5—6 mm, metasomatic Pb—Ag—Cu deposits originate in the neighbouring limestones. Metasomatic Fe deposits are formed in limestones when the grain-size of the igneous rock ranges from 6 to 6.5 mm.

e) The relationship between ore deposits and the size of intrusions

Pegmatites and pneumatolytic and hydrothermal deposits are associated with large plutons. Various sized deposits of similar composition are related to the hypabyssal stocks of small intrusions, since the magma and ore solutions are intruded along the same zones. They form by the differentiation at depth of the same parental magma (Fig. 42). In practice it is important to distinguish these small intrusions from the apical parts of large plutons (Table 39), since more deposit types are associated with the former in a smaller area. Tin mineralization, for example, is related to the apical parts of small granite cupolas. In the deeper parts of the massif, it follows chiefly

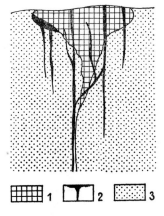

42. A small intrusion and the veins related to it. 1 — intrusive stock; 2 — ore body; 3 — country rock.

100

the moderately dipping contact zone. The relationship between dykes and endogenic deposits is shown in Fig. 43 and has been discussed in detail by Abdullayev (1957).

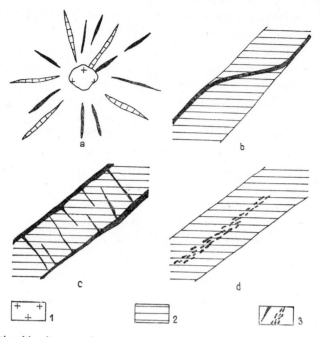

43. Spatial relationships between dykes and mineralization (horizontal sections). a — dykes and veins fill radial fractures around the mother intrusion; the fractures formed during upward movement of magma and as a result of its volume reduction after solidification; b — ore vein alternately following the upper and lower vein/rock contact (see the "dyke theory"); c — ladder vein; d — ore galls and impregnations in dyke (modified after Smirnov, 1957); 1 — mother intrusion; 2 — dyke; 3 — ore vein and ore impregnations.

TABLE 39

Distinction of small intrusive stocks from the apical parts of large plutons

	Small intrusive stocks	Apical parts of plutons
penetration	along faults	stoping
composition	granosyenite, quartz diorite, gabbrodiorite, gabbro	centre: granite or granodiorite
alteration of country rocks	narrow contact zone	broad contact zone, granitization, pegmatitization
deposits	hydrothermal	streaky syngenetic pegmatites, quartz veins

f) The relationship between deposits and the internal structure of intrusions

The flow structures in basic and ultrabasic rocks determine the (streaky) form of liquation and segregation deposits. The other internal structural features of intrusions (Fig. 44) can affect the localization of hydrothermal mineralization.

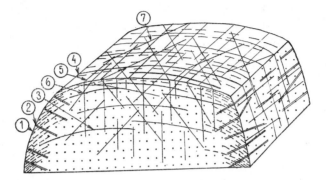

44. Diagram showing structural elements of an intrusive massif (modified after Smirnov, 1957). 1 — flow structure; 2 — marginal fractures; 3 — joints subparallel to the intrusion surface (*L*, sheet joints); 4 — vertical longitudinal joints (*S*, rift joints); 5 — vertical transverse joints (*Q*, grain joints); 6 — inclined joints; 7 — diagonal joints.

g) The relationship between deposits and the depth of magma cooling

Plutonic intrusions congeal at depths of over 2000–3000 m and often give rise to pegmatite fields. Hypabyssal intrusions are accompanied by deposits of various types. Effusive facies are divided into old facies (spilites, diabases and others of Palaeozoic age) with sulphidic deposits and young (Tertiary) facies, which contain deposits of sulpho-arsenides, sulpho-antimonides and gold.

5. METAMORPHOGENIC CRITERIA

Metamorphic deposits form as a result of the mobilization and concentration of elements during regional metamorphism. Metamorphosed deposits originate by contact or regional metamorphism of primary sedimentary or magmatogenic deposits. Both types are included under the term metamorphogenic deposits. Many deposits in metamorphic complexes whose spatial connection with magmatites is uncertain are probably metamorphic. It is postulated that they have formed under the action of water of non-magmatic derivation (rutile and ilmenite deposits in crystalline rock complexes, veins of crystallized quartz in quartzites, and deposits of phlogopite and graphite are probably of this type). Metamorphosed deposit types are numerous, especially among pre-Palaeozoic and Palaeozoic sedimentary or magmatogenic deposits (e.g. the regionally metamorphosed Krivoi Rog Fe-deposit, U.S.S.R.; the Broken Hill Pb–Zn deposit, Australia; the Witwatersrand gold deposit,

Rep. South Africa; the contact-metamorphosed Kowary Fe deposit, Poland; the Sonora graphite deposit, Mexico; and the Naxos emery deposit, Greece). Prospecting for these deposits is based on the same principles as those used in the search for sedimentary and magmatogenic deposits, but the relationship between the mineral composition of the wall rocks, before and after metamorphism, must also be considered. *Ultrametamorphism* associated with anatexis and palingenesis at a depth of 10—20 km is preceded by the mobilization of metals (in addition to other substances), which are carried upwards to form a front of ore metamorphism. These metal deposits have the character of hydrothermal deposits.

The metamorphic facies are a criterion for metamorphic deposits. Deposits of native copper are associated with zeolite facies (Lake Superior, U.S.A.); magnetite-hematitic quartzites and Au, U, sulphides, emery, amorphous graphite and asbestos deposits with greenschist facies; silicate Mn and Zn ores and magnetite-amphibole ores with the glaucophane facies; taconites and itabirites, Fe-ores, cyanite, sillimanite, andalusite, corundum, emery, crystalline graphite and ilmenite with hornblende facies; amphibole-pyroxene-magnetite quartzites, garnet, rutile with granulite facies; and rutile with eclogite facies.

6. GEOCHEMICAL CRITERIA

The behaviour of elements in the earth's crust is governed by certain laws; some are typical of basic rocks and other of acid igneous rocks or sediments (Rankama, Sahama, 1950—Tables of average element contents in various rocks). Geochemical studies are mainly of intrusive, sedimentary and metamorphic rocks which have a higher than average content of ore elements. However, intrusions accompanied by ore deposits are sometimes impoverished in certain ore elements, as a result of their separation and concentration in ore bodies.

An increased amount of accessory ore minerals in intrusive rocks indicates the potential occurrence of deposits of these minerals. The presence of some ore minerals in the heavy mineral concentrate is further evidence of such deposits in the area. An ore element that accompanies deposits of various types (e.g. in polymetallic deposits) almost invariably forms separate deposits. In contrast, other elements never occur together in the same ore province (e.g. tin and copper). All these conclusions are derived from the laws governing the behaviour of elements in the earth's crust and from the metallogenesis of the area. The geochemistry will be dealt with in greater detail in the chapter on geochemical prospecting methods.

7. GEOMORPHOLOGICAL CRITERIA

Geomorphological criteria are particularly important in the prospecting for placer deposits, which should be carried out first in the areas that have been poorly investigated geologically, for example, in developing countries. Even so, successful results of the study of tin placers in the Krušné hory Mts. (Czechoslovakia) have shown that this method can be advantageous even in countries whose mining history is about 2,500 years old.

Indirect geomorphological criteria such as tectonic steps, hogbacks and cuestas, reveal the tectonic structure of the area. Direct criteria concern the surface features of the deposit, which can be either positive or negative (see the chapter on prospecting methods and prospecting indications).

Deposits can be related to the topographic evolution of the area (exogenic deposits) or formed irrespective of it (endogenic deposits). The former can be related to the recent topography (placers, residual deposits, some deposits of bauxite, clay, sand and gravel) or to fossil relief (marine sedimentary Fe and Mn deposits, deposits of bauxite, phosphorite or carbonate rocks); prospecting for the latter utilizes palaeogeographic criteria.

Contour maps and aerial photographs are useful in the study of recent relief and deposits associated with it, such as placer deposits of gold, platinum, cassiterite, wolframite, diamonds, monazite, zircon and other resistant minerals as well as sand, gravel and some types of clay and loam deposits.

Initially it is necessary to distinguish areas of recent denudation and accumulation and determine the intensity of neotectonic movements. Uplifted areas are dissected by a network of deep meandering V-shaped valleys; the rivers have a constant course and rock terraces develop. Depressed areas on the other hand are characterized by aggradation, closed basins and rivers which often shift their channels.

In areas that have undergone block movements, headward erosion and rapids which produce an uneven stream gradient appear in the uplifted blocks. The ancient stream pattern to which placers could be related has been preserved only in patches. In the sunken blocks, alluvium has accumulated and the streams have developed meanders. Neotectonic movements can be deciphered from the terraces which are likely to contain alluvial deposits.

If stream reaches of the same evolution stage are connected, several zones that are characterized by particular types of placers can be delimited (Fig. 45). Bilibin (1948) distinguished four evolutionary stages of a river valley associated with particular placer types (Fig. 46): 1. senile stage—a broad valley of the old erosion cycle with valley placers (at the base of the alluvium); 2. youth stage—valley incision with new placers in the river beds and old placers in the terraces (i.e. in the remnants of valley sediments of the old erosion cycle); 3. mature stage—widening of the valley during which the placers in the river bed become valley placers and the terrace placers are destroyed; 4. senile stage—broad valleys of the new erosion cycle with valley placers. All four zones (Fig. 45) migrate upstream. Zone IV increases gradually until it predominates completely. Prospecting using the heavy-mineral concentrate method is especially useful in stages 2. and 3. In sampling the river valleys in stages 1. and 4., test pits (up to several tens of metres in depth) must be sunk to examine the sediment at the base of the alluvium.

104

Placers are completely absent from canyon-like valleys, although they are concentrated in open valleys downstream of canyons. Buried fossil placers in sunken
blocks can be covered by effusive bodies or by marine, lacustrine and glacial sediments.
If they contain magnetite, they give a positive magnetic anomaly in acid rocks and

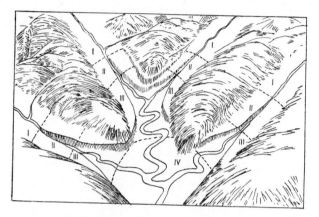

45. Zonal division of stream placers (after Bilibin, 1955b). Zones I and IV — valley placers; zone
II — terrace and river-bed placers; zone III — conversion of river-bed placers into valley placers
and destruction of terrace placers.

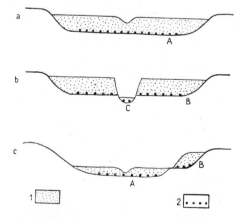

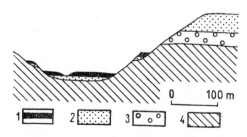

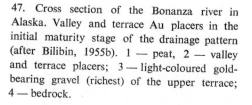

46. Dependence of placers formation on the
evolution stage of a drainage pattern. a —
senile stage—valley placers (A); b — youth
stage—river-bed (C) and terrace (B) placers;
c — mature stage—conversion of river-bed
placers (C) into valley placers (A), destruction of
terrace placers (B); 1 — alluvium; 2 — placers.

47. Cross section of the Bonanza river in
Alaska. Valley and terrace Au placers in the
initial maturity stage of the drainage pattern
(after Bilibin, 1955b). 1 — peat, 2 — valley
and terrace placers; 3 — light-coloured gold-
bearing gravel (richest) of the upper terrace;
4 — bedrock.

a negative anomaly in basic and ultrabasic rocks. Tertiary and Mesozoic buried placers are known from the Urals, West Siberia and Kazakhstan among other places. In addition to alluvium (valley, terrace and river-bed deposits, Fig. 47), placers with heavy ore minerals also occur in glacial moraines, lacustrine, deltaic, littoral (Fig. 48) and eluvial (in the weathering crust) sediments. Colluvial placers, developed on the slopes and at foot-hills, supply material to all other types. It is of interest to note that the drainage pattern can be affected by a mineral deposit. Figure 49 shows the evolution of a valley which does not fit into the normal pattern of development, as a result of the differential weathering of a sulphide body.

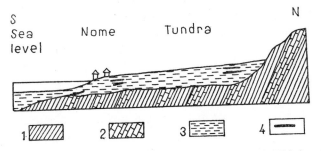

48. Marine placers near Nome in Alaska (after Bogdanovich in Bilibin, 1955b). 1 — crystalline schists; 2 — crystalline limestone; 3 — marine and fluviatile sediments; 4 — marine placers, including littoral and foreshore (on sea floor) placers, and fossil coastal placers in marine and fluviatile sediments.

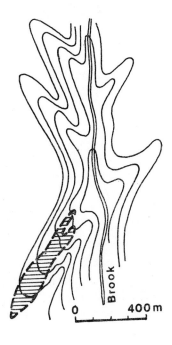

49. Anomalous course of a river valley cut in the oxidation zone of a sulphide body (shaded) (after Smirnov, 1957)

8. PALAEOGEOGRAPHICAL CRITERIA

Mineral deposits related to ancient relief forms developed in peneplains (residual deposits of the weathering crust), in lakes (often with through-drainage), swamps and rivers (platform coal deposits with a small number of very thick seams, slightly affected tectonically), on coastal plains (U and Cu-bearing sandstones), in coastal lagoons and swamps (transitional type of coal deposits), lagoons (deposits of salt, sulphur, sedimentary copper and uranium ore, bituminous shales), in the littoral of shallow seas (phosphattes, sedimentary iron and manganese ores) and near deltas (intensively folded and metamorphosed geosynclinal coal deposits with a large number of extensive but thin seams, from the late stage of geosynclinal evolution).

Placers (alluvial deposits) are formed predominantly by physical weathering in areas of dissected relief and composed of igneous rocks, and may be found in fossil valleys unrelated to the present-day drainage pattern.

Laterites constitute the uppermost layers on plateaus and plains separated by shallow valleys, whilst red kaolins originate in hilly areas. Bauxites are confined to depressions, often of karstic origin. Where they have been deposited in fresh-water basins developing into peat bogs, they contain pyrite and siderite and may grade laterally into coal seams. Marine bauxites usually occur at the base of a transgressive complex after a long hiatus.

The mineral composition of sedimentary iron ores depends on how far from the shore they have been deposited (Fig. 50). The surrounding land has usually been peneplaned and exposed to tropical weathering. A particularly large amount of Fe is liberated during the weathering of basic rocks. Rich deposits have been formed in embayments at the mouths of streams of the Amazon river type. Continental (lacustrine) iron ores develop under identical climatic and topographic conditions. Arid zones are unfavourable to the development of iron deposits in the contiguous seas. Since the Precambrian the focus of sedimentary iron ore formation has moved from the deep-sea facies (Precambrian Fe-bearing quartzites — reserves over 3000 milliard tons) to the littoral and continental facies (bog ores). Simultaneously there has been a decrease in the total amount of ore (the post-Proterozoic Fe-ore reserves make only about 120 milliard tons) as a result of the growth of platforms and the rising importance of landmasses.

Sedimentary manganese ores are formed in shallow seas and their mineral composition changes with depth. A typical sequence from shallow to deep water is: concretionary psilomelane and pyrolusite — oxides of trivalent Mn — carbonate ores — carbonate or siliceous rocks. This regular zoning was utilized in prospecting in the Urals Mountains. Carbonate Mn ores were found initially, then oxide ores were located by drilling between the ancient shore-line and the area of carbonate ores. The width of the zone containing Mn deposits ranges from several hundred metres on the steeply inclined, diversified shelf, to 8 — 10 km on the moderately sloping shelf. The thickness of oxide ores varies considerably and depends on the morphology of the

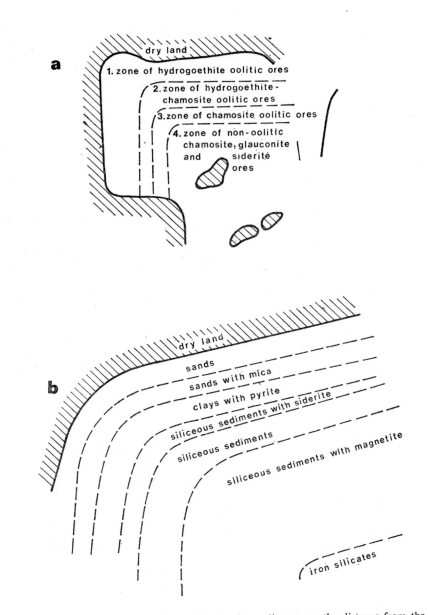

a — zone of hydrogoethite oolitic ores

50. Dependence of mineral composition of Fe-bearing sediments on the distance from the shore. a — Cretaceous in the Urals (after Krotov); b — Upper Proterozoic jaspilites in Minnesota (after White in Rukhin, 1962).

sea floor before the transgression. As with iron ores, sedimentary Mn ores originate close to a landmass with a tropical climate.

Copper ores, associated with variegated sandy-clayey sediments (of the *red-bed type*), generally follow the periphery of the red beds rimming areas of continental uplift. The ores were formed at the mouths of rivers in shallow lagoons, bays and lakes constituting belts even many hundreds of kilometres in length. Occasionally copper has migrated and been concentrated in sandstones containing plant remains. Another type of sedimentary Cu ore is *black argillite*, in which copper is accompanied by the ores of U, Mo, Pb, Zn, V, Ni, Au, Pt, Pd, Rh, and sometimes Ag as, for example, in the Mansfeld shales in Germany and the Lubin–Polkówice deposit in Poland. The individual metals attain their maximum concentration in different horizons of the same bed complex. Mansfeld shales originated at mouths of rivers in non-aërated sea gulfs. Metals were first absorbed by organic substances and clay colloids and, therefore, they only occur near the shores of embayments even after the ore minerals have been recrystallized. Polymetallic deposits may also be associated with dolomites (Triassic dolomites in Upper Silesia and in the Alps). Sedimentary polymetallic deposits were formed in warm, periodically dry climates near land with surface outcrops of primary deposits.

Sedimentary uranium ores, related to weathering crusts, phosphates, carbonate rocks, caustobioliths, and clays and sand, have been laid down in seas, lagoons, deltas and alluvial plains ever since the Proterozoic. *Residual uranium* deposits develop in arid climate. Uranium is bound in vanadates, sulphate-carbonates and others. Infiltration uranium deposits develop along the margins of platforms and intermontane depressions in sandstones and other porous sediments. The ores are

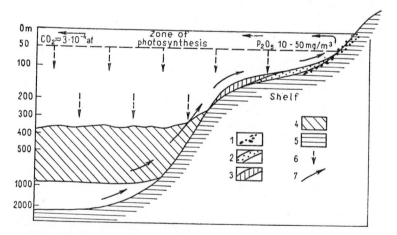

51. Diagram showing the formation of phosphates (after Kazakov in Strakhov, 1962). 1 — facies of littoral gravel and sand; 2 — phosphate facies; 3 — facies of calcareous sediments; 4 — zone of maximum CO_2 and organic P_2O_5 content (partial pressure of CO_2 up to 12×10^{-4} at, P_2O_5 — 300 to 600 mg/m^3); 5 — landmass; 6 — sedimentation of plankton remains; 7 — current directions.

precipitated in places where an oxidizing environment changes to a reducing one and the readily soluble U^{6+} compounds convert into the rather less soluble U^{4+}.

Phosphates precipitate from sea-water at a depth of $50-80$ m and organic P_2O_5 is of deep-sea origin. The solubility of P_2O_5 is directly proportional to the partial pressure of CO_2 in the sea water. CO_2 escapes near the water surface because of the decrease in pressure and P_2O_5 precipitates from the solution and forms calcium phosphate concretions in a zone parallel to the shore (Fig. 51). The width of this zone depends on the inclination of the sea bed; thus in epicontinental seas phosphates are dispersed over a 1,000 km wide belt, whilst they are concentrated in several layers of greater thickness within a narrower belt in geosynclinal seas. Ascending sea currents which raise cool water with CO_2 and P_2O_5 to the shelf, are caused by offshore winds from an arid landmass.

Bituminous shales were laid down in fresh-water basins (e.g. in the Palaeocene — Green River, Colorado), lagoons (e.g. in the Late Carboniferous — Irati, Brazil), or sea embayments (e.g. in the Late Jurassic — Russian Platform). They usually formed at the peak of a transgression (e.g. Ordovician kuckersite). The landmass around the basins in which the bituminous shales were formed, had the character of a peneplain.

Oil deposits are not significantly controlled by their palaeogeographical position, since oil can be produced in diverse parent rocks and does not remain in situ but migrates into porous reservoir rocks. It is generally accepted that oil forms in the foredeeps of geosynclines; from the Alpine foredeep, for example, two thirds of the world's oil production is extracted. The porous reservoir rocks are all deposited in virtually the same regions, since the greater part of the clastic material is deposited in the sea close to the developing fold mountain range at the contact with the subsiding platform margin. The formation of oil is encouraged by a warm and humid climate.

Coal deposits were formed in extensive shallow littoral and terrestrial depressions of tectonic or erosional origin, which were filled with deltaic sediments.

Refractory clays are associated with continental, generally lacustrine sediments on platforms and are often accompanied by coaly sediments. The neighbouring land generally formed a peneplain and was exposed to lateritic or at least kaolinitic weathering. If present, the amount of free hydrated Al-oxides decreases with distance from the shore. Sedimentation occurred in the presence of organic matter and in a reducing environment, which caused the conversion of Fe^{3+} into Fe^{2+}, that is, the change of relatively insoluble Fe compounds into more soluble ones, which were subsequently leached from the clays.

Glass quartz sand owes its almost pure silica constitution to the decomposition of other minerals in the weathering crust or to their removal during redeposition. Glass sand deposits are formed in shallow littoral seas, close to the mouths of large streams, which have traversed plains mainly composed of easily eroded sedimentary rocks, that is, previously sorted quartz material. Any mafic minerals which remain are separated by prolonged washing of river alluvium by wave action, or are winnowed out.

Foundry sand contains an admixture of clay and is deposited in rivers, lakes and shallow littoral seas.

The palaeogeographical factors that are significant in prospecting for mineral resources can be placed in several categories, each of which demands special attention. These are: 1. the relief of the source and accumulation areas, 2. the climate, 3. the ancient drainage pattern, 4. the form of the shoreline, 5. the direction of currents in the accumulation area (in the river, sea or lake), and 6. the presence of volcanic centres. All these factors are affected by tectonism.

Particular attention should be directed to the peripheries of peneplain areas since these are potential sites of sedimentary chemogenic and biogenic deposits (bauxite, iron and manganese ores, salts, refractory clays, coal). Also of importance is the topography of the sedimentation area; for example, refractory clays and diatomites often form at the centre of a lacustrine basin, whereas bauxite and iron and manganese ores are confined to its margins. Karstic depressions are particularly favourable for the accumulation of iron and bauxite.

The genesis of sedimentary deposits is supported by a tropical or subtropical climate, which makes the reconstruction of climatic zones in various geological epochs particularly desirable. Only placers and quartz sand deposits are relatively independent of climate.

The determination of the course of old river valleys is especially useful in the search for placers and Cu and U deposits. Other types of deposits occur in areas where these valleys enter the sea, mainly in embayments. The importance of establishing the position of the shoreline can be demonstrated in the case of manganese deposits. In oxidic ores the P content increases with distance from the shore, as does the amount of Fe and S in carbonate ores, as well as the Ca and Mg content. The direction of sea currents can be determined from the angle of diagonal stratification, as it corresponds to the direction of dip.

Submarine volcanic centres produce ore elements (Fe, Mn, Pb, Hg, Cu, Zn and Sb), which help to form effusive-sedimentary deposits, especially during geosynclinal subsidence.

Tectonic quiescence enables residual deposits and deposits of chemogenic rocks to develop. Weathering crusts and sedimentary deposits are preserved in rapidly sunken blocks. Since the fold regions are composed of petrographically variable rocks, they provide a wider spectrum of residual deposits and sedimentary cycles after peneplanation. Tectonic movements leading to regressions favour the formation of a weathering crust on land and salt formation in lagoons. Younger sediments deposited by transgressing seas protect the weathering crust from erosion. Transgression also promotes the ascent of water from the depths and thus the formation of phosphates. Deposits of bauxite, Fe and Mn deposits, and placers are usually located just above the transgression surface.

The palaeogeographical conditions also affect the mineral deposits after their origin, for example, the uplift of deposits results in their oxidation and leaching and occasionally in their partial or complete destruction.

9. PALAEOCLIMATIC CRITERIA

Palaeoclimatic criteria are particularly important in prospecting for deposits related to weathering crusts. The residues of some rocks are enriched through weathering with poorly migrating elements to form economically valuable accumulations, such as Ni-hydrosilicates on serpentinites, Al-rich laterites on rocks poor in Fe, kaolins on feldspathic rocks, Mn-oxides on Mn-rich rocks (e.g. gondites) and gossans with Au, Pb and Fe on the corresponding primary deposits. This type of weathering has occurred since the oldest geological periods; kaolin, for example, has been formed in post-orogenic phases since the Carboniferous. Residual and many sedimentary deposits form in a tropical climate. If we know the approximate position of the equator in individual geological periods, we can infer where intensive weathering occurred. Conclusions on the Late Carboniferous and Cretaceous to Palaeogene dates of kaolinization in Europe (Kužvart and Konta 1968), will have practical significance if the occurrence of pre-Upper Carboniferous or pre-Cretaceous feldspathic rocks on either side of the equator are established. Where these rocks had a slightly dissected surface in the Late Carboniferous or in the Cretaceous to Palaeogene, kaolinic or lateritic weathering crusts may be preserved in sunken blocks and beneath younger sediments. In addition to kaolins and laterites, placers with Au, Pt, Ti, diamonds and other heavy minerals also formed in humid zones. The redeposition of a weathering crust rich in Al minerals gives rise to bauxites. Sedimentary Mn- and coal deposits also formed in a humid climate. The formation of the majority of coal deposits during three principal periods – the Carboniferous, Permian to Jurassic, and the Late Cretaceous to Tertiary – can be explained, like the formation of thick weathering crusts, by extraterrestrial effects (e.g. an increase in solar activity) or, as is more probable, by planetary events (e.g. orogenies accompanied by intensive volcanism and changes in the configuration of the earth's surface). In periods of slight arid weathering fewer deposits of sedimentary iron ores were formed, since the small quantities of iron liberated by exogenic agents remained in the weathering crust and were not transferred to sedimentary basins (Devonian, Permian). Surface and ground waters in the dry savannah of Ghana, where laterites are presently being formed, contain only 1.6 ppm R_2O_3, whereas waters in the rain-forest region in southern Ghana contain 2.1 ppm R_2O_3. Precambrian ferric quartzites of the Krivoi Rog type formed as a result of Fe mobility (in the form of $Fe(HCO_3)_2$) in an atmosphere containing CO_2, CH_4, NH_3, H_2O, that is, in a reducing alkalic atmosphere. The alternation of Fe-bearing laminae with quartz laminae may have resulted from seasonal changes in weathering conditions.

Other deposits of arid zones include dolomite, Cu-bearing sandstones, sedimentary Pb and Zn ores, gypsum, halite, potassium salts, celestite, borates and bromine.

10. HISTORICAL CRITERIA

Historical criteria include written reports on ancient mining, old mine maps, archaeological finds (hammers, chisels, lamps, remnants of timbering), traces of old mine workings, relics of old dressing and smelting plants, slag heaps and local names. Until modern times only gold, copper, tin, silver, iron, lead, antimony and mercury deposits were exploited. Deposits of other metals are usually not affected by ancient mining. The study of historical criteria in countries with old mining tradition helps very substantially in later prospecting (e.g. for uranium in the Erzgebirge – Krušné hory – and for gold in Czechoslovakia). It is very important that the material from old dumps and slag heaps should not be used, for example, as road aggregate, since the gangue material and slag may contain important elements formerly considered useless. A legal basis for such protection is given in the mine law. The oldest reports on the protection of dumps date from the sixteenth century (Kutná Hora, Czechoslovakia).

The importance of theory in prospecting

A negative theory stating that mineral deposits cannot occur in a certain area or under certain geological conditions may cause greater damage than an erroneous positive theory. In some cases a deposit may be discovered using an incorrect theory inferred from a correct observation. It is wrong to develop a theory, prove its accuracy even if the results of observations have to be distorted, and stick to it irrespective of contrary findings. The American geologist Ira B. Joralemon (in H. E. McKinstry, 1948) says that "the geologist must not devote himself to any one theory or to the facts that support it long enough to fall hopelessly in love with it. He must make each theory the object of a summer flirtation and not a wife – he must be ready to throw each one over the moment a more attractive mental maiden comes along." Washington Irving mentions a man who "finding that the world would not accommodate itself to the theory, he very wisely determined to accommodate the theory to the world" (in our case to facts observed in nature).

NATURAL FACTORS CONTROLLING
THE CHOICE OF PROSPECTING METHODS

On the basis of structural-geological conditions two types of areas can be distinguished: 1. lower structural layers of platforms and uplifted fold areas, both of them characterized by strong faulting, metamorphism and intense magmatism;

2. platforms with a sunken lower structural layer, and sunken fold areas overlain by a sedimentary cover which has undergone only weak tectonic disturbance. Uncovered and covered terrains are differentiated according to the nature of the Quaternary mantle. The weathering crust and its disintegration products transported a short distance (colluvium) are developed predominantly in uncovered areas. In covered areas, alluvial, glacial, marine, lacustrine or eolian sediments, transported a great distance, mantle the bedrock. Geomorphological conditions particularly affect prospecting procedures in uncovered areas. From the bioclimatic point of view the following can be distinguished: a tropical forest zone with acid soils and waters, and savannah, steppe and desert zones with alkalic or neutral soils and waters. The desert zone can be divided into a subzone free from permafrost, a subzone with patches of permafrost and a subzone with continuous permafrost.

Prospecting methods are selected on the basis of the natural factors and the criteria mentioned above. In covered areas, for example, reliable results are obtained by combining geophysical measurements with drilling, but not by geological methods. In uncovered fold mountain areas, on the other hand, geological prospecting methods can be used successfully. Geochemical prospecting provides the most reliable results in folded, slightly dissected areas, covered with eluvium and colluvium. The depth of sampling depends on the particular bioclimatic zone.

PROSPECTING METHODS AND INDICATIONS

All prospecting work is based on geological maps. Prospecting criteria determine the location of this work and prospecting methods determine how the prospecting is to be carried out. At present, mining-historical, field and aerial geological, ground and aerial geophysical and geochemical methods are used.

MINING-HISTORICAL METHODS AND INDICATIONS

In areas to be investigated by geological and mining work, archives and scientific publications should first be studied. On *old mining maps* of Central Europe both underground and opencast mines as well as outcrops of deposits are plotted. Drifts are shown either by a single line or two parallel lines as is done today. The directions of the drifts (N–S, E–W) and the levels are depicted in different colours. Ore material and gangue are differentiated on some old mine maps. For every mine or group of mines a separate surveying network was constructed; this network was usually oriented according to the geographic or magnetic north, but in some maps the south is at the top.

The dip of veins can be determined from the course of two levels which follow the same vein but at different elevations (Fig. 52).

The geology of the deposit can be estimated from the course and termination of drifts even without outlining the geological situation on the map. A fault with a small horizontal displacement is reflected in a moderate bend of the drift (Fig. 53a). The interruption of a vein by a fault can be inferred from numerous branches and short cross-cuts driven from the drift, which indicate futile attempts by former miners to master the complicated tectonic situation (Fig.. 53b). Exploration in such places is promising.

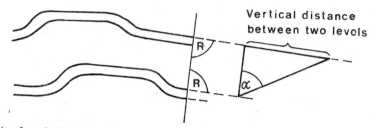

52. The dip of a vein (α) followed by two galleries at different levels can be determined from a mine map, if the approximate distance between levels is known (R — right angle).

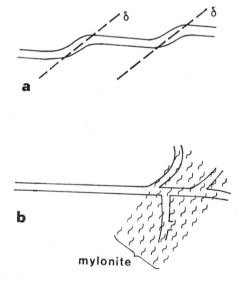

53. Geological conditions of ore deposit inferred from a mine map. a — slight bends in a gallery indicate that the deposit is disrupted by faults with a small horizontal displacement; b — termination of a gallery in several branches reflects vain effort of former miners to find a vein disturbed by complicated faulting.

The distribution of "prospecting circles" and claims can also provide valuable information. In Central Europe everybody had a right to stake out a "prospecting circle" after indication of mineralization had been found. This circle, having a radius of 425 m, could be laid out anywhere outside existing "prospecting circles" and fenced-in land. Its centre was designated by a numbered post and its position surveyed; the tower of the nearest church was used as the datum point and the distance of the post from it was given in metres and the direction in degrees.

If a deposit was found, it was requested that a claim be granted. From the beginning of the nineteenth century a claim in Central Europe was 448×112 m in size. In the U.S.A. the claims on vein deposits are 1500 feet long and 600 feet wide. An individual claimant can own 20 acres on a placer deposit and an association can be alloted a placer claim of up to 160 acres. Individual states and territories have their own statutory rulings concerning the size and location of claims. The shape and extent of a deposit can be estimated from the *arrangement of claims* on a map (on Fig. 54 the claims follow a vein and its offshoot). The significance of the distribution of "prospecting circles" must be assessed very carefully, since they were also placed between geologically unrelated deposits in order to join the lots of one owner.

54. The grouping of claims suggests the shape of the deposit (vein with offshoot).

After studying the archives and literature, field evidence for old mining should be examined. The relics of open-pit mining such as pits and old quarries are most easily traceable. *Pits* were produced by the opencast working of deposit outcrops down to the ground-water table. Figure 55 shows a section through (a) an opencast during mining and (b) after caving-in. Pits are invariably concentrated in elongated groups or narrow rows following the outcrop of a deposit. By mapping these rows, faults (Fig. 56a), the local loss of ore or pinching out of the deposit (greater distances between trial pits Fig. 56b) and the abrupt (tectonic?) termination of the deposit can be recognized. In the last case, prospecting is particularly promising (similar to the example shown in Fig. 53b). If the row of pits ends where a rock complex is covered by younger transgressive sediments, it is advisable to search for the continuation of the deposit beneath the younger sediments where it has not been worked. The tectonic structure of the bedrock should be extrapolated beneath the younger cover (Fig. 57). The structure of a vein and its primary lateral changes can also be determined by following the trend of the row of pits (Fig. 58); in hilly terrain its course indicates the general direction and the angle of dip of the vein (Fig. 59).

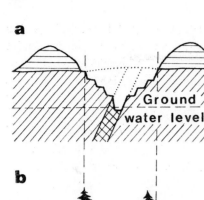

a

Ground
water level

b

c

55. Vertical section of a pit in operation (a) and after partial caving-in (b); ground plan of a pit (c). Horizontal ruling—material extracted, cross-hatching—vein.

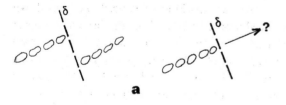

a

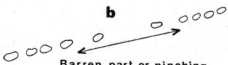

b

Barren part or pinching
out of the vein

56. Lines of pits indicate disruption of the deposit by a fault (a) and by pinching out or local barren zone (b).

57. Line of pits (4) ends where the younger formation (3) transgresses the crystalline complex (1). In following the unextracted vein segment (5), tectonic disturbance of the amphibolite interlayer (2) must be taken into account.

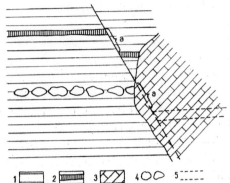

58. Diagram showing the lateral changes in vein material, as determined by following the line of pits. Quartz, pyrite and sphalerite found in pit heaps.

Exploited part of the deposit

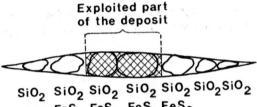

SiO_2 SiO_2 SiO_2 SiO_2 SiO_2 $SiO_2 SiO_2$
FeS_2 FeS_2 FeS_2 FeS_2
ZnS ZnS

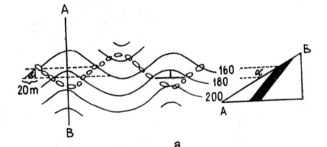

a

59. The strike and dip (α) of a vein can be inferred from the line of pits in a diversified terrain.

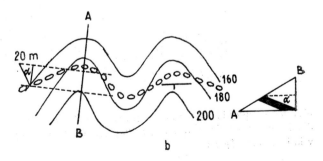

b

Old underground mining has left depressions on the surface as a result of subsidence at the mouths of shafts, caved-in openings to galleries, and waste heaps. The *sinks* formed by subsidence are similar to the pits produced by opencast mining and they must be distinguished strictly from the latter, as they can also develop above cross-cuts which do not warrant following. Surface subsidences often continue long after the mine has been abandoned. In the neighbourhood of sinks no tipped waste material is found (Fig. 60a). The width of the sinks above the worked-out ore indicates the dip of the vein (Fig. 60b, c).

The study of archives, the precise survey of all traces of old mining, and petrographical, mineralogical and chemical analyses of waste material, provide data which can replace metallometric mapping, and even geophysical survey, in locating outcrops and aureoles. In favourable cases information is also furnished on the economic value of the deposit at depth, the mode of deposition and the hydrogeology, among other data.

The openings of old galleries sometimes remain undisturbed, especially where the gallery was opened in firm rock. The timbering at a gallery opening lasts at most

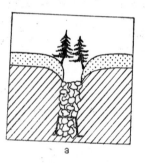

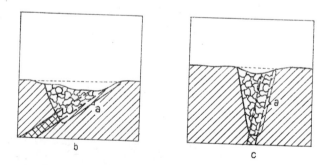

60. Surface depressions (a) formed after caving-in of mine works; these are broad above a moderately dipping vein (b) and narrow above a steep vein (c). Sector a represents the worked out part of the vein in all three figures.

several dozen years. The entry to the gallery may be marked at the surface by an elongated depression running nearly perpendicular to the slope. The caved-in mouths can also be pinpointed by the presence of fresh green vegetation which results from water discharge from the old gallery; this water is sometimes coloured by Fe-hydroxide.

An old *waste heap* provides information on the mineralogy and petrography of the deposit and indicates the position of the particular gallery. The length of drifts can be inferred from the size of the heap. Rich ore samples are rarely found on the heap of an abandoned mining work and samples with a low metal content predominate. Ores should be sought in large boulders in the mid-slope of the dump. On a dump of an exploration work rich ore samples can occasionally be found near the gallery entrance. The succession of rocks through which the gallery was driven is reflected in a reverse position in the material of the waste heap (Fig. 61).

The position of a gallery can usually be established quite easily. However, if a thorough examination of the slope at the level of the heap proves unsuccessful, the gallery must be sought higher up. It is possible that gangue material from the gallery had to be hauled across a bridge to be deposited at a lower site in order to avoid burying a road or other important structure.

The length of galleries and drifts can be determined either by geophysical methods or from the volume of the waste heap. This is either calculated mathematically if the

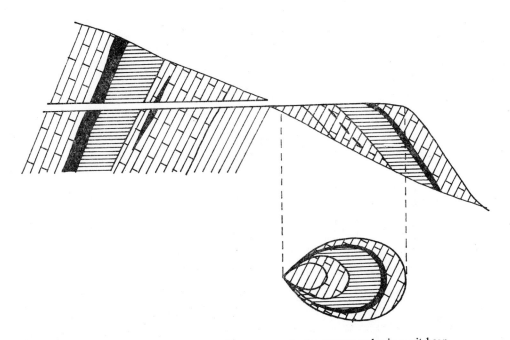

61. The rock succession traversed by a gallery appears in the reverse order in a pit heap.

120

heap is of a regular shape, or by surface levelling along several lines of reference points. From these data several longitudinal sections of the heap are constructed. Their areas are measured, summed and multiplied by the distance between the sections (in metres). As the waste material has a larger volume than the in situ rock, the volume is divided by the coefficient of bulking, an average value of 1.4. The resulting value is further divided by 2, because on driving 1 m of gallery about 2 m^3 of rock is excavated.

The gross length of a gallery can be calculated from the formula $D = \dfrac{P \times a}{2.8}$, where D — the length of the gallery, P — the sum of all the section areas, and a — spacing of the sections.

If several galleries have been driven to work one vein, its course can be inferred from their distribution (Fig. 62); the dip of the vein is estimated from the length of cross-cuts and this in turn from the size of the pit heaps (Fig. 63). In some cases the faulting of a deposit can be assumed with a certain degree of probability (Fig. 64) and, in exceptional cases, also the primary and secondary differences in the vein (Fig. 65 — example from the Spišsko-gemerské rudohorie, Slovakia). The prerequisites for determining the original depth differences are a large vertical distance between the highest and lowest galleries (of the order of hundreds of metres) and rapid ore differentiation.

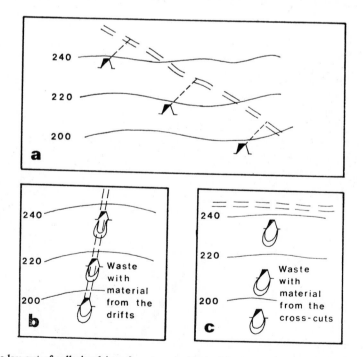

62. From the lay-out of galleries driven for one vein (dashed line) it can be deduced whether the vein runs along the slope (a), at right angle to the slope (b), or parallel to the contours (c).

Waste with material from the cross-cuts

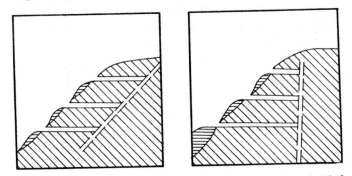

63. The size of pit heaps lying above one another indicates the dip of the vein (dashed lines).

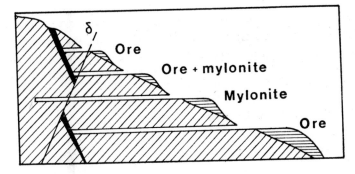

64. Material of pit heaps from galleries driven above one another for the same vein can show the tectonic disturbance of the deposit.

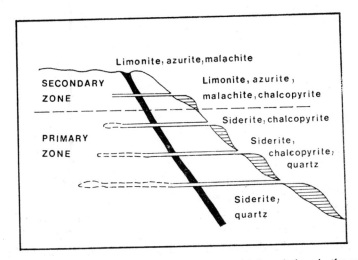

65. Material of pit heaps can show primary and secondary depth variations in the vein.

Ancient mines were usually connected by *mine paths,* and these can also lead to galleries and waste heaps. Pebbles of resistant gangue minerals (barite, vein quartz) in the brooks also point the way to old dumps.

Old placers must first be plotted on a detailed map and then an approximate estimate of the volume of the washed alluvium and the output of useful minerals can be made. By sinking pits to the bedrock it is possible to determine whether the richest and least accessible part of the deposit, that is, material from grooves in the bedrock has been exploited. Care must be taken not to mistake large boulders in the alluvium for the bedrock. Enriched alluvial bodies, elongated parallel to the valley (old channels) should be examined by means of a series of cross sections, which are constructed on the basis of a quantitative evaluation of heavy mineral concentrates from large-diameter drill holes. The primary deposit should then be sought on one of the valley sides near the upstream end of the heaps of washed material.

GEOLOGICAL GROUND PROSPECTING METHODS

Ground prospecting methods, which are the oldest and simplest, involve searching for outcrops of deposits or fragments of ore material. The fragments may form aureoles near the deposit or may have been transported by a stream or glacier, often for a considerable distance. In ground prospecting, the geologist proceeds in the following way: first ore minerals in river alluvium or moraine are sought, then the aureole of ore fragments is examined and finally the outcrop of the deposit is located.

Prospecting for deposit outcrops

The outcrops of deposits may be topographically conspicuous even at a distance. Resistant veins form long elevated ridges (*positive relief*), the most striking of which, however, are formed by barren (for example quartz) veins. If the mineralized parts of veins have been leached, these appear as depressions in the relief. There are, of course, gossans of rich deposits that form prominent elevations because of a firm wall rock. This was the case of the Broken Hill deposit of Pb−Zn ores in Australia, which was discovered by the first European to enter the area by its morphological peculiarity.

Fragments of a resistant vein are usually dispersed for some distance beyond the outcrop and this may lead the explorer to overestimate the true width of the vein. A positive relief and wide scatter of fragments are distinctive features of outcropping zones of silicification, barite, pegmatite, Fe-quartzite, corundum-bearing rocks, skarn, magnetite (in Sweden), baked beds in the vicinity of burnt coal seams, and serpentinites. The bodies of the Vermilion Fe-deposit near Lake Superior (U.S.A.) because of their topographic elevation are locally called the "ranges". The Fe-ore deposits of the Minas Gerais State in Brazil also appear as elevations in the topography.

123

An elongated depression (*negative relief*) develops along the outcrop of veins of low resistance as, for example, carbonate gangue, limonite and graphite deposits, in zones of intensive tectonics and at the outcrops of hydrothermally altered rocks (sericitized, chloritized, dolomitized, talcized, kaolinized or carbonatized). Depressions may also form above concealed deposits, the upper parts of which have suffered oxidation and a consequent volume change (Fig. 66), as occurred in the Bisbee deposit in Arizona. The depressions are direct indicators of the occurrence of Ni-silicate ores and bauxites.

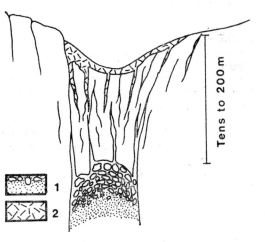

66. Surface caving-in above the oxidized part of a concealed sulphidic deposit (after Smirnov, 1957). 1 — oxidized part of the ore deposit; 2 — surface sediments.

The "*ore karst*" features which originate near ore veins as a result of the dissolution of neighbouring limestones by CO_2-bearing hydrothermal ore solutions, also show a negative relief. Similarly, sulphuric acid liberated during the weathering of sulphides in the oxidation zone can dissolve limestones to such an extent that underground caverns, several thousand cubic metres in size can be formed (e.g. Karatau and Karamazar in the U.S.S.R.). Some supergene alterations of ore minerals are accompanied by volume changes; the conversion of arsenopyrite into scorodite causes a twofold increase in the mineral's volume. Such a process results in the extension of the deposit outcrop, and the fracturing and preferential denudation of the wall rock.

An outcropping deposit undergoes certain physical and chemical changes. Thus *coal* in outcrops has a higher ash and water content, is poorer in carbon and does not sinter. Such alterations affect coal seams to a depth of 10—20 m. Seams and veins are of a smaller width in outcrops since the terminal curvature of beds results in their dip being closer to the slope inclination. Cases have been recorded in the literature of a 2 m wide pyrrhotite vein and a polymetallic vein 15 m wide appearing at the surface as a limonite "lead" only several millimetres thick. *An asphalt or ozokerite plug* formed over the outcrops of reservoir oil beds prevents the deposits from being further destroyed. *A gypsum cap* is typical of salt domes and an *alum cap* characterizes sulphur deposits; *caliche* (fine-grained calcite) is distinctive of some ore deposits in arid regions and *jasperoid* of others, for example, the supergene Ground Hog chalcocite deposits in New Mexico (U.S.A.).

Alterations affecting the ore outcrops are particularly diverse. A gossan develops not only on iron ores but also on polymetallic deposits, although it is most typically found on pyrite and pyrrhotite deposits.

The formation of a gossan and cementation zone and their preservation are controlled by the chemistry of the deposit and the ground water, and by the climatic conditions, i.e. temperature, humidity and vegetation. These relationships are shown in Table 40. Calcite gangue and carbonate wall rocks extend the gossans.

TABLE 40

Climate / Zone	humid tropical	dry tropical	humid mild	dry mild	humid polar	dry polar
oxidized	prominent, often leached	very prominent	occasionally prominent	prominent	very slight	very slight
cementation (enriched)	missing or faint	often very prominent	faint	often prominent	missing	missing

In prospecting for mineral deposits, the study of redeposited gossans is important; they are detected by the rusty or ochreous colour of the sediment. A layer with washed gossan material at the base of a younger transgressive sedimentary formation suggests that the advancing sea met an ore outcrop, which was destroyed by wave action and the material spread by the currents over a large area. In this case it is necessary to study, among other things, the palaeogeography of the area during the transgression, especially the position of the shoreline at certain periods, the direction of currents as derived from the facies and palaeontological content of the sediments, and the areal extent of rusty sediments as determined by boring. With favourable conditions such an integrated investigation can detect deposits even if they lie buried by a thick layer of younger sediments.

An example of *redeposited gossan material* preserved in the Plzeň basin (West Bohemia) was described by Z. Pouba (1955). During the Tertiary, Algonkian pyritic shales were strongly decomposed by sulphuric acid produced by the weathering of pyrite. The Proterozoic shales were kaolinized near the surface and an Fe-bearing horizon, 30 m thick, developed below the leached layer. During the Oligo-Miocene period, this material was re-washed and deposited in depressions to produce an inverted profile. White beds, originally forming the surface layer, were deposited at the base and the ochres that initially formed the underlying bed, were deposited on top of them.

If the direction of streams transporting the gossan material is known, its original location can be determined. The stream direction can be deduced from the disposition of flat pebbles in the river bed, in a conglomerate or a river terrace; they are generally moderately tilted upstream.

Normally a gossan has a mineralogical composition similar to that of a typical bog iron ore. The two types of limonite ore are distinguishable on the basis of shape, structure and accessory minerals. The limonite gossan deposit shows a linear form, a cellular structure and an admixture of secondary Cu, Pb and Zn minerals. A deposit of bog ore on the other hand usually covers a considerable area, the limonite may cement breccia or gravel, and indications of stratification are visible in places.

The minerals of the weathering crust of ore deposits (gossan, oxidation zone) can be divided into:

1. relict sulphides,
2. insoluble sulphates, basic sulphates, phosphates, arsenates, vanadates, hydrosilicates,
3. carbonates and basic carbonates of Cu, Pb and Zn ,
4. clays, particularly montmorillonite clays, which sorb Zn and to a certain degree Pb, Cu and other elements,
5. oxides of manganese, iron, lead and other minerals.

The oxidation zone of sulphidic deposits was studied in great detail by S. S. Smirnov (1951). A distinctive feature of this zone is the gradual impoverishment in sulphur and heavy metals. The same compounds (SiO_2, Fe_2O_3, Mn_2O_3 and Al_2O_3) as in the weathering of silicates remain in situ and combine with one another and with water. Leaching of metals is prevented by the carbonate wall rocks, which neutralize acid solutions and induce precipitation of secondary minerals. The surface layer, several decimetres thick, is depleted in even those elements with a small migration capacity (e.g. Pb, Sb, Bi), and it must therefore be removed before exploration.

In terms of their *migration capacity*, elements can be classified as very mobile (e.g. K, Na, Ca, Mg), mobile (e.g. Cu, Ni, Co, Mo, U, Ra, Zn) and poorly mobile (Fe, Ti, Al, Zr, Pt, Au, Sn, W, Hg); elements of the last group are often present in relict minerals of the primary zone. In studying the outcrops of ore deposits attention should be directed to stable minerals, typomorphic secondary minerals, the structure of limonite and the presence of a small amount of metals.

The *following minerals are not altered in the oxidation zone:* primary oxides and hydrates such as hematite, limonite, pyrolusite, Al-minerals of bauxite, chromite (in New Caledonia this is leached to a depth of 10 m as a result of intensive weathering), cassiterite, corundum, rutile, diamond, fluorite, barite (which dissolves in association with sulphides and fluorite!), topaz, monazite, wolframite, scheelite (sometimes covered by tungstite), cinnabar (the most stable sulphide of the oxidation zone), gold in quartz veins, and platinum. The chemical analysis of samples from the oxidation zone is also valid for the primary ore. The oxidation zone of Au, Ag, Sr, Ta, W and

Pb deposits may occasionally have an enriched mineral content. Vanadinite and wulfenite are concentrated only in the oxidation zone.

The alteration of minerals in the oxidation zone of Pb, As, Bi and Sb deposits, and those of carbonate Fe and Mn ores, occurs without substantial leaching of the metal. The principal ore mineral of lead – galena – is unstable in the oxidation zone and is oxidized to anglesite which in turn is converted into stable cerussite; this may still contain relics of galena. The lead may form compounds with Mo and V (wulfenite, vanadinite), which are present in minute quantities in the primary ore (thousandths of one per cent). In the Mammoth-St. Anthony deposit (Arizona, U.S.A.) the Mo content in the lead ore is workable in the oxidation zone. Only under desert conditions can lead be completely leached (Mianda deposit in Tanganyika). *Arsenopyrite* is usually altered into stable scorodite or beudantite, and *bismuthine* is converted to bismite or bismuthite. *Antimony oxides* (such as valentinite and cervantite) are the end products of the decomposition of antimonite. *The carbonate Fe and Mn ores* are converted into their relevant oxides in the outcropping parts. In this group of deposits only the mineral composition, not the metal content, changes in the oxidation zone. Ore analyses from the oxidation zone are also valid for the deeper parts of the deposit.

Zinc is *intensively leached in the oxidation zone* and its principal mineral – *sphalerite* – alters into the readily soluble sulphate. Zinc remains close to the deposit only when the Zn sulphate encounters carbonate rocks. The differences in the history of individual metals in the oxidation zone is demonstrated particularly well with zinc and lead, which often occur together in the primary ore. The lead remains in situ as cerussite (see above) whilst the zinc migrates a considerable distance unless it meets a carbonate rock. A prospector can thus draw an important conclusion: if Pb ores are found near the surface in carbonate rocks, smithsonite ores should be sought at a greater depth nearby.

Copper minerals also decompose into a soluble sulphate, which migrates from the deposit, especially in rich sulphidic ore bodies. The oxidation zone of Cu deposits with malachite, azurite and cuprite is underlain by a leached zone with loose quartz, and this in turn by an enriched zone with chalcocite and covellite, which contains 2 – 3 times more copper than the primary zone lower down. In prospecting, it is important to assess the existing stage of the oxidation zone. If a large amount of relict sulphides has been preserved at the surface and the subsurface waters are acidic, the gossan is in the initial evolution stage and the cementation zone has not had enough time to develop and exploration related to it will be useless. A "dead" gossan, from which metals have been leached, and a neutral reaction of subsurface waters suggest the presence of a well-developed cementation zone. Limonite with pitch lustre, and Cu carbonates and oxides in the gossan indicate that the alterations occurred in the absence of acid, this being either neutralized or originally present only in small quantities due to the lack of pyrite. In this case copper has remained in the oxidation zone.

The Ni sulphides decompose into soluble sulphate, which is either removed from the deposit or reacts with As (if present) to form insoluble annabergite. Cobalt shows a similar behaviour and the end-product of its alteration in the oxidation zone is erythrite, heterogenite or asbolite.

Molybdenite, the principal Mo mineral, decomposes slowly into a sulphate with a high migration capacity or into an acid, which reacts rapidly to form ferrimolybdite or powellite (Fe and Ca molybdates), both highly soluble. However, in the oxidation zone of the Hall deposit (Nevada, U.S.A.), 30−40 % of the molybdenite has been preserved whilst in the Climax deposit (Colorado, U.S.A.) as much as 80−90 % remains. In arid regions Mo can be completely leached out as a soluble sulphate.

Uranium deposits often show zoning in the oxidation zone. In descending order the sequence is: hyalite with adsorbed uranium, uranium sulphates (e.g. uranophane), yellow micas (e.g. autunite), green micas (e.g. torbernite), and sulphates (e.g. zippeite). In the absence of phosphoric, arsenic and vanadic acids in the deposit, the soluble compounds of U^{6+} (produced by weathering of uranite) can be completely removed by acid waters.

Submicroscopic *gold* (below 0.1 mm) occurs in sulphidic ores and in the oxidation zone passes into solution and precipitates in loose quartz and barite (zone of leaching), in concentrations 50−100times higher than in the primary ore. Silver and selenium sometimes behave in the same way. If the Au content is too low in the oxidation zone, the exploration of the primary ore is useless. In arid regions, gold accumulates in the cementation zone and is transported in the form of poorly soluble chloride (chlorine is formed through the reaction $H_2SO_4 + MnO_2 + NaCl$). The Cu, Ag and Cd deposits become significantly enriched in the cementation zone.

When solution of Cu, Ag, Zn and Pb sulphates from the oxidation zone come into contact with pyrite of the primary ore, they react to form sulphides which constitute the zone of secondary enrichment. Solutions of salts of any metal from the Hg, Ag, Cu, Bi, Cd, Pb, Zn, Ni, Fe, Mn series will be decomposed by the sulphide of the following metal and precipitated as sulphide. On Ag, Cu, Pb and Zn deposits, for example, secondary sulphides (argentite, chalcocite, galena, sphalerite) will precipitate on reaction with pyrite (FeS_2), since Fe is almost at the end of this series. This *Schürmann rule* is generally valid but there are some restrictions since the origin of cementation minerals is also considerably affected by the affinity of individual minerals, the hydrolitic and coagulation properties of the originating transient compounds, and other factors.

Oxidation products can penetrate deeply along fissures. On the Santa Eulalia Pb−Ag deposit (Chihuahua, Mexico), new bodies of oxidized ore near the surface were detected using overhand stoping, by following limonite streaks along the joints in limestones. On the Bisbee deposit, the solutions rich in Fe-sulphate reacted at depth with limestone to form siderite bodies, which are indicative of the presence of copper ores in the overlying rock complex.

The non-oxidized ore can outcrop at the surface where the oxidation zone has been removed, for example, by glacial erosion, and if the outcrop was subsequently covered by an impermeable clay. The oxidation zone on the Noranda chalcopyrite deposit (Quebec, Canada) was only 0.5 cm thick. In some places, large bodies of non-oxidized ore have preserved in gossans, such as sphalerite in the Hanover deposit gossan (New Mexico, U. S.A.). If a primary deposit is unworkable at the surface, it is usually superfluous to explore it to a greater depth.

The technological properties of ores from oxidation zones have been the least explored, and dressing procedures for oxidic Zn, Ni, As, Sb and Mo ores have not so far been developed.

The structure of gossan limonites (i.e. a mixture of lepidocrocite, goethite and hematite) may help to determine the minerals from which the gossan originated (Fig. 67). The cell walls are formed of silicified limonite, which was the first to fill the fissures in the sulphides. Later the cells were filled with soft limonite and other secondary minerals. Their outline indicates the boundaries and cleavage systems of primary minerals, along which replacement by limonite or occasionally jasper occurred. The dimensions of diagnostic limonite sections should be determined with the aid of a magnifying glass. If pyrite represents more than a quarter of the sulphides in the ore, this method cannot be used because a large amount of soft limonite is formed. Soft limonite precipitates from very acid solutions at a variable distance from the site of the pyrite decomposition, which produces the acid reaction of the solutions. Since a less acid medium is produced by the decomposition of Cu, Zn and Pb sulphides, Fe precipitates from acid solutions almost in situ (indigenous limonite). On replacing pyrite by chalcocite, for example, copper neutralizes the greater part of the acid formed by the pyrite decomposition. Most of the iron soluble in an acid medium therefore precipitates in situ as small porous grains forming a cellular framework (relief limonite) to the soft limonite supplied by solutions from elsewhere. In the period 1925—1935 Blanchard and Boswell published the systematics of limonite types from the oxidation zone of ore deposits in the journal Economic Geology. According to these authors, limonite from porphyry copper ore deposited in situ (relief limonite) is craggy and cavernous when formed after chalcocite or covellite and pyrite, but has a pitch lustre when formed after bornite. With a higher proportion of pyrite (in chalcocite or covellite ore) a nodular or botryoidal crust lines the small voids. Coarsely cellular firm limonite develops in situ from chalcopyrite. Transported limonite forms the haloes and rims of voids after pyrite when the solutions are of low alkalinity (e.g. in a kaolinized feldspathic rock).

In the gossans of sulphidic deposits cellular or spongy limonite usually occurs. Limonitic stalactites and stalagmites form from the transported decomposition products of pyrite. Every sulphidic mineral has a limonite with a typical structure, but their description cannot give a true picture. The method of approximate determination of primary sulphidic minerals from the limonite structure can provide adequate results only after a thorough study of samples from the field and collections, and after long experience of various deposits from different climates. The character of the primary ore can sometimes be inferred from the mineral composition of the limonites,

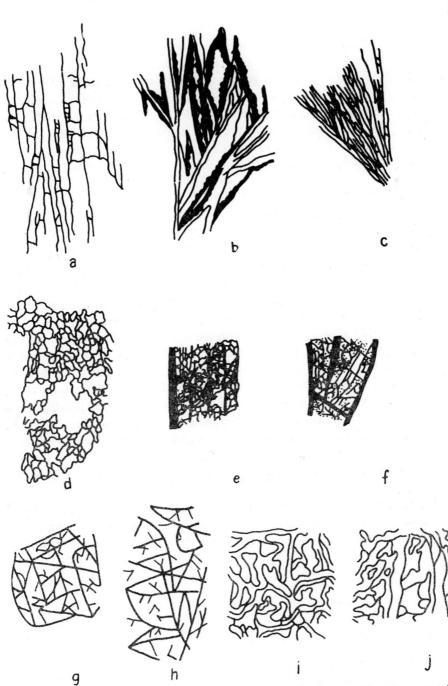

67. Structures of gossan limonites (according to Blanchard and Boswell, 1934). Primary ore minerals were: a, b, c — galena (a — cleavage boxwork, cell walls 0.005—0.05 mm, b — mesh boxwork, 0.2—0.5 mm, c — radiate structure); d, e — sphalerite (d — sponge structure, e — cellular boxwork, thick quartz walls); f — chalcopyrite, cell walls coated with limonite of pitch lustre; g, h — bornite (triangular cellular structure); i, j — tetrahedrite (contour boxwork).

particularly from the lepidocrocite/goethite ratio. Goethite, for example, predominates in the limonite from the oxidation zone of Pb—Zn deposits.

Minerals of ore gangue may be a good guide in prospecting. White vein quartz present in a large amount is usually barren. The quartz gangue of ore deposits contains cavities after weathered sulphides; these can be distinguished from the voids after carbonates by the shape and colour of the limonite (the carbonate neutralized the acid solutions containing Fe and caused their precipitation). Rhodochrosite gangue often accompanies silver ores, whilst fluorite may be indicative or Pb—Zn deposits.

Prospecting for the mechanical aureoles (haloes) of ore fragments

An ore body ourcropping at the surface or in a weathering zone undergoes chemical decomposition (see the section on geochemical methods and outcrops) and its chemically resistant components disintegrate. Along outcrops of quartz, barite, pegmatite or magnetite veins, fragments of these resistant veins accumulate in the form of elongated ridges. If the veins outcrop on a slope, their fragments move slowly downhill by gravity or solifluction. Solifluction causes the movement of fragments on slopes with an inclination as little as 3°. On steeper slopes the fragments slide. Ore fragments invariably move at right angles to the contour lines. If the vein ends in colluvium, fragments will accumulate near the outcrop, at the base of the waste layer, but on moving downslope they will get towards the surface due to the differential movement of individual layers in the waste mantle (Fig. 68). The form of an *aureole* (*areal or fan-shaped*) depends on the course of the vein and the slope morphology (Fig. 69). *Linear continuous or discontinuous talus aprons* are formed in gorges (Fig. 70), whilst discontinuous linear talus aprons occur on mountain slopes. Ore boulders may come to a standstill on moderately sloping valley sides. It is therefore erroneous to sink exploratory pits higher than the first ore talus above the valley floor.

On finding ore fragments on the surface of a slope, prospecting in the colluvial layer should proceed upslope and below the surface. As the deposit outcrop is approached, the fragments will occur deeper below the ground surface and the surface

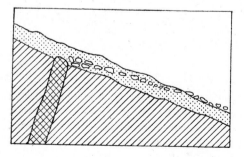

68. Fragments of vein material approach the ground surface as a result of creep.

fan will become narrower (Fig. 69a). In J. London's story "Pocket" there is a very good description of the location of an auriferous pocket in a river terrace by tracing an aureole of ore fragments.

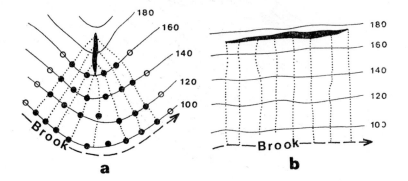

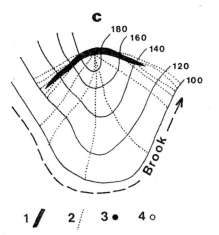

69. Formation of fanwise (a, c) and areal (b) ore debris aureoles on the slope below the outcrop of a deposit. 1 — outcrop of deposit; 2 — fragments of vein material; 3 — positive exploration works; 4 — negative exploration works.

Prospecting based on erratic ore boulders

Fragments of ore minerals transported by glaciers are rounded and deposited in moraines as a particular type of secondary aureole of erratic ore boulders. This aureole usually has the form of a triangle up to a 20 km distance, farther away its shape becomes indistinct. It is most promising to follow the basal moraine. By tracing the route of erratic boulders indicated by the striation of the bedrock, the outcrop of the primary deposit can be found; in the final stage of exploration, geophysical methods and drilling must, of course, be used. In Scandinavia important mineral deposits have been discovered using this procedure.

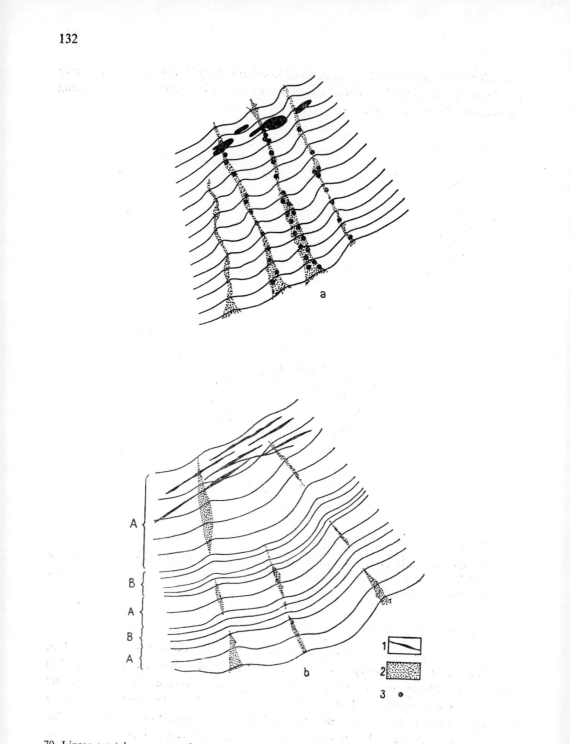

70. Linear ore talus. a — continuous; b — discontinuous (after Smirnov, 1957): 1 — outcrop of deposit; 2 — debris with fragments of vein material; 3 — ore boulders; A — moderate slope with a linear talus; B — steep slope without talus.

The first successful example of prospecting based on erratic boulders was the discovery of the Cu deposit in Outukumu in Finland. A boulder of brecciated quartzite with an ore cement, 5 cu. m in size was found near Kivisalmi, close to boulders of an ultrabasic olivine-bearing rock. By tracing the glacial striation and studying the geology of the area, the Outukumpu deposit was discovered 50 km NNW of the first boulder. Here, prospecting was facilitated by the fact that the common occurrence of brecciated quartzites and olivine-bearing rocks had been mapped only in the Outukumpu area.

The Takkejaur pyrite and sphalerite deposit in northern Sweden was found by following a large number of erratic ore boulders weighing from 1 kg to more than 1 ton (Grip, 1953). The distribution of the boulders was consistent with the direction of glacial striation. The boulders are angular near the deposit and rounded farther away; poor or apparently burnt vegetation reveals their presence even below ground surface.

The reliability limits of this method can be demonstrated in the case of the Näsliden Zn deposit in northern Sweden. One deposit was discovered by following erratic ore boulders and another, nearby, was found using geophysical methods. The latter, however, was not a source of erratic boulders, probably because of its flat surface at the time of ice movement. The erratic boulders from the Ag−Pb−Zn deposit in central Sweden had been dragged out to the surface 700 m from the deposit; closer to the deposit test pits had to be dug. The maximum distance so far recorded between the first boulder found and the deposit was 125 km, on the Ultevis manganese deposit in Lapland.

Prospecting based on heavy mineral concentrates

After the weathering of ore outcrops some resistant and heavy ore minerals are transferred into the ore-fragment halo, and from there into surface streams, which deposit them as placers with other alluvial material. Gold, platinum, diamond, cassiterite, ilmenite, rutile, monazite, zircon, wolframite, scheelite, cinnabar, chromite, columbite and tantalite are among the minerals found in alluvial deposits. The prerequisites for the formation of rich placers are a long period of weathering of rocks containing heavy minerals on a peneplain, and uplift of the area with erosion, to liberate the heavy minerals from the eluvium and enable their deposition in favourable sites. The placers must then be protected from contemporary or subsequent erosion by a cover of lava or fluviatile sediments. Very rich deposits occur where the placers were rewashed during a rejuvenation of the drainage system. Their presence in recent and fossil alluvium (river terraces) can be determined by panning. Search for placers by panning is one of the prospecting procedures used in uninvestigated areas, unless the region has been glaciated. Panning makes it possible to find 1. a primary deposit (if we proceed from the site of the first appearance of an ore mineral upstream to the aureole and the outcrop of the deposit) (Fig. 71); 2. a secondary deposit of the placer type (if the concentration of the ore mineral is so high in the alluvium that it can be exploited); 3. the association of heavy (non-ore minerals included) minerals which is characteristic of the petrographic composition of the area and a guide to further prospecting. Areas composed of acid and intermediate intrusive rocks, for example, are characterized by orthite and anatase; basic intrusives by chromite, olivine and picotite; alkaline intrusives by sphene and perovskite; metamorphic rocks by sillimanite, disthene, staurolite and cordierite; pegmatites by cassiterite, tantalite, columbite, beryl and xenotime; the contacts between acid intrusives and carbonates by vesuvian, wollastonite, spinel and scheelite, and the contacts between acid intrusives and argillaceous rocks by corundum. Some heavy minerals occur in several of the mineral assemblages mentioned above and their importance is apparent only when the whole concentrate has been studied.

Samples for panning are taken from the river, convex banks (Figs. 72, 73), shallow and deep pits sunk into the alluvium of broad valleys, river terraces and occasionally from colluvial and eluvial deposits. The site and density of sampling is determined by 1. the scale of the geological mapping, 2. the type of drainage pattern and the nature of the river bed, and 3. the petrographical character of the alluvium.

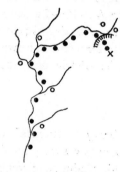

71. Localization of deposit outcrop by tracing ore mineral concentrates upstream and by verifying ore aureole.

1 — concentrate with heavy minerals; 2 — concentrate lacking ore minerals; 3 — ore debris aureole; 4 — outcrop of deposit.

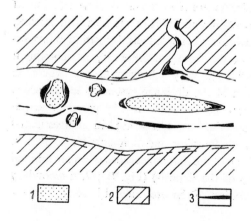

72. Enrichment of alluvium in heavy minerals (according to Zeschke, 1964). 1 — sand and gravel; 2 — valley slope, 3 — heavy mineral concentrate.

With mapping at scales from 1 : 1,000,000 to 1 : 500,000 samples are taken only in those valleys being mapped. At larger scale mapping, 1 : 200,000 to 1 : 50,000, a more detailed areal sampling is carried out. Samples are taken at intervals of 1 to 0.2 km. If this sampling has proved favourable ore concentrations, detailed sampling to localize the ore halo and mapping at 1 : 25,000 are carried out. All tributaries are also sampled back to their sources at a density of approximately eight samples per kilometre.

A mature stream pattern is characterized by broad valleys, meanders and a low gradient. When taking samples along mapping traverses, the sites should be widely spaced and concentrated downstream and upstream of junctions with tributaries (Fig. 74a). In areal and detailed sampling, samples are taken from pits dug in the alluvium. The pits are arranged in lines (Fig. 74b) and the distance between the pits and lines is controlled by the map scale. Surface sampling is possible in the river bed, on convex banks and at the base of undercut terraces (Fig. 74a). Both these procedures enable bands of alluvium enriched in ore minerals (so-called pay streaks) to be delimited.

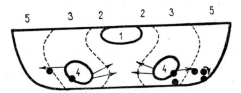

73. Cross section of a straight stream reach (according to Routhier, 1963). 1 — zone of greatest velocity; 2, 3 — belts of moderate flow and strong turbulence; 4 — zone of strongest turbulence (heavy minerals pulled into belt 2 are carried downstream); 5 — belts of slow flow and weak turbulence (heavy minerals from zone 4 remain in belt 5).

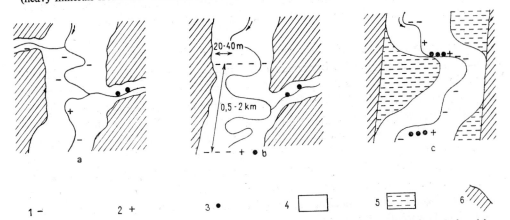

74. Sampling of alluvium in a mature stream; samples taken in trial pits at widely spaced sites (a), arranged in lines (b), and directly from the river bed (c) (after Itsikson, 1953). 1 — concentrate lacking ore minerals; 2 — concentrate with sporadic ore grains; 3 — concentrate with weight amount of ore minerals; 4 — alluvium; 5 — terrace; 6 — valley slope.

Youthful streams are commonly of steep gradient, with numerous rapids and rocky banks in a narrow valley. A general example of sampling in a young drainage system is shown on Fig. 71. Samples are taken particularly where the flow velocity slackens as the resulting fall in transporting capacity of the stream allows the load material, especially the heavy minerals, to settle (see Figs. 72, 73). Such places include the points of abrupt valley widening, the concave sides of sharp bends and below hard-rock ledges in the river bed. Depressions in the river bed due to the selective erosion of poorly resistant rocks, convex banks of meanders, and islands (in those parts facing upstream samples can be taken above the ground-water table), sites upstream of obstacles in the river bed (boulders, tree stumps and trunks) and the bases of river terraces are all favourable sites for ore mineral concentration. The best time for sampling is after floods, when the level and transporting capacity of a river fall rapidly.

All other conditions being equal, a river bed formed of limestone or shale is more favourable for the origin of placers than one consisting of granitoids. It is very important to examine places where the stream-flow direction is consistent with the strike of the schistosity, since longitudinal grooves are more effective in catching heavy minerals than transverse grooves. Panning almost invariably proceeds in an upstream direction.

Panning itself involves taking samples of unconsolidated material and its reduction in a water-filled washing pan. This method utilizes the different densities of ore (heavy, dark) and gangue (light-coloured) minerals and thus their different ability to float in water. Samples are taken directly from the valley floor or from concave meander banks, but in alluvium the upper humus layer has to be removed first. From the base of a river terrace samples are taken most easily from the undercut bank, by the channel method. Heavy minerals tend to concentrate in and are thus most readily obtained from unsorted coarse sand with pebbles and a small admixture of clay.

In order to obtain an adequate sample of heavy minerals (10—15 g) to provide not only reliable qualitative information but also some knowledge of the mineral ratio in the alluvium, it is necessary to take a sample at least 30—32 kg in weight from each bed of the alluvial deposit (in several places at least). The sample is weighed or its volume is determined (e.g. by pouring it into a vessel of known volume) so that the results can be assessed quantitatively. First, the large cobbles are picked out and then the clay is removed by washing the sample in a pan. The heavy minerals are separated from the light by continuously shaking and rotating the pan. The heavy fraction will gradually concentrate at the bottom, whilst the light fraction will remain at the surface. The latter is either removed by hand or by splashing water over the pan edge. The smaller the quantity of material remaining in the pan, the more careful must the separation be so that the heavy minerals are not removed with the light fraction. The presence of garnet in the heavy concentrate indicates that minerals heavier than garnet have not escaped. The concentrate which is black in colour is then dried over a small fire and packed in a bag or glass. Some minerals can be determined in the field, at least broadly by their colour: magnetite is black, pyrite greyish-green, cinnabar red, garnet pinkish, monazite yellow-brown, zircon white, sphene yellowish and gold yellow. However an exact determination must still be made in the laboratory.

On the basis of field and laboratory studies, *a map of ore concentrates* is constructed using the circle, stripe or normal method, or by means of isolines. The circle method consists of plotting each heavy mineral in a particular quadrant of the circle (Fig. 75). It also enables the ratio of minerals in the concentrate to be expressed, as the filled-up area of the quadrant is proportional to the percentage of a particular mineral. The diagram using the stripe method is constructed on the map by plotting the proportions of a mineral at right angles to the stream at the sampling locality and by connecting the ends of the abscissae (Fig. 76). The universal method also shows the geological structure at the sampling locality (Fig. 77). The circle and universal methods are suitable for small-scale mapping, whilst the stripe method is most useful during the initial stage of field mapping. The method using isolines (Fig. 78) is possible only when one or two minerals are to be plotted.

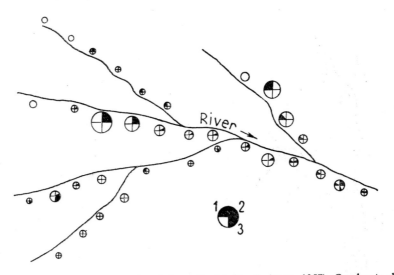

75. Plotting of panning results by the circle method (after Smirnov, 1957). Quadrants show the gold (1), cassiterite (2) and scheelite (3) contents. The radius is proportional to the total weight of the heavy mineral concentrate.

Samples for panning can also be taken from colluvium or eluvium. In West Australia grid sampling of a desert soil and the construction of isolines of metal content enabled a subsurface gold deposit to be located, its shape being delineated by the isoline of the highest Au content. In arid regions where water is not available in sufficient quantity for washing, an air current can be used to remove the light minerals (*dry washing*).

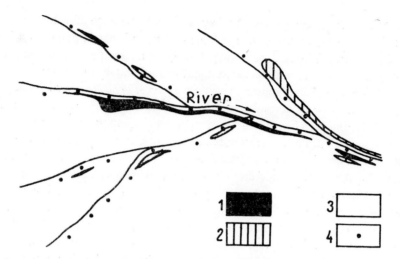

76. Panning results plotted using the band method (after Smirnov, 1957). 1 — gold; 2 — cassiterite; 3 — scheelite; 4 — sampling localities.

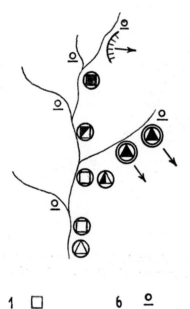

1 □   6 ⚲

2 △   7 ○

3 □ △   8 ◎

4 ◪ ◭   9 ⌇

5 ■ ▲   10 ↘

77. Panning results plotted using the universal method (after Itsikson, 1953). 1 — cassiterite; 2 — wolframite; 3 — individual grains; 4 — tens of grains; 5 — weight amounts; 6 — samples without ore minerals; 7 — sample taken from alluvium; 8 — sample taken from colluvium or eluvium; 9 — aureole of ore fragments; 10 — direction of prospection for the primary deposit.

The distance between primary deposits and placers depends on the mineral type. Gold can be transported for several hundred kilometres but the richest placers are located near the primary deposit. Colloidal cassiterite (wood-tin) showing concentric zoning can be found more than 100 kilometres from the deposit. This contrasts with wolframite, which succumbs to rapid mechanical destruction as a result of its easy cleavage; 5—8 km downstream from a deposit no trace of it remains in the concentrate. Tin placers are found at most 5—6 km from the deposit, although occasionally this figure may increase to 15 km. In general, rich placers are no farther than 2—3 km from their primary deposits. A small amount of ore mineral can be transported a greater distance (diamond as far as 500 km) and in the washed placer material at least a few grains may be collected (Fig. 79).

The distance to the primary deposit can also be inferred from the crystal form of the fragments. Beryl twins are not usually found more than 4 km from the outcrop of the deposit. Small prismatic crystals of zircon, rutile, cassiterite and kyanite, flakes and grains of gold, and fibres and wires of platinum can be transported for 10—30 km without any change in form. Large crystals (above 1 mm) with preserved edges in a concentrate suggest that the primary deposit is in close proximity.

The crystal habit of minerals may be indicative of the genetic type of their primary deposits. Thus the cassiterite of sulphidic deposits has a long columnar form and the

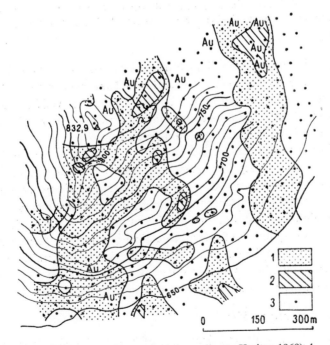

78. Sampling results ploted by the isoline method (according to Kreiter, 1960). 1 — scheelite content ranging from individual grains to 0.000 01 %; 2 — scheelite content from 0.000 01 to 0.0001 %; 3 — sampling localities.

crystals terminate in sharp pyramids, whilst cassiterite from pegmatites has a di-pyramidal form and contains an admixture of Nb and Ta. Zircons from granites are clear, but larger brownish-pink zircons are derived from pegmatites. Corundum from plagioclasites and marundites has a pyramidal habit but that from sphene-corundum rocks is of a thick tabular form.

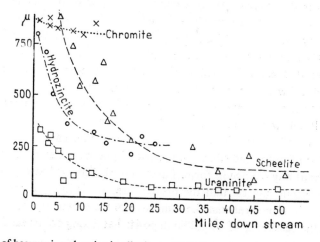

79. Reduction of heavy mineral grains in alluvium with distance from the primary deposit: chromite—Hindubagh, Pakistan; uraninite—Indus river, Pakistan; hydrozincite—Rio Ter, Spain (after Zeschke, 1961).

Prospecting using heavy mineral analysis has enabled hundreds of deposits to be detected in Siberia, especially in Primorie and the Khabarovsk district, and diamonds to be found on the Siberian platform. Diamond is accompanied in concentrates by pyrope (up to about 100 km downstream from the deposit), chrome-diopside (a small distance from the deposit) and magnesium-rich ilmenite.

In developing countries prospecting for placers should be given special attention since deposits of this type are very economic. The working of placers can begin as early as six months to two years after the commencement of exploration (by means of shallow pits or large-diameter hand driling), whereas the interval between the start of exploration and exploitation of primary deposits is about five years. The working of placers may directly follow the detailed exploration. The technical equipment required is very simple: for small deposits wooden compartmentalized washing troughs with an installation to supply water to the trough, and for large deposits, a drag shovel excavator and large sluicing unit or a bucket ladder dredge.

Prospecting based on ore pebbles

In addition to small crystals and fragments of heavy minerals, streams also carry larger ore fragments and pebbles. The size of fragments transported by water is a function of the stream velocity. At a speed of 0.5 m/sec, (i.e. in plains) sand grains

1.7 mm in diameter can be transported. Streams in piedmont areas with a velocity of 1.5–2 m/sec can carry pebbles up to 50–100 mm in diameter, and torrential streams with a velocity above 3 m/sec can even transport rock fragments more than 300 mm in diameter. The weight of the largest fragments that can be moved is proportional to the sixth power of the stream velocity (by doubling the velocity fragments 64 times heavier can be transported). There is no relationship between the stream velocity and the rounding of pebbles; at the end of a river reach, pebbles will be worn to almost the same extent regardless of the speed of their travel through the particular river sector. No rock is hard enough to withstand continual water transport and after a certain distance fragments are comminuted by impact to 0.001–0.1 mm particles, which are carried in suspension. Sandstone pebbles weighing 40 g can be carried 15 km at most, fragments of argillaceous shales (24 g)–42 km, limestone (61 g)–64 km and granite (36 g)–278 km. Large, hard and light fragments are carried farther than small, soft and heavy ones. Pebbles of ore minerals chemically and mechanically unstable in water (e.g. sulphides) are destroyed in one tenth of the distance that pebbles of stable rocks (e.g. granite) remain intact, all other things being equal. The last remnants of oxidized sulphidic minerals are found at most 3.5–5 km (exceptionally 7 km) from the primary deposit.

The distance from the primary deposit or the rock outcrop can be calculated approximately from *the pebble rounding*. For this purpose typical pebbles of a rock or gangue are gathered fom river gravel at distances of 0.5, 1, and 2 km from the outcrop. Their outline is traced on paper and their coefficient of roundness (K) is calculated by dividing the radius of minimum curvature R (in mm) by the radius of maximum curvature $r\left(K = \dfrac{R}{r}\right)$. A curve showing the relation of the K coefficient to the distance from the outcrop can be constructed (Fig. 80). The distance of other pebbles of the same rock (or rock of similar resistance) from the outcrop in the same area is then readily determined by computing the K coefficient and using the curve.

Fragments occurring 0.5–1 km from the outcrop are unworn, those 1–4 km away are roughly worn ($K = 3$–1.5) and fragments of homogeneous rocks at a still greater distance are spherical ($K = 1$).

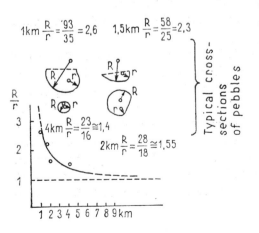

80. Pebble rounding as a function of distance from the outcrop (according to Smirnov, 1967).

USE OF AERIAL PHOTOGRAPHY AND SATELLITE IMAGERY IN PROSPECTING

Aerial and satellite prospecting methods use various portions of the electromagnetic spectrum, within and beyond the visible range, to take photographs of an area and for spectrometric survey and microwave and radar imagery. All these methods are usually preceded by aerovisual observations.

The observer in a slow-flying light aircraft or helicopter notes *features distinguishable by colour* which may indicate the presence of deposits (e.g. rusty gossans, outcrops of ochre deposits, kaolinized zones around ore deposits in lateritic terrains, black outcrops of coal, graphite and Mn-ores, and green minerals on outcropping Cu deposits), any *topographic features indicative of deposits* (positive and negative relief of vein fillings and altered wall rocks — see the chapter on the outcrops of deposits), as well as faults, cuestas, folded zones, domes, small intrusions, traces of the old drainage pattern and terrace system which may be the sites of placers, spring lines, particular vegetation in mineralized zones and on the outcrops of salt, oil and gas deposits, karstic terrains, and evidence of old mining and salt lands. Aerial photographs are taken of these features (and even those unobserved), for later interpretation and evaluation. Prospecting criteria, especially stratigraphical, magmatogenic and geomorphological, are distinguishable on photographs at a scale of 1 : 50,000 and prospecting indications can be discerned on photos at scales of 1 : 15,000 to 1 : 20,000. Diamond-bearing pipes and copper-bearing sandstones in the U.S.S.R. have proved to be most easily located on photographs of 1 : 5,000 scale.

Aerial cameras are provided with filters and emulsions sensitive to different wave ranges. *Infrared thermal scanning* is used to identify exothermal reaction zones, for example, around sulphidic deposits. *Microwave imagery* registers radioactive radiation within the range of micro- to centimetre and decimetre waves. *Microwave radiometers* capable of accurately distinguishing bodies with approximately the same temperature are used to complement the infrared method. *Ultraviolet radiometers* are applicable to the study of geological structures and deposits.

Radar and colour TV are suitable remote sensing tools in the airborne investigation of the geological structure of an area.

Aerial prospecting methods are economic in both time and money. The use of helicopters is at least three times more productive than ground exploration and aerial radiometric prospecting at a scale of 1 : 50,000 is about 50 times more rapid. The success of aerial methods depends on the amount of geological data available, the topography, the lithology of rocks, the vegetation cover, the presence of overburden, the scale of reconnaissance works and the height of the flight. Helicopter with a carrying capacity of 2,000 kg can fly 5 — 10 m above the ground at a velocity of 15 km/hour. The latest Soviet MI-1 (carrying capacity 300 kg, flight radius 267 km) and K 15 (five persons, flight radius 175 km) helicopters are even more suitable for observation and photography.

The advantage of airborne methods in prospecting has been particularly appreciated in the sparsely inhabited parts of Canada (Williams 1959 – 1960, Wright 1959 – Fig. 81 shows an example of helicopter flight passes) and in Australia (Fitzpatrick, Wilson and Hartman in Lawrence, 1965).

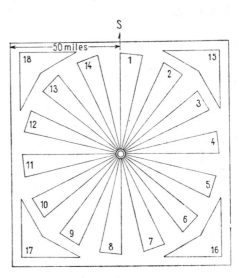

81. Flight lines of a helicopter from the camp in prospecting a sector in Northwest Canada (after Wright, 1959).

Satellite imagery is employed to accelerate the prospecting of extensive areas. A map of the whole of the U.S.A. would require one million air photographs, while it can be constructed from only 400 satellite images obtained in a much shorter time. The data provided by the first specialized satellite Landsat-1 (previously ERTS-1 – Earth Resources Technology Satellite) launched in the U.S.A. in 1972 helped to locate large salt deposits in the Andes, the so far unknown branches of the Amazon River, which may contain gold placers, and potential areas for oil and porphyry-copper deposits during the first seven months of its flight. Manned Earth satellites will facilitate still further prospecting for new mineral resources.

GEOCHEMICAL PROSPECTING METHODS AND INDICATIONS

Geochemical methods have been extensively used for prospecting for about 30 years. Areas with concentrations of a particular metal (*regional geochemical anomalies*) which warrant further exploration can be established even in mapping at a scale of 1 : 200,000. Geochemical methods should be employed during mapping at scales of 1 : 50,000 to 1 : 5,000, enabling primary or secondary aureoles of microscopic or submicroscopic ore impregnations to be discovered. Primary aureoles (contemporaneous with a deposit) accompany endogenic and exogenic deposits and are sought in the wall rocks (Fig. 82). They generally extend along the strike of the deposit rather than laterally (Fig. 39). The supply channels of primary impregnations were faults (especially barren fractures where the deposit pinched out), shattered and fractured zones and dykes, as is reflected in their metal contents. In one case, the metal content of a sample taken from the fault was 30 times that in a sample taken 1 m away.

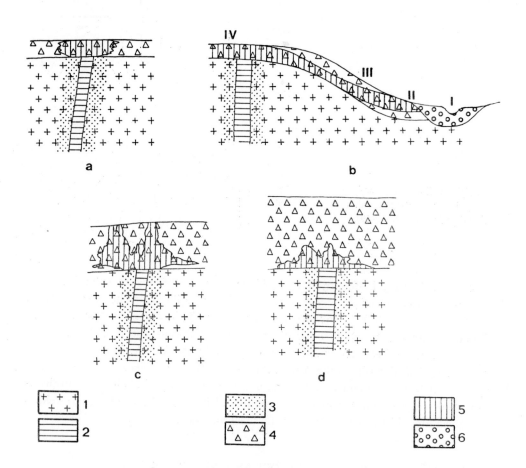

82. Primary and secondary aureoles of ore impregnations at a small thickness of overburden (a — in a plain, b — in undulating terrain), at a greater thickness (c) and very great thickness (d) of overburden (according to Smirnov, 1957, modified). 1 — wall rock, 2 — deposit, 3 — primary aureole of ore impregnation, 4 — unconsolidated overburden, 5 — secondary aureole of ore impregnation, 6 — fluviatile sediments. I — collection of hydrochemical samples (from watercourses), sampling by "stream-sediments" method (clayey sediments from the stream bottom), collection of heavy mineral samples; II — collection of samples from clayey rocks at the alluvium-slope contact, III — collection of samples from secondary aureole in slope material (fragments of ore and gangue, soil, ground water), IV — areal sampling of secondary or primary aureole of ore impregnations (soil from shallow holes or rock from deeper test pits).

Uneven distribution of ore impregnations is also related to the petrography and chemistry of the rock. The boundary of ore elements aureole in individual rocks varies according to the content of ore elements in non-impregnated rocks (*background metal level*). In the neighbourhood of one deposit, for example, limestones contain 0.0067 % heavy metals, dolomites 0.01 %, hornfels 0.015 % and dolomites with siliceous concretions 0.05 %. In porous rocks impregnation spreads as readily as on faults and shattered zones and extends to a greater distance from the deposit. Impermeable beds exert the opposite effect.

The vertical dimensions of primary aureoles vary considerably. Values of 300–350 m were found for the Hg deposits of Central Asia, in the polymetallic deposits of the Altai the Cu aureole extended upwards for 250 m, Pb and Ba — 400 m and Zn — 500 m; data recorded on American deposits vary from 120 to 170 m. This implies that concealed deposits at these depths will show themselves on the surface by an increased concentration of ore elements. The area of impregnation greatly exceeds the projection of the deposit on a horizontal plane. Primary aureoles can extend up to 1 km along tectonic lines; their width is 100 m at the most.

The chemical character of primary aureoles largely corresponds to the composition of ore bodies but it differs in detail, owing to the various migration capacities of the elements. For example, Ba, Pb, Cr and Te form an aureole close to the deposit, but As, Sb, Cu, Ag and Zn are located very far from it. Hg and Ta aureoles will be found the farthest away. The Pb–Zn ratio increases approaching the deposit (mainly from above). It has been established on a polymetallic deposit (U.S.S.R.), which was located in a shattered zone and covered by as much as 100 m of sediments, that Cu, Sr, Mo and As penetrate upwards 100–120 m, zinc, 200 m, Pb, Ba and Ag, 50–60 m and Bi, 20–25 m. The aureole is even more extensive in the shattered rocks, reaching up to 1 km from the deposit along the strike.

Pyrite impregnations around a deposit are accumulators of trace elements. Pyrite above sulphidic Cu ores contains increased amounts of Cu, Pb, Ni and Co (occasionally Ag and Au); increased contents of Pb, Zn and Cd above polymetallic deposits and Sn and Cu above tin ores have also been observed. This zonal aureole can extend up to several hundred metres from the deposit.

An example from the sulphidic deposits in the Ural Mts.: Pb migrated to a distance of 20 m from the deposit, Ni and Co to 27 m, As to 30 m, Cu and Ag to 40 m and Zn to 70 m. The aureole of polymetallic deposits in the Altai shows the following zonation: Pb, Sb, Cu, Ba to 100–150 m; As, Bi, Mo, occasionally Zn and Ag, to 200–500 m.

Metallometric mapping has recently been used successfully in the U.S.S.R. even in gold prospecting. Samples were taken from a humus layer from a depth of 15–30 cm.

The migration properties of elements are controlled by their valence, ionic radius (the greater the ionic radius, the smaller the aureole), ionization potential, atomic and ionic polarizability, aggregate state, radioactivity caused by the presence of radioactive elements in minerals, structure of the crystal lattice and the density. The principal external factors in migration are temperature, pressure, concentrations of various substances, oxidation-reduction conditions, water pH and volcanic and biological processes.

The ratio of metallic elements in the ore body cannot always be determined from their proportions in the primary aureole. Besides ore-forming elements *accessory elements can also be indicators of concealed deposits.* Examples: On the Tintic Pb − Zn deposit, not only Pb and Zn but also Mn and Ba indicate concealed bodies; As and Zn are indicative of the Sierra Mojada Pb − Ag deposit; Co, Bi, Ag and Sn of the Goldfields Au bodies; Zn of Hg deposits in Central Asia. On the Pb − Zn deposits in the U.S.S.R., As, Cd, Cu, Ag, Sb, Br and occasionally Mn and B are accessory elements, as are Pb and Mo with uranium deposits. Hg aureoles occur above Cu-sulphidic deposits, As aureoles above Sn and Hg deposits, and Ge aureoles, in topaz and axinite, above tin deposits.

Long circulation of hydrothermal solutions before mineralization causes leaching of the background (clarke) contents of metallic elements from the rocks and thus the formation of a *"negative aureole"*, which can also be indicative of mineralization at depth.

Generally, primary aureoles are only discovered during detailed mapping, geophysical, geochemical or exploration work. Secondary aureoles form by weathering and disintegration of the vein filling and by impregnation in the wall rock (primary aureole) near the surface, displacement of fragments by solifluction and transport of ore elements by ground-water in the eluvium, colluvium and proluvium, and by the action of surface water-courses.

The content of ore elements in a weathered bed does not differ in order of magnitude from that in the mother rocks, i.e. in the primary aureole. The normal Cu, Pb and Zn content in rocks varies between 0.002 and 0.02 per cent and in soils between 0.001 and 0.01 per cent. Eluvium above a polymetallic vein contains about 170 times more lead, 100 times more copper and 10 times more zinc than the weathered material from non-mineralized rocks. Such unusual concentrations of ore minerals in weathered rock (or soil) suggests the presence of a deposit. With the Tintic deposit the following migration capacities of metal elements in ground-water (secondary aureole) were determined: Pb virtually does not migrate; Au − maximum migration distance is 1 dm; Cu − up to 10 m; Zn − hundreds of metres; Ag − more than zinc. In chemically active rocks (e.g. in carbonate), the aureole is narrower (0.5 m in dolomites on the Tintic deposit) than in barren (silicate) rocks (50 m in monzonites of Tintic deposit).

If an old formation containing a mineral deposit, e.g. crystalline complex, is covered with younger sediments, the secondary aureole extends upwards several dozen metres into them and lengthwise up to 400 m from the deposit.

The dimensions of secondary and primary aureoles are also controlled by the permeability of the rocks: they are larger in permeable rocks, but are soon leached on account of rapid ground-water flow. In contrast, the aureoles are less extensive but more resistant in poorly permeable rocks. Secondary aureoles extend along tectonic zones as far as 0.5—3 km from the deposit (Milyaev—Fokin, 1963).

If the deposit is overlain by poorly permeable rocks, a hydrogeochemical anomaly can be identified on the surface as much as several hundred metres from the projection of the deposit onto the surface. The results of hydrogeochemical research and metallometric mapping can be distorted when ore elements are retained in the overlying rocks with a high sorption capacity, e.g. in montmorillonitic clays.

Macroscopic mineralization is the most reliable guide in prospecting for endogenic deposits, even if they are concealed. It is manifested by macroscopic ore impregnations in altered country rocks. Polymetallic deposits in the Altai Mts., for example, are fringed by an aureole of pyrite impregnations extending several hundred metres from the ore bodies. Tin deposits in the Far East are accompanied by an aureole of carbonate and quartz veins. Fluorite and barite veins on the surface can be indicative of polymetallic mineralization at depth. Such associations of certain ore minerals with certain gangue minerals and their spatial relations are due to primary differences in vein composition, depending on depth.

Secondary aureoles of microscopic impregnations are traced during metallometric mapping and sampling and during hydrogeochemical, gas, biochemical and geobotanic prospecting. Samples are examined mainly by spectral analysis.

For metallometric mapping, soil samples 20—50 g in weight are collected from a depth of several centimetres up to 1 m, usually beginning with the humus layer (up to 10 cm), since large amounts of ore elements occur in humic and podzolic soils above a deposit (Fig. 83). It is advisable to study the soil profiles and to establish

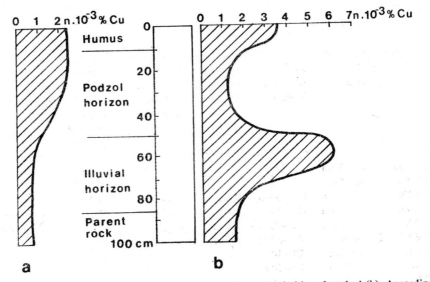

83. Copper content in a soil profile of chernozem and grey earth (a) and podzol (b). According to Kreiter (1960).

the sampling depth for each soil type and thus to determine the layer richest in ore minerals (0.5—0.7 m in podzols). This is then sampled systematically.

A deposit or its primary aureole is indicated by an increased metal content near the surface to the maximum thickness of the residual cover of 2—3 m (rarely 5—10 m). Below the much thinner alluvial, eolian or glacial sediments, anomalies are identifiable only by samples from the base of the overburden. Linear and areal metallometric mapping are suitable in following known deposits along their strike, in searching for new ore bodies near a known deposit and for deposits in promising areas. In geological mapping at scales of 1 : 200,000 or less, linear metallometric mapping across structures is carried out. The density of sampling should be chosen so that one sample is taken from an area corresponding to 1 cm² on the map. Against the background of regional anomalies, deposit anomalies can be established by areal sampling. The results of chemical (spectral) analyses plotted on the axis corresponding to a line of drill holes will furnish the curve shown in Fig. 84. Its highest point will depend on the highest ore mineral concentration encountered (as seen from Fig. 84, the concentration will be greater at depth in segment a, and near the surface in segment b). The form and symmetry of the curve will be governed by the topography of the area, by the degree of impregnation of the wall rock (heavily impregnated wall rocks produce a flat and moderate curve) and by the solubility of metal compounds (easily soluble compounds produce maxima which are elongated downslope). Anomalies above ore fields can merge together to reach a size of several tens of sq. kilometres. Examples showing the dimensions of a network for areal sampling are listed in Table 41.

TABLE 41

Examples of network dimensions (according to Kreiter, 1964)

For metallometric sampling			For sampling of clays from river-beds (stream sediments)			
scale	spacing profiles	spacing of points in lines	spacing of samples	minimum length of floodplain to be sampled	number of samples per 1 km of the drainage pattern	number of samples per 1 km²
1 : 1,000,000	12—8 km	100 m				
1 : 500,000	6—4 km	100 m				
1 : 200,000	2 km	100—50 m	800 m	0.8 km	1.25	1.7—2.1
1 : 100,000	1 km	100—50 m	400 m	0.4 km	2.5	4.0—5.0
1 : 50,000	0.5 km	50—40 m	200 m	0.2 km	5	8.5—14.0
1 : 25,000	250—200 m	40—20 m	100 m	0.1 km	10	18.0—32.0
1 : 10,000	100 m	20—10 m				
1 : 5000	50 m	20—10 m				
1 : 2000	20 m	10—5 m				
1 : 1000	10 m	5 m				

If the size of the anomaly above one deposit is known, spacing of lines equal to 0.8—0.9 of the strike length of the anomaly and spacing of sampling points equal to one half its width are chosen for further mapping. The results of detailed metallometric mapping are plotted on maps of anomalies which delimit the aureoles. This map together with the geological map warrant further research using surface assessment works and drill holes to a small depth.

A special kind of metallometric mapping consists of sampling of clayey sediments from the bottom of watercourses (*stream sediments*). The clay minerals, mainly montmorillonite and hydromicas, and humic substances, SiO_2 and Al_2O_3 colloids and Fe and Mn hydroxides, absorb and concentrate cations of some metals so that they can be identified several hundred metres (in forest areas) and several kilometres (in steppes) from the deposit. Limonite concretions or infiltrations in clay and coating of pebbles can contain Cu, Zn, Pb, Ni, Co, Mo, W, Sb and Bi; Mn oxides contain Co. The aureole of rock fragments near the deposit can be localized further by determining the metal elements in clays, in limonite and manganese concretions and in beds at the contact of slope and alluvial sediments. In the final phase, sampling thus moves to the *foot of slopes*. In Ghana, sampling by these two methods proved to be successful; samples were taken at points 200 m apart and at the foot of slopes they were collected at a depth of 5 to 20 cm below the surface.

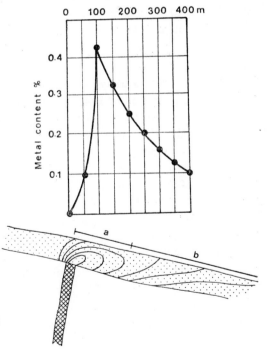

84. Geochemical anomaly above a deposit outcropping in slope.

In the initial prospecting stage, the stream-sediment method is to be preferred to other geochemical methods, particularly in geological mapping at scales of 1 : 200,000 or 1 : 500,000, since anomalies determined by this method are prominent and sampling is simple and rapid (50 g from one point in the alluvium near the centre of the stream is sufficient; for the spacing of points see Table 41). The content of heavy minerals, hydrochemical conditions, secondary aureoles in the eluvium and aureoles of rock fragments, which are usually not far from the primary deposit, can be studied simultaneously. The method is essential in areas with evanescent streams, where the lack of water makes the use of hydrochemical methods and panning impossible, but its application is limited by geomorphological conditions, as it cannot be employed on peneplains and in areas with thick overburden.

Hydrogeochemical methods have made an appreciable contribution to the discovery of U, Mo, Zn, Cu, Pb–Zn, Ni, V, Cr, Sn and Au deposits. Surface waters percolating to great depths dissolve ore and gangue minerals in the deposits they encounter. Chemical analysis indicates the type of environment the waters flowed through. In sulphidic deposits, the content of SO_4'' ions increases (the aureoles extend up to several kilometres from the deposit) as do the SO_4''/Cl ratio ($= 2.5 - 6$) and metal ion content (Cu and Ni content $= 0.0000n$ to $0.00n$ g/l compared to the background amount $0.000\,00n$ to $0.0000n$ g/l in waters that have not come into contact with the deposit); the pH value is low. Metals can be arranged according to their decreasing migration capacities: Zn (aureole of several kilometres), Mo, Ag, U, Ni, Cu (aureole – 0.5 km under rapid flow, 1.5 km with slow flow), Cr, Pb (aureole – several hundred metres), Sn and W. The presence of fluorine in the water suggests the existence of fluorite deposits, iodine and bromine indicate oil deposits and bodies of potash salts manifest themselves by a $\dfrac{\text{NaCl}\,\%}{\text{KCl}\,\%}$ ratio below 37. These data should be complemented by a study of the flow of ground-water. The distance to which the aureole of metals in ground-water can be determined is controlled by the background value, the size of the deposit, the velocity of the water flow, the content of dissolved solids (the higher this value, the less additional metal ions can be absorbed; almost no metals are absorbed when the salt content exceeds $2-3$ g/l), and the composition of the rocks through which the water containing the metal ions flows (carbonate rocks and waters cause neutralization of acid solutions and precipitation of metals). The zone of metal-enriched waters is extended along the ground-water flow and is deformed by the structures in its path. Before beginning hydrogeochemical prospecting it is advisable to determine the content of metals in the ground-water at about $1-2$ km from the known deposit. In normal, unenriched waters it is necessary to determine the background metal content, so that anomalies can be identified. Water is taken from springs, spring lines, wells, drill holes, test pits, surface waters or marshes and the following measurements are taken: temperature of the water and the air, water level, depth and yield (if possible), pH, content of sulphate and chlorine ions, total metal content (by the dithizone method) in the field, if possible. The hydrogeology and geology of the surroundings of the sampling site are studied. Chemical and spectral analyses are made in the laboratory and the results are plotted on a geological map or an overlay sheet on a geological base map.

For sampling of waters with a higher evaporation residue than 500 mg/l, 0.1 litre is usually sufficient; if the evaporation residue is below 100 mg/l, the sample must be larger (1 litre). In humid regions, the dry season is most suitable for sample collecting, whereas the period with the highest ground-water levels is to be recommended for sampling in arid regions.

In mapping at a scale of 1 : 200,000 or 1 : 100,000, water samples are taken at the intersections of valleys with important rock contacts, faults, etc. Depending on the map scale and the complexity of the tectonic structure, 0.09 to 0.6 samples should be taken for each sq. kilometre. In mapping at scales of 1 : 50,000 and 1 : 25,000, streams are sampled over their entire length from the mouths of major valleys to the heads of tributary valleys. 0.7 to 3.6 samples are collected per sq. kilometre.

Samples of water from rivers and streams flowing over an outcrop or halo of a mineral deposit contain very small amounts of metals and highly sensitive analytical methods are therefore necessary for their determination. A positive anomaly has 5—10 times the metal content of an unenriched stream. Samples are collected near the mouths of streams, which in the positive case are examined in an upstream direction using other methods.

Many deposits in the Rudnyi Altai Mts. (U.S.S.R.), Canada, the U.S.A. and Japan were discovered by hydrogeochemical prospecting. The copper deposits in Zambia were also found partly because of a hydrogeochemical anomaly. They were discovered close to unpotable water. In gas mapping, gaseous products resulting from the decay of radioactive elements (emanations), helium, Hg vapours, oxygen and carbon dioxide (in the oxidation zone of deposits), SO_3 and H_2S (in sulphur deposits), and gaseous hydrocarbons accompanying oil deposits are entrapped.

In emanation mapping, the products of Ra and (partly) Th decay are studied: radon (half-life = 3.8 days) and thoron (half-life = only 54 seconds). Emanations are chiefly concentrated in soil air, whose background varies between 3.7 and $37 \, s^{-1} \times 1^{-1}$. The suitability of this method depends on the thickness of the overburden, which should be 0.5—10 m; 1.5—2 m are most suitable. Amounts regarded as anomalous should be at least 3times the background value. Soil air samples are collected during field traverses at 10—20 m intervals. Using this method, not only can uranium deposits and their primary aureoles be identified in areas where the gamma-ray method cannot be employed because of a thick overburden, but also major faults, boundaries of rocks and geological units, mineral waters and even bodies containing a small amount of radioactive elements (placers of Li and Nb minerals, beryl, monazite and phosphorites) can be found.

Helium is determined in samples of soil air taken from a depth of 20—30 m in order to eliminate the effects of the atmosphere. Anomalous He contents indicate deep-seated faults which can be associated with kimberlites and carbonatites.

Gas prospecting for oil deposits is based on the assumption that small amounts of hydrocarbons penetrate the impermeable overlying rocks over a geologically long period (of the order of millions of years). These microindications may be determined by soil air analysis or by a bacteriological method. Geochemical oil-gas prospecting (according to Sokolov) makes possible the discovery of oil deposits in common structural types and, what is particularly important, those of lithological oil traps that cannot be found by geophysical methods. The origin of lithological traps is due to different facies development of a bed at different places. A well-cemented sediment may pass laterally into unconsolidated porous rock, which is a favourable reservoir rock. Oil deposits of this type have the shape of lenses, troughs or stringers. Microindications of hydrocarbons are of greater extent so that they can be more easily identified.

Samples of soil air are collected in 1.5—2 m deep test pits or drill holes. Air samples are taken by a flat bell sampler driven into the bottom of a hole. The sampler is connected by a tube with a vessel filled with water; when water flows out, the air is drawn from the sampler into the vessel due to underpressure. Samples are collected either randomly or at small intervals within a grid of $100-300 \times 200-500$ m. They are analysed in the laboratory using a low temperature method (liquid or solid hydrocarbons are obtained) or in the field by chemical microanalysis. The latter consists of burning the hydrocarbons contained in soil air; the amount of CO_2 thus formed is proportional to the original hydrocarbon content. The amount of CO_2 is determined by passing it through water containing barium ions, in which the CO_2 reacts with Ba to form crystalline $BaCO_3$. The rate of separation of the crystals is proportional to the CO_2 concentration in the soil air. It is important to record the time of formation of the first $BaCO_3$ crystals, as the concentration of CO_2 can be determined by correlation with a calibration curve.

Curves showing the shape of a deposit and partly also its position are obtained by plotting the results of soil air analysis in profiles, in which the pits or drill holes

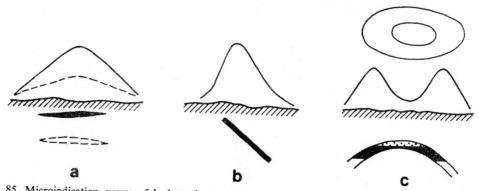

85. Microindication curves of hydrocarbons above oil deposits: a — at depth (dashed), near the surface (solid); b — above inclined deposit; c — above domal deposit (gas—dotted); after Sokolov in Šuf (1957).

are indicated. The curve is symmetrical above a horizontal linear deposit of a trough or string shape. The height of the curve is related to the depth of the mineral source (Fig. 85a). The dip of a linear body is expressed by the asymmetry of the curve (Fig. 85b). The so-called "ring effect" forms above a typical domal structure (Fig. 85c).

By evaluating the results of analyses from an area of 10–15 sq. km we obtain a number of curves, from which parts of the same dip and sectors of the same intensity of microindications can be determined. In order to determine correctly anomalous concentrations, the background microindications should be known beforehand. The gas method has been experimentally used for coal deposits.

In *bacteriological investigation of soil samples* from a shallow drill hole, a sample is submerged in water and exposed to the action of some hydrocarbon (e.g. methane) for several days. If methane bacteria are present, they multiply because of improved living conditions and form a yellow-brown film on the water surface.

The biogeochemical method is based on the fact that an increased content of metal elements in ores, primary aureoles and soils leads to an increased content of these metals in plants growing above them. Some metals, particularly nickel, copper, chromium, lead and molybdenum, form soluble compounds in the soil and accumulate in plants far more intensively than poorly soluble (even if more abundant) compounds of, for example, aluminum or titanium. Plants with the deepest root systems are the most suitable for sampling. The largest amount of metal elements accumulates in leaves, less in the bark and wood-pulp of trees and shrubs. Some plants possess an outstanding capacity to preferentially concentrate a definite metal in their tissues (e.g. maize–Zea Mays–concentrates gold in its stems). Anomalies over metal deposits may attain values many times that of the background (Zn — 100 times, Ni — 30 times, Fe — 6 times, Pb — 2 to 5 times). A mineral deposit often produces an increase in the metal content of plant ash even if it occurs at a considerable depth (Cu — 50 m, Zn, Ni, Cr, Co — 30 m, As — 10 m, U — 10 to 25 m, Mo — 3 m). Samples of tree foliage (the weight of one sample is 15–20 g, i.e. 20–100 mg ash) are collected in rectangular grids of various dimensions. Fallen leaves can also be analysed, contents of non-ferrous metals being relatively concentrated after alkalis, alkaline earths, iron and manganese have been leached.

The biogeochemical method is more effective than metallometric mapping as it provides information on geochemical conditions at a depth of up to 20–30 m, which is attained by roots of some trees. It can be employed where metallometric mapping does not furnish reliable results, if mineral deposits are covered by thick alluvial, coarse colluvial, eolian, glacial or marine sediments. It has also proved successful in areas with perenially frozen ground (the sampling of dwarfed birch and sallow leaves and mosses yield good results), in sandy deserts and in semideserts. The results of this method must be checked against geological study of the area to avoid serious errors. In Michigan (U.S.A.), for example, sampling and analysis of plant samples revealed a copper smelting plant!

A new biochemical "*air-trace*" method has recently been developed (Barringer,

Fifth International Symposium on Geochemical Prospection, Montreal 1974). It is based on the assumption that the Earth's surface (woods, tundras, steppes, deserts, oceans) continuously produces gaseous organic substances which are bound with metals. The organic fraction containing metals accumulates in the atmosphere, increasing the geochemical contrast between mineralized and non-mineralized areas. The metal content in metalorganic compounds is determined by emission spectroscopy or mass spectrometry of air samples. These are collected by airplanes, ships or, in ground investigations, from automobiles. Provided the atmospheric conditions during sample collecting are controlled meteorologically, anomalies can be established and mineralization, particularly of Hg, Cu, Pb, Zn, Ni and Ag, and manifestations of CH_4, CO_2, SO_2, H_2S and other gases can be detected.

The geobotanical method is a useful tool in prospecting for outcropping deposits and for concealed deposits whose ore impregnations extend to the surface. G. Agricola (1556, Book 2, p. 31) regarded "sick" grass and trees whose leaves are greyish in the spring as indicative of a nearby ore vein. Some plants flourish on soils rich in certain elements or fail to flourish in their absence. There are some specific plant accumulators of Cu, Zn, Li, Mn and other metals, which are good indicators of mineral deposits. Examples (from Ginzburg, 1957): lead — *Amorpha canescens*, America; galena — *Rhus*, England; tin — *Trientalis europea*, Bohemia; zinc — *Viola lutea* var. *calaminaria*, Central and Western Europe; copper — *Polycarpaea spirostylis*, Australia; mosses — *Gymnocoleae, Acutiloba, Cephaloziella*, and *Melandryum rubrum*, Czechoslovakia; serpentinite (chromium) — *Asplenium adulterinum*; nickel — *Dianthus cajullifrons, Sempervivum pittonicus;* manganese — *Digitalis purpurea, Zostera nape, Fucus vesiculosus;* uranium and vanadium — *Astragalus pattersonii;* selenium — *Stanleya, Onopsis* and *Xylorrhiza* genera; phosphates — *Convolvulus tricolor*, Spain; salt — *Salicornia herbacea;* clay — *Cirsium arvense, Tussilago farfara;* sand — pine tree, birch, bamboo.

The presence of metals in the soil causes a change in the colour of some plants. *Emolcia californica* has bluish-grey blossoms in the presence of copper and lemon-yellow where zinc is present. Uranium and thorium produce pathological changes in plant growth: nanno- or giant forms originate. Manganese causes a grey spottedness on oats. An increased bitumen content also has a pathological influence on vegetation and may lead to the origin of new species.

Sulphuric acid formed by weathering of sulphides has a deleterious effect on plant growth, so that outcrops of sulphidic deposits are often free of vegetation. Boron and arsenic have the same effect. Of all plants, birch trees are most resistant to an acid habitat, so that not only deposit outcrops but also old waste dumps can be identified by their occurrence. Thermal energy liberated in the decomposition of sulphides results in the snow cover appearing later and disappearing sooner on their outcrops than elsewhere.

Outcrops of kaolin deposits are covered with thin vegetation since alkalis necessary for its nourishment have been leached out.

Geobotanical relationships of a definite kind are valid only for one climatic zone. They are useful in compiling lithological maps, for determination of subsurface water, in the search for salt domes and in prospecting for bitumens, boron, sulphur and metal deposits.

Geochemical methods useful in tropical and subtropical countries differ from those used in moderate climatic zones (see above). The most important method in arid and semiarid regions with outcrops of primary rocks is the investigation of primary aureoles accompanied by geological mapping, by the study of heavy minerals in weathered rocks and trace elements in some primary minerals (e.g. biotite) and by the application of geophysical methods. If the primary rocks are covered by a weathering crust, metallometric investigation of soil samples and biogeochemical and hydrogeochemical methods are carried out. Areas with thick eolian cover are not suitable for geochemical prospecting methods.

In humid tropical regions with deep weathering and lateritic hardpan on the surface, only the biogeochemical method can be used. Where the hardpan is missing, metallometric soil sampling as well as hydrogeochemical investigation is useful. Sampling of clay sediments (stream sediments) and heavy minerals in river beds can be used in humid tropics with or without lateritic hardpan cover.

PROSPECTING FOR CONCEALED DEPOSITS

Concealed deposits occur both in areas whose geological and mining history is well known and in unexplored areas. A search for them is especially relevant in areas which have already been thoroughly investigated geologically, where exhausted deposits make it necessary to find new mineral resources; the availability of the necessary geological data facilitate prospecting.

According to Smirnov (1955), in prospecting for concealed deposits three problems should be studied: 1. the theoretical possibility of the existence of concealed deposits, 2. surface indications of concealed deposits (see geochemical indications) and 3. prospecting methods.

Theoretical possibility of the existence of concealed deposits. Theoretically, potential reserves of concealed deposits are larger than those of deposits cropping out on the surface. According to calculations by Soviet geologists, only 20 per cent of all deposits occurring to a depth of 1,000 m crop out on the earth's surface. The importance of the remaining 80 per cent of deposits which are hidden at depth will steadily increase, but precise methods for their exploration have not been developed so far. Their discovery and exploration are bound to be more time-consuming and more expensive than is the case with other deposits. The methods used should attempt to ensure that the deposits provide mineral materials of suitable quality for the future.

The existence of concealed deposits is governed 1. by the vertical range of the ore complex, 2. by the vertical dimensions of ore bodies and their dip, 3. by the depth of erosion of the ore complex, 4. by the character of ore-bearing structures, and 5. by the composition of the wall rock.

The vertical range of an ore complex or ore-bearing interval. An ore complex involves deposits of one type, i.e. deposits which originate within a certain temperature range (e.g. polymetallic, tungsten-molybdenum, antimony-mercury deposits). These deposits occur within a given space interval in the mantle of the ore-bearing intrusion. The interval is bounded by two isotherms, which form the upper and lower limits for the origin of deposits of a given type. The vertical range of mineralization of the same type is an *ore-bearing interval* (Fig. 86), which is limited either by more or less horizontal surfaces (apomagmatic deposits) or by distorted surfaces (perimagmatic deposits). The form of these surfaces is related to the nature of the contact of the intrusion with the country rock, to the structure of the country rock and to the distance of the ore-bearing interval from the contact. For simplicity's sake, the limits of the ore-bearing interval will be regarded as horizontal in the text below.

Endogenic (especially hydrothermal) deposits could originate down to the depth of 10–12 km below the surface. Distribution of deposits of all types within this range and of deposits of one type in the contingent ore interval is of particular importance for the potential discovery of concealed deposits.

In Fig. 87 the lengths of individual vertical ore bodies are shown in relation to the ore-bearing interval, assuming that the interval is halved by the present surface. Three cases are considered. First, in a relatively rare instance (Fig. 87a), the length of the ore body is equal to the ore-bearing interval. The Sn–W deposits in the Krušné hory Mts. (Czechoslovakia) belong to this group; according to recent investigations (J. Janečka) they end at a depth of more than 300 m. The composition of the deposit, geochemistry of the cassiterite, and the alterations of the wall rock change with depth. Secondly (Fig. 87b), a situation occurs in which it is assumed that the deposits are distributed from the ore-bearing magma up to the former surface; the length of one ore body equals one tenth of the ore-bearing interval. The third case (Fig. 87c) assumes the length of the ore body to be one half that of the ore-bearing interval. The second situation is most frequent. Possibilites for the preservation of concealed deposits can be seen in the figures.

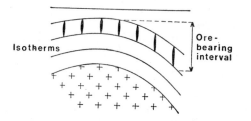

Isotherms

Ore-bearing interval

86. An ore-bearing interval is a vertical range of a one-type mineralization (e.g. polymetallic).

The extent of an ore-bearing interval has not yet been established for individual geological conditions. On the basis of empirical data, Graton determined the vertical range of hydrothermal zones (not ore-bearing intervals!) at 8—15 km for hypothermal conditions, 3—6 km for mesothermal and 2 km, at the most, for epithermal conditions. So far, it is known only that ore-bearing intervals are larger for deposits formed in abyssal conditions than for those derived from hypabyssal conditions (compare with the chapter on the crystallinity of rocks).

Vertical dimensions of ore bodies and the angles of their dips are also decisive for the preservation of concealed deposits. The smaller the vertical dimension of an ore body and the smaller its dip, the more probable is the preservation of the concealed deposit. *The depth of erosion* of the ore-bearing interval containing vertical bodies is indirectly proportional to the number of preserved concealed bodies.

The character of ore-bearing structures. Greatly simplified situations for concealed deposits have been considered above, irrespective of the geology, the character of ore-bearing structures and the composition of wall rocks. Ore-bearing structures (Fig. 88) can be divided into open, half-closed and closed structures. *Open structures* are, for example, fractures and steeply inclined beds of favourable rocks, which sometimes pass through the whole ore-bearing interval. An ore field of open

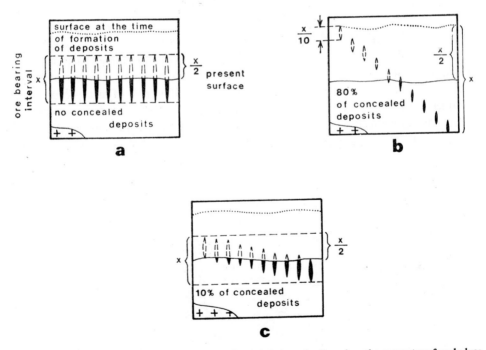

87. Examples showing potential preservation of concealed ore bodies when the present surface halves the ore-bearing interval and the lengths of ore bodies is equal to (a) or is a tenth (b) or one half (c) of this interval (according to Smirnov, 1955).

structures is generally devoid of concealed deposits. *Closed structures* are associated with fold deformations and mineralization is limited to moderately inclined beds. Examples of closed structures are bedding surfaces, especially in folded beds, crushed zones on the inner sides of fold bends, interformational breccias, subhorizontal contacts of rock complexes, and slightly dipping fault zones, particularly if they were warped before mineralization. Closed structures are the most promising sites for concealed deposits. On the other hand, prospecting is complicated by the absence of both primary and secondary aureoles. A primary aureole in a horizontal direction is the only exception.

Half-closed structures show a transitional character. Their importance in prospecting for concealed deposits is appreciable.

Another factor determining the location of mineralization is the petrography of wall rocks. If rocks favourable to mineralization (e.g. certain types of limestone) occur near the lower boundary of the ore-bearing interval, the presence of concealed deposits is more probable than if these rocks lie near the upper boundary.

It can be concluded that the number of concealed deposits will be larger 1. the greater the ore-bearing interval, 2. the smaller the vertical dimensions of the ore bodies, 3. the smaller the dip of the ore bodies, 4. the higher the erosion level in the ore-bearing interval, 5. the more numerous the closed structures and 6. the lower in the ore-bearing interval rocks favourable to mineralization are located.

Concealed deposits are divided into *covered* (Fig. 89a), *deep-seated* (unexposed) (Fig. 89b) and *buried* (Fig. 89c). With covered deposits the thickness of the younger formation is decisive for the application of geophysical prospecting methods and subsequent drilling at the sites of established anomalies. Small deposits of polymetallic, rare and precious metals at a depth of tens to hundreds of metres cannot be identified by present-day geophysical methods. The possibility of the preservation of deposits as yet unexposed by erosion has been discussed above. Prospecting for them is facilitated by structural criteria, as they are often grouped in ore zones.

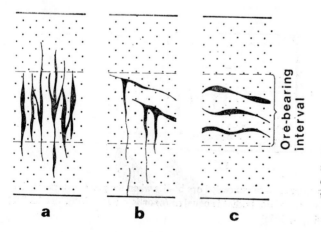

a   b   c

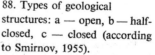

88. Types of geological structures: a — open, b — half-closed, c — closed (according to Smirnov, 1955).

As far as the prospecting methods are concerned, prospecting in poorly explored areas is based on study of the metallogenesis of large geological units. It is aimed at delineating areas with deposits of a given type such as, for example, parts of ore zones. Applying experience gained on deposits already investigated, the sectors showing promise are determined using a geological map at a scale of 1 : 50,000. These sectors are then mapped geologically at scales of 1 : 5,000 of 1 : 10,000 and, on this basis, vertical sections are constructed. At the same time, metallometric mapping is carried out, maps of heavy-mineral concentrates are prepared and a geophysical survey is conducted. Detailed metallometric mapping should be preceded by determination of the background metallic content in barren rock, especially granitoids, so that anomalies (primary aureoles of ore impregnations) can be identified. The background content of metallic elements is established by taking samples during several reconnaissance field trips. Only then is detailed sampling in a promising sector possible; it is achieved by closely spaced traverses through the area. All outcrops of faults should be first sampled. Favourable anomalies are assessed by drilling. Since core recovery is very costly, the presence, character and intensity of primary aureoles are often determined without coring, using radioactive nuclear resonance methods to analyse the chemistry of aureoles of Cu, Pb, Hg, Mn, B, Be and other elements. The methods employed are gamma-gamma density log, selective gamma-gamma log, induced radioactivity log, neutron-neutron log and others. The utilization of both geophysical and geochemical methods in the search for concealed ore deposits can reduce the cost of drilling by one half.

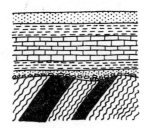
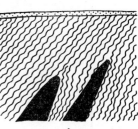

a   b

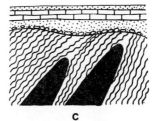

89. Concealed deposits are divided into covered (a), unexposed (b) and buried (after Smirnov and Khrushchev, 1961, modified).

c

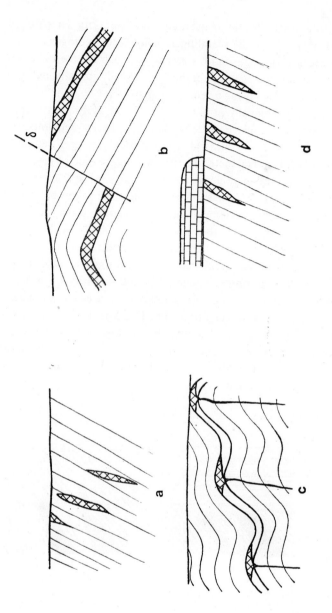

90. Ore fields with concealed and outcropping deposits.

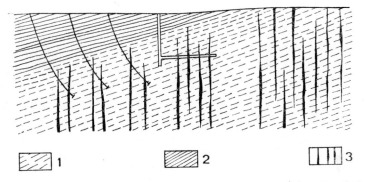

91. Prospecting for concealed deposits related to a layer of "favourable rock" (after Smirnov, 1957). 1 — sandstone, 2 — shale, 3 — ore veins.

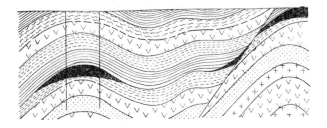

92. Prospecting for a concealed deposit confined to the contact of two rock types (after Smirnov, 1957).

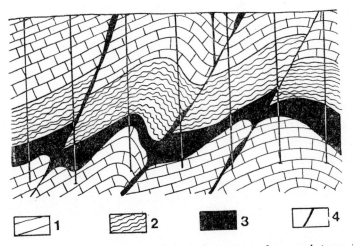

93. Prospecting for a concealed deposit confined to the contact of two rock types, based on ore indications on faults (after Smirnov, 1957). 1 — limestone, 2 — shale, 3 — ore, 4 — fault.

Prospecting in worked ore fields is easier as it can be based on regularities established on known deposits cropping out on the surface. Several examples of ore fields with both outcropping and concealed deposits are shown in Figs. 90—94. The following instances may be encountered: 1. some sectors of the ore field are covered with a younger formation (Fig. 90d), 2. ore bodies are limited to a rock complex which extends to a greater depth (Figs. 90c, 91, 92) and may be indicated on the surface by mineralization along faults (Fig. 93), 3. sunken parts of the deposit are preserved at depth (Fig. 90b), 4. mineralization occurred at depth in a horizon parallel to the productive layer cropping out on the surface (Fig. 94), and 5. ore bodies are part of a steeply dipping ore-bearing interval (Fig. 90a).

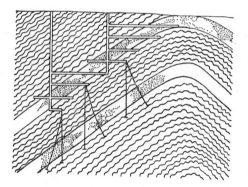

94. Prospecting for concealed deposit related to a layer of "favourable rock".

PROSPECTING FOR ORE DEPOSITS IN VARIOUS GEOGRAPHICAL AREAS

The prospecting methods applicable in little known or uninvestigated areas differ considerably from those used in areas whose geology and mining history are well known.

At present, only Central and West Europe and some parts of the United States can be regarded as thoroughly investigated geologically, but even there new, especially concealed, deposits may be discovered, as was the case with Pb—Zn deposits in Ireland. It is, however, improbable that they will substantially affect world mineral production. The need to provide a supply of minerals for continually expanding industry leads governments and mining companies to carry out intensive, scientifically directed and organized prospecting, especially in unexplored regions. These include the vast areas of the Asian part of the U.S.S.R., inland areas of China, a large part of South and South-East Asia, Africa, and inland areas of South America and Australia. Although some geological data concerning mineral deposits are available in these regions, deposits cropping out at the surface have also been discovered there quite recently.

Prospecting in little investigated areas

In high-mountain areas (above 3,000 m a.s.l.), exposed bedrock, low oxidation of mineral deposits and the great scatter of mineral fragments on mountain slopes and in valleys facilitate prospecting. A prospecting route leads along the major river valley and continues upstream along the nearest tributary to its source and then back to the major river, if possible along the ridge separating the tributary from the neighbouring valley, which is inspected subsequently. The prospector examines pebbles and heavy mineral concentrates and takes samples of clay from the river bed.

Mountainous areas (up to 3,000 m a.s.l.) are often extensively covered with eluvial, colluvial, proluvial and river deposits. The oxidation zone is prominent. The scatter of mineral fragments is large. Prospecting routes are planned as in the previous case or only along the rivers if the headwater and divide areas are covered, or along the dividing ridges if the river valleys are muddy and devoid of outcrops. In mountainous areas metallometric methods can be used in addition to those employed in high mountains.

Elevated mountainous plateaus of steppe character. The topographic forms vary from rounded hills to plateaus ending in steep mountain sides. The valleys on elevated plateaus can be as much as 10 km wide (they are called "ova" in Anatolia). Sparse vegetation cover is characteristic. In dry summers only a few grass tufts or spiny shrubs are found per sq. metre. This provides ideal conditions for prospecting and geological mapping; if the rocks differ in colour and resistance, the course of strata and their disturbances can be followed from a higher place or from an airplane even over several kilometres. Strikingly coloured outcropping deposits (for example ochres, gossans, graphite, coal seams) are detectable in this way.

Steppes and plains are almost completely covered with surface deposits, the outcrops occurring only in valleys and gullies. The oxidation zone is shallow but prominently developed. The scatter of minerals is small. In prospecting, the methods used in the search for concealed deposits should be employed.

E. Ackermann (1937) described a case of prospecting in southern Central Africa, carried out during 1929—1930, which is today of historical interest. The whole area covered by a virgin forest was divided by rides into strips, 20—30 km broad, striking E—W. Sixty-two geologists were divided into pairs. Each pair was to explore and map one of these strips. Their camp was in the midst of the belt from which the geologists undertook separate mapping tours, each of them being accompanied by about ten bearers. Traverses were laid so that the two geologists could check on each other. The geologist proceeded along a N—S traverse accompanied by several bearers and in the midst of a line of bearers, posted about 80 paces apart, who picked up stones for him. Distances were measured by a cyclometer. Since watercourses, paths and villages were plotted in addition to geological features, a topographic map could be constructed together with a geological map and a map of ore occurrences. Reports and samples were dispatched monthly to the central camp of the expedition. On the approval of the chief geologist, the ore prospects could be examined by trial pits. Every geologist walked 300—500 km a month and mapped an area of 200—300 km².

At the present time, helicopters and geophysical methods provide good services in prospecting in tropical and other terrains difficult of access.

Prospecting in African rain or semideciduous forests and in bush is greatly hindered by dense vegetation and a thick cover of red soil (up to 30 m). In mapping and prospecting work, forest rides must be cut; a group of eight men can cut a ride through bush at a speed sufficient for the mapping geologist. Outcrops of eluvium whose nature can be determined, are rare and therefore shallow drill holes and test pits must be sunk. Somewhat different conditions exist in tropical savannah, where the vegetation is rather sparse. The trees are mostly low (4—5 m at the most), only baobabs (in Africa) reach a height of up to 15 m. A jeep can penetrate everywhere, even into thickets if the tree trunks are not more than 5 cm thick. Geological work in savannah is facilitated by the accessibility of the area and relatively good exposures in the slopes, but it is hampered by a lack of water in the dry season and by floods in the rainy seasons. In the rainy season, field work in the whole tropical zone is retarded by downpours lasting several hours or days. But in Ghana, for example, geologists work in the field even in the rainy season, provided it has not rained since early morning. During long rains, they write mapping reports at their field stations. In this season trial pits should be logged immediately after excavation, as they are filled with water within several hours.

In sandy deserts prospecting is made difficult or impossible by a sand layer or sand dunes. In such an environment geophysical methods and especially those which allow at least reconnaissance of a large area are used.

Stony deserts provide better conditions for prospecting. The bedrock is particularly well exposed in steep slopes of inselbergs and plateaus, which present a good survey of the succession of strata. Talus deposited at the foot of an inselberg can readily be removed by excavation. Broad shallow valleys (wadis) are often filled with a layer of gravel, sand and clay up to several tens of metres thick. Plateaus, on the other hand, are covered with only a few decimetres of overburden, often containing gypsum "desert roses". Salt efflorescences are abundant. The oxidation zone is thin but very marked. Prospecting routes are oriented at right angles to the structures. Metallometric mapping is advisable and the study of geomorphological conditions yields valuable information. Lack of water and insufficient base maps are the main disadvantages of prospecting in deserts. Since water for drilling must be transported by a tank truck, for economic reasons the holes cannot be drilled too far from water sources. Base maps are sometimes lacking altogether, as no industrial or military institution is interested in mapping a desert which apart from mineral resources, is of no economic value. The geologist must therefore prepare at least a rough topographic base for geological mapping.

Tundras and polar regions are covered by tundra vegetation in low-lying areas and by rock fragments at low elevations. The recent oxidation zone is faint, and the fossil oxidation zones in the permafrost, which developed in a warmer climate, can be thick and well marked. Fragments of ore minerals are scattered over a small area, mostly due to solifluction. The mapping traverses on elevations are oriented at right angles to the structures. A geologist looks for ore fragments and pebbles, heavy mineral concentrates and metallometric anomalies; erratic ore boulders are sought in moraines.

In conclusion it should be mentioned that the prospector should invariably collect and critically assess information from local inhabitants. Even in sparsely inhabited areas, valuable reports may be obtained from indigenous or nomadic tribes who know their country well. One must also not forget that some economically backward countries, today considered to be unexplored, were centres of high civilization in ancient or prehistoric times. This is true of India, China, Iran, Mesopotamia, Anatolia, the Arabian Peninsula, Egypt and North Africa. For example, in Oman, copper was mined by the Sumerian population as early as three thousand B. C. In prospecting in countries with ancient cultures it is therefore necessary to study historical reports and to look for traces of old gold and copper mining in the field with the aid of local inhabitants.

The following natural and economic features are important for prospecting in developing countries: 1. differences in the degree of topographic and geological investigation, 2. lack of transportation routes, 3. in many places a thick weathering crust, which necessitates prospecting by drilling, 4. dense vegetation cover (rain forest and humid semi-deciduous forest), 5. a vegetation-free surface (desert, semi-desert) and 6. parcelling of the area into concessions of foreign companies. Prospecting is carried out 1. for the compilation of a general metallogenetic and prognostic map of the country (duration of work 1 to 2 years for two to three workers), 2. during multidisciplinary geological expeditions (more than 3 years for dozens of workers), 3. during specialized expeditions organized for prospecting for one or several raw materials (5 to 10 years, at least dozens of workers), 4. during short consulting missions (2 to 6 months for 2 to 3 workers), 5. during employment in the local Geological Survey (1 to 3 years, individual contracts) or 6. in the service of United Nations (1 to 2 years, individual contracts).

The most complete form of prospecting in developing countries is assessment of the territory of the whole state on the basis of published and archive reports complemented by field work. A general metallogenetic and prognostic map is constructed from the data thus obtained.

At the same time, other economic, natural and social factors are studied in preparing a feasibility study, and proposed improvements in the national economy that are feasible in the next few years are set out. Such studies are elaborated, for example, by the experts of the International Bank for Reconstruction and Development in Washington, where they are also published. A geologist or mineralogist is invariably a member of the advisory team (e.g. the Report on the Economic Development of

Libya, 1960, 524 pp., was elaborated by a medical doctor, a hydrologist, pedagogue, an agronomist, three economists, a mineralogist and one expert in each tourism, transportation, industry and trade).

One to three geologists usually cooperate in compiling a metallogenetic and prognostic map from published and archive reports. Multidisciplinary geological expeditions are also conducted, members working more in the field than in archives and offices. It is desirable not to overspend on the exploration of the deposits discovered (in any case funds are usually limited beforehand). Thus, a cycle of geological works on a deposit in a developing country does not usually last longer than three to six months. The geologist must be prepared to carry out simple dressing tests (e.g. gravity enrichment on a vibrating table, washing for a concentrate of heavy minerals) in order to save expenses on transport.

In developing countries specialized prospecting works are conducted for one or more mineral materials, most frequently for oil and uranium. These expeditions consist of hundreds or thousands of workers of various professions and the area to be explored may be hundreds to thousands of square kilometres in size. In recent years, expeditions of American oil companies have been very successful in Libya and southern Nigeria (Port Harcourt) as were the teams of French geologists who discovered uranium deposits in the Sahara (Ahaggar, Algeria). Expeditions prospecting for other ores have fewer personnel (only several geologists) and the methods used are similar to those employed during short consulting missions.

The purpose of short-term missions in developing countries is usually the elaboration of a feasibility study (whether, for example, construction of an industrial plant, a cement, glass or ceramic factory will be profitable). A geologist evaluates all published and archive reports and after a field study advises on the most promising deposits for exploration and technological sampling.

Field work is controlled by many factors so that it differs in many respects from one country to another. It is usually conducted in two phases. First, the deposits are assessed quantitatively either in collaboration with the local geologist or after reconnaissance drilling (with a light, possibly hand-operated drill rig) and small samples are taken to the base (which is usually in the capital). Secondly, large technological samples weighing several tons are taken from those deposits which have satisfactory grade and reserves. The samples are stored near the sampling site and are treated in local pilot plants or sent abroad.

The final report on the study should contain a description of the objective of prospecting, of the localities examined and of prospecting works performed, theoretical considerations on the origin of the raw material and their application to local geological conditions, correlation of geological, technological and mining conditions with those of world deposit types, conclusions on the economic importance of the mineral reserves, recommendations for further geological and technological exploration, including requirements on the quantity and quality of the reserves that should be verified, and a list of references, archive reports and acknowledgements. For his own use, the geologist calculates the cost of one ton of exploited ore, taking into account the volume of mineral reserves and of overburden (in case the deposit is to be opened by open pit), local wages, the cost of explosives and distance to the place of treatment.

The report should be accompanied by as many figures, charts and graphs as possible, as they reduce the length of the text and make the results comprehensible even to non-geologists (a map of the country with locations of the mineral occurrences plotted; sketches of the most important deposits, showing sites of exploration and sampling; and graphs representing the preliminary results of laboratory tests performed in the field). In addition, the report should be accompanied by a list of storehouses where samples were temporarily deposited, by instructions for their shipment, and all technological data determined in the field. A report on sampling and instructions for pilot-plant testing of large technological samples should be completed at the same time.

Prospecting in a developing country can also be carried out for a local geological institution. In this role the prospector is sometimes charged with assessing the finds of local inhabitants. This demands very careful assessment and considerable diplomacy, as it is risky to arouse vain hopes in local dignitaries.

In the service of United Nations, geologists take part in the work of technical aid to developing countries and cooperate on economic commissions, in the Commission for the Peaceful Utilization of Atomic Energy and in UNESCO. Generally, a geologist acts as a consultant, a reviewer of the projects and final reports and instructs his successors. His work as a prospector is usually restricted to the project and preparation of the metallogenetic and prognostic map.

A prospector working in developing countries should also be acquainted with demands on the dressability of ore and with the basic principles of the treatment of ore and industrial minerals, especially with methods applicable to primitive conditions. It must be borne in mind e.g. that, as a result of *martitization* (alteration of magnetite into hematite, easily discernible in a microscope), the dressability of iron ores by magnetic separation appreciably decreases. On the other hand, hematite can be converted by heating into magnetite (maghemite), thus improving its magnetic dressability. *Intergrowths of minerals and the size of their grains*, observable under an ore microscope, control ore dressability; generally, coarse-grained ore minerals intergrown with accompanying minerals along even planes can be dressed more easily than fine-grained interlocking minerals. In quantitative determination of the composition of a simple ore by *planimetric analysis*, the length of the line measured must be greater than a hundred times the diameter of the largest grain. *Pb−Zn ores* must be assessed mainly with regard to their amenability to flotation. Electrolytic treatment of Pb−Zn concentrates is impaired by the presence of Co (usually in linneite). Gold grains covered, for example by Fe minerals, must be deprived of this coating by crushing or dissolution in an acid before amalgamation or cyanidation. *Gold* in solid solution in pyrite or arsenopyrite cannot be recovered by a normal treatment. Dressing of telluridic gold ores is practicable by flotation or smelting, amalgamation and cyanidation are ineffective. In gravity concentration of gold-bearing alluvium it should be borne in mind that gold grains below 0.1 mm in diameter cannot be caught. *Metals of Ag — Co — Ni — Bi — U* paragenesis can usually be separated only by smelting. Low-grade *Cu ores* should be concentrated by flotation before smelting. Lixiviation is more suitable for oxidic, carbonate, silicate ores and strongly fractured ores; copper is extracted by electrolysis. Ores with intergrowths of chalcopyrite and cubanite and with magnetite cannot be concentrated by magnetic separation since cubanite is strongly magnetic. *Zinc* bonded to tetrahedrite in chalcocite ore cannot be removed from the concentrate. *Molybdenite ore* (as far as possible without chalcopyrite) can be treated by flotation and wulfenite by gravity concentration. By flotation of *Ni — Cu ores of the Sudbury type* only a copper concentrate with Ni or a nickel concentrate containing Cu are obtainable. *Cinnabar* can be flotated only when very fine-grained. Generally it is more advantageous to roast an ore containing at least 0.25 % Hg to 500 — 600 °C and condense the mercury vapours. The presence of freibergite in galena Ag-ore often causes great losses of *silver,* because the susceptibility of the two minerals to flotation is different owing to the dissimilar properties of their surfaces. *Pitchblende* is concentrated most easily by gravity separation. Intergrowths with other minerals cause losses. Oxidic uranium ores can be concentrated by hand picking and lixiviation. *Ferberite* is separable from cassiterite by a magnetic process; hübnerite ore can be concentrated by flotation as can scheelite, the high density of which also makes it amenable to gravity concentration. With regard to the excellent cleavage of wolframite and scheelite, excessive grinding might cause considerable losses. *Cassiterite* is concentrated by gravity process. Magnetite, garnet, wolframite, epidote and others are removable from the cassiterite concentrate by a magnetic process, sulphides by flotation, pyrite by roasting followed by magnetic separation, sulphur and arsenic by roasting, and Fe-oxides by lixiviation.

It should be noted that the tailings often contain a large amount of finely crushed cassiterite. Fine-grained cassiterite (below 300 mesh) and cassiterite intimately intergrown with As minerals cannot be utilized. Some kinds of barite, fluorite, phosphorite, magnesite, talc, graphite and feldspar can be concentrated by hand picking; clays, kaolin and sand by flotation; sandy gravel by washing and sieving; talc by crushing and dry separation; and gypsum by sieving. As with ores, the contact of an economic mineral with accompanying minerals should also be examined in industrial minerals.

In primitive conditions, hand picking and sieving is practicable. Ore fragments can be cleared of clay and sand by washing in an inclined trough in which running water is whirled by a rotating log provided with forward curved blades. Ore material is poured into the trough at the lower end while water flows along the whole trough; the ore is shifted slowly by the whirling blades towards the upper end and the clay is carried down to an overflow (Bernewitz 1943).

Crushing is done either by hand with a hammer, or by a piston hung on a spring or a thin flexible tree trunk, or by a small mobile jaw crusher. Concentration by gravity separation demands either a washing installation (see the chapter on the exploration and mining of placers) combined with, for example, corduroy covering of the bottom of the sluice which catches gold flakes, or simple vibration tables.

Prospecting in industrial countries

Central Europe can be cited as an example of an area with long mining traditions. It has been an important mining centre since the Celtic settlement, 2,000 — 2,500 years ago. Czech and Saxon medieval mining is a connecting link between ancient and modern mining techniques. The mineral deposits of Bohemia were the object of intense prospecting in the mid-13th century and at later times, when even Saxon and Italian prospectors arrived there. The majority of ore deposits cropping out on the surface were discovered then.

Chrt (1955) described a number of methods and criteria for prospecting in the Bohemian Massif, which has possessed the character of a platform since the Permian. These methods, directed mainly at outcropping deposits (many of them known since the Middle Ages), and published and archive reports were employed during 1950 to 1965.

Medieval mining works were not invariably abandoned after depletion of the deposit. Mining was often brought to an end by wars (e.g. Hussite wars) or persecutions (e.g. emigration of evangelic miners), by ignorance of tectonics or by a large water inflow. In these cases it may reasonably be hoped that part of the deposit has remained intact.

Every *oxidation limonite zone* was examined by trial trenches and pits to determine the extent and character of the primary mineralization.

The character of the deposit can be estimated by reference to *fragments of accompanying minerals*. Fluorite and barite deposits are accompanied by quartz veins with hematite, Mn oxides and amethyst. Quartz is often banded. Smoky quartz and quartz of pegmatitic type with tourmaline occur on Sn and W deposits. Cavernous quartz (with cavities after decomposed sulphides) coloured by ochre and limonite is characteristic of Pb—Zn—Cu deposits. Pink and purple barite is typical of deposits of greenish to violet fluorite. Milky white barite accompanies Pb—Cu deposits.

Mapping of the outcrops and course of *thick quartz veins* provides a picture of ancient mineralized faults and fractures. Barren quartz veins sometimes occur together with veins of Pb—Zn ores, Cu ores, barite, fluorite, pyrite and antimony ores. These veins are usually almost parallel with quartz veins.

Important results can be deduced from the study of *mineralizing pulsation*. In the Krušné hory Mts. (Bohemia) the sequence of hematitic quartz—purple barite—fluorite—quartz II is frequent. If purple barite is found, the occurrence of fluorite can be expected at greater depths. The following deposits are associated either structurally or genetically with hypabyssal intrusions: Sn, Cu deposits with quartz and granite porphyry (Krušné hory Mts.), Pb—Zn—Ag ores with diabase (Příbram), Cu—Ni ores with proterobase, Sb—Au ores with kersantite and minette, Sn ore (greisen) with aplite (Krušné hory Mts.), Sn—Li, Zn—Cu ores with pegmatite and Pb—Zn—Cu ores with bostonite.

The *contact haloes* are occasionally indicative of Sn, As, W and Mo mineralization.

Inspecting *temporary excavations* (for pipe lines, cables, highways, railway lines) may prove profitable. Barite fragments were discovered, for example, in this way and a fluorite-barite vein was found by tracing them. Information obtained from collectors, foresters, teachers, and other laymen is often useful.

At present, great emphasis is laid on the *study of the relationships between deep structures and mineralization*, especially in the search for concealed deposits (e.g. fluorite); geophysical (mainly electrical) and geochemical (including hydrogeochemical) methods are used more extensively, and heavy mineral concentrates are studied. Expensive exploratory works can be started only after the whole areal extent of the mineralization has been determined, so that investments can be directed to the most promising sectors.

Today, prospecting and exploration demand that prospectors have a more thorough geological education than previously. At the beginning of prospecting, the amenability of ore to treatment, the influence of impurities, variations in the material and technical conditions of mining have to be established. Particular attention should be paid to conflicting interests of the mining industry and agriculture (it is not economic to occupy good arable soil with a sand pit), health resorts, transport, telecommunications (e.g. a high-power radar station can produce a spontaneous firing of charges in a quarry or opencast to a distance of up to 15 km), gas or oil pipe lines, water supply and urban planning. It should be also kept in mind that the use of geophysical methods in industrial countries is limited by such features of civilization as railways, pipe lines, power circuits, reinforced concrete etc. Industrial pollution makes all geochemical methods liable to error, especially in the neighbourhood of refineries.

GEOLOGICAL MAPPING, PHASES AND TYPES OF PROSPECTING

We conclude this chapter on the general principles of prospecting with a short survey of the types and the proper sequence of prospecting works.

Geological mapping should be carried out at least one season before prospecting. The objective of geological mapping is to determine the structural units and to outline the geological history of an area. On the basis of the assembled data, criteria for prospecting and potential indications of mineralization under the relevant geological conditions are established. Their correct establishment is the only standard by which to evaluate geological mapping. It cannot be expected that new deposits will be discovered by reconnaissance geological mapping.

Prospecting takes place in four successive stages: general, preliminary, detailed and prospecting-exploratory. Special and deep prospectings are of a particular character.

As a result of geological mapping on a 1 : 1,000,000 or 1 : 500,000 scale combined with general prospecting conducted simultaneously or subsequently, broad prospecting criteria and indications of mineralization are determined.

Preliminary prospecting is based on geological maps on scales of 1 : 200,000 or 1 : 100,000. At this stage, the stratigraphy, tectonics, petrography, and the activity of geological agents and hydrogeology should be studied so that the prospecting criteria are determined precisely and indications of mineralization located as accurately as possible. Following general and preliminary prospecting and mapping on scales of 1 : 1,000,000 to 1 : 100,000 a map showing the distribution of mineral resources is compiled.

Detailed prospecting based on a geological map of a scale of 1 : 50,000 or 1 : 25,000 should enable evaluation of indications of mineralization and a decision as to whether the occurrence is of mineralogical interest only or a prospecting-exploratory stage is justified. For the construction of the geological map at least one outcrop or man-made exposure per 1 cm^2 of the map is required. If necessary, a few deep drill holes are sunk.

During geological prospecting and mapping on scales of 1 : 200,000 to 1 : 50,000 (i.e. during preliminary and detailed prospecting) it is advisable to carry out aero-radiometric and metallometric mapping to the same scale, sampling of heavy mineral concentrates and clays from stream floors and hydrogeochemical and radioactivity surveys. The use of satellite imagery at this stage will accelerate prospecting. In concealed areas aerial photogrammetric and aeromagnetic mapping combined with seismic profiling, gravimetric survey and deep drilling are useful. In some areas (the Far East, eastern Siberia, U.S.S.R.) detailed prospecting is possible with a map on a scale of 1 : 200,000. Detailed prospecting based on a map of a scale of 1 : 50,000 or 1 : 25,000 is the first stage enabling mineral reserves to be calculated. Calculations are made on the basis of a well-documented map and two profiles perpendicular to each other (reserves Δ_2 in the German Dem. Rep., or Δ_1 if they are sampled for technological assessment on one outcrop). Technological investigation provides general information. For limestones, for example, chemical analysis, DTA curve, X-ray analysis, description of thin sections and reference to analogous stone under exploitation are sufficient. The volume weight is taken from Tables. The result of all stages of prospecting is the construction of metallogenetic maps summarizing information obtained by geological, geophysical, geochemical, hydrogeochemical and other methods.

The prospecting-exploratory stage is the first to provide detailed data on the deposits. Geophysical and structural-geological prospecting works are conducted on the basis of geological maps of scales of 1 : 10,000 and 1 : 5000 (of 1 : 25,000 or 1 : 50,000 for sedimentary deposits) and outcrops of deposits are assessed; prospecting for concealed deposits is aimed at distinguishing industrial and non-industrial deposits, determining the industrial type of the deposit and its gross evaluation. The prospecting-exploratory stage lies on the boundary between prospecting and exploration; a drill hole which encounters a concealed deposit is a "prospecting hole" but the the following drill holes will be "exploratory". Mapping at a scale of 1 : 10,000 or 1 : 5000 is carried out in areas where some deposits have been found in order to discover concealed deposits and in areas where indications of mineralization have been established by prospecting. Prospecting is usually directed at a certain genetic type of deposit, so that it would be superfluous to study all geological features in the same detail.

Special prospecting is related to a definite metal or mineral material. It is carried out on the instructions of governmental planning institutions when a sudden economic need emerges. It is not usually necessary to map the whole territory in question; the sum of all the knowledge concerning the given territory and geological maps that are available are evaluated from the point of view of prospecting for a new metal or mineral.

Deep prospecting is important in areas where geological and mineralization conditions are well known. Its objective is to find concealed deposits which were not discovered by medieval prospectors. This is the most difficult kind of prospecting.

THE PROSPECTING-EXPLORATORY WORKS ON MINERAL DEPOSITS

At this stage detailed geological mapping, geophysical and geochemical methods and structural studies are employed; outcrops and man-made exposures are described thoroughly. The purpose of the prospecting-exploratory stage is economic evaluation of the mineral deposit.

DETAILED GEOLOGICAL MAPPING

Detailed geological mapping (usually at a scale of 1 : 25,000; 1 : 10,000 or 1 : 5000) is carried out in areas that were mapped at scales of 1 : 50,000, 1 : 100,000 and less often 1 : 200,000. Before such mapping is begun, the results of previous mapping, of the analyses of heavy mineral concentrates and of metallometry and geophysics are evaluated. Air photos are very helpful. In areas bearing endogenous deposits, detailed geological mapping is related especially to *structural features*; in areas containing exogenous deposits *facies-lithological conditions* are chiefly considered, and in metamorphic areas attention is concentrated on the petrography of the rocks; mapping according to metamorphic series or geological units is not considered to be adequate. In deciphering the structural pattern of an area it is necessary to understand the time sequence of geological events. In addition to structural elements, *contacts* of any type must be determined. The overburden need not be plotted on the map. Prospecting is undertaken before, at the same time as, or after detailed geological mapping. The principal objectives are discovery of all mineral deposits cropping out on the surface in the area to be mapped, establishment of their relationship to the lithology and structure, determination of the limits of ore fields and deposits, delineation of segments geologically favourable to the location of concealed deposits and assembling the data needed for exploration.

Hydrogeological mapping is aimed at assessing the effect of hydrogeological conditions on exploration and mining and the provision of a water supply for any future mining plant. *Geomorphological investigation* is recommended for topographically controlled deposits (e.g. placers). In detailed geological mapping, any occurrences of building and other (especially slag-forming) materials should not be omitted.

Detailed geological maps of extensive deposits (e.g. phosphates, building materials) are drawn during detailed or production exploration. A less detailed scale, for example, 1 : 25,000, can be employed for mapping sedimentary deposits such as coal or oil.

Detailed geological mapping includes the study and description of outcrops, construction of a schematic geological map on this basis and its verification by artificial exposures. These are of several types. *Cut-offs* are excavated on slopes to expose the bedrock. In outcrops of deposits large samples of ore material can be obtained by *blasting*. The thickness of overburden to a depth of 2 m and the course of the deposit can be determined by *hand augering*. *Trial trenches* should be 0.6 – 3 m broad and several metres deep (at most 10 m if they are provided with casing), their length varying according to their purpose. *Central trenches* are excavated throughout the ore zone (1 – 3 km long). *Transversal trenches*, up to 30 m long, accurately determine the strike of a deposit located by the central trench. *Trenches following the outcrop* of the deposit or a fault reveal lateral changes in the deposit. Trenches excavated parallel to the contours on a steep slope cave-in easily; it is therefore better to drive them along the maximum inclination of the slope if the geological structure permits. *Trial pits* with a cross-section of 0.75 × 1 m to 1.5 × 2 m properly cased, can be sunk to a depth of 10 – 30 m. They are excavated to expose the surface of the bedrock and to determine the thickness and quality of poorly-consolidated mineral materials. Material is lifted by a winch. In kaolin and similar rocks, pits of circular cross-section and a relatively small diameter prove adequate. In some, especially water-bearing terrains, mobile, *light drilling rigs* and *short galleries* can be used to advantage. Where the overburden is of considerable thickness, mineral occurrences are assessed by exploratory works (see the chapter on exploration). At present, there is a trend to substitute the largest possible number of artificial exposures by a geophysical survey, where it is economically and geologically warranted.

In an unknown area, central trenches are employed. Vein deposits are traced by trenches driven at right angles to their course and spaced 10 – 100 m apart. Trenches following the strike of the deposit are seldom excavated. Where the overburden is relatively thin, trenches and pits provide more reliable results than geophysical and geochemical methods. Ore pipes and nest-shaped deposits are first examined by two trenches perpendicular to each other. Man-made exposures are inspected visually and sampled immediately after excavation. One outcrop or artificial exposure per cm² of map of any scale and construction of cross-sections spaced 10 cm apart and connected by a longitudinal section are usually necessary, but not so essential for very detailed maps. The density of datum points is greater in the neighbourhood of mineral deposits and major geological units than over large, geologically monotonous areas. Geophysical and geochemical surveys are carried out before or simultaneously with the construction of the synoptic map. They are employed, in consultation with a geologist, to determine contacts beneath the overburden, faults, veins and rock and ore bodies. The survey begins at the most exposed sector in order to obtain information on the physical properties of the known geological units. Electrical profiling, the mise-a-la-masse method, spontaneous polarization, magnetometry and other techniques are usually suitable. The effectiveness of geophysical methods is reduced, for example, by the fact that anomalies established by composite profiling can be produced by flood plain, buried valleys or elongated depressions and not by mineral deposits.

The map of rocks and structures with which the deposit is spatially or genetically connected shows the contours of ore zones, deposit outcrops, lithology, pre- and post-mineralization metamorphism and indications of mineralization. Rock strips 1 mm wide, depending on the scale, should be plotted on the map; smaller units apart from mineral deposits are not shown. Discharge of springs, yield and composition of water in wells and drill holes, and water content of individual rock types are also marked in the map. One group of field geologists can map $5-10$ km^2 at a scale of 1 : 10,000 or $2-5$ km^2 at a scale of 1 : 5,000 in a month.

Detailed geological mapping usually follows detailed topographic mapping. Only exploratory works are accurately surveyed topographically. Outcrops and geological boundaries should be plotted on the map using a compass, pacing or a cyclometer. Their trends and distances are taken from the numbered datum points arranged in a precisely surveyed square network (10×10, 20×20 or 40×40 m). This can be prepared without additional effort if a geophysical survey precedes geological mapping. The construction of such a network can save considerable expense on detailed topographic mapping if the mineralization turns out to be economically unimportant. Topographic and geological mapping can only be carried out simultaneously in well-exposed, non-forested areas. After outcrops have been studied, they are surveyed along with faults, contacts and intersections of veins, which are plotted on a topographic map. Air photos prove very useful in detailed geological mapping. They help to decipher the structures, the strikes of beds, buried relief, granite tectonics of massifs and the trend of dyke rocks (which are often emphasized by weathering), and belts of altered rocks, where the alteration was connected with decoloration (sericitization, kaolinization, silicification, mylonitization). Coloured air photos provide even better results.

Detailed geological mapping is supplemented by geophysical methods of metallometry, especially a search for secondary and later for primary aureoles of ore dispersion, and by study of heavy mineral concentrates from samples taken on slopes and divides. These two methods are often combined and augmented by hydrogeochemical sampling if the overburden is thicker than 5 m. Samples taken during metallometric mapping are analysed in a spectrographic laboratory or in a field laboratory using the dithizone method. This is a simple, rapid ($15-25$ determinations per day per worker) and relatively precise method because only metal from ore minerals and not that forming an isomorphous admixture of rock-forming minerals is extracted. It can be applied to the analysis of Zn, Pb, Cu, Ag, W and Mo ores.

In mapping the environs of mineral deposits other methods are employed to obtain more precise results; for example, in mapping true bauxite deposits a map of the thickness isolines of the deposit and a contour map of its substrate are constructed.

DETAILED METALLOMETRIC MAPPING

Detailed metallometric mapping carried out in the prospecting-exploratory stage is carried out on a larger scale than prospecting metallometric mapping (see above). On occasions samples are also taken from rocks in man-made exposures, not from the weathered materials or soil near the surface. This lithogeochemical method is applicable in prospecting for ore deposits of nearly all types and is most suitable for deposits of a large areal extent, even if their metal content is low (e.g. impregnated ores). Samples are taken in trial pits or trenches, either in a grid or as loose pieces of ore (see Chapter on sampling) and are examined by semiquantitative or quantitative spectral analysis. Relict primary and newly-formed supergenic minerals, including pseudomorphs, the structures of limonites and the shapes of cavities left after leaching of primary minerals are also studied in the excavations.

Uranometric mapping is facilitated by the high migration capacity of uranium. It is usually performed after measurement of emanations (in areas covered with eluvium and colluvium) or after a gamma-ray survey (in exposed areas). Samples are studied by quantitative luminescence analysis. In mapping at scales of 1 : 10,000 to 1 : 2,000, samples are taken in a square grid, the side of which is only a few metres long.

In both the American and Soviet literature cuprometric mapping has been described in greater detail than any other metallometric method (Fig. 95 — example from South Rhodesia). Kreiter (1960) described its application on the Almalyk deposit in the U.S.S.R. Samples are taken in outcrops or excavations in a 10×10 m or 20×20 m square grid and the sampling locations are plotted on a topographic map of a scale of 1 : 2,000. Before sampling, the surface layer of weathered rock should be removed to a depth of $0.2 - 0.3$ m. The quantitative representation of supergenic minerals and limonite is initially assessed macroscopically during sampling. The Cu-content is determined by the coloration of a blowpipe flame, which is played on a small part of the sample moistened by hydrochloric acid. A Cu content of up to 0.2% produces a faint light-blue ring on the flame, up to 2 mm broad; $0.2 - 0.5 \%$ Cu — a bright light-blue ring, $2 - 5$ mm broad; $0.5 - 1 \%$ Cu — a $10 - 12$ mm broad ring of the same colour and green coloration of the flame interior; $1 - 3 \%$ Cu — a $15 - 30$ mm broad green ring. The metal content can also be determined by comparing the macroscopic habit of the sample with a set of standards, the composition of which has been determined by accurate chemical analyses. The Cu content ranges of the standards are $0.0 - 0.2 \%$, $0.2 - 0.5 \%$, $0.5 - 1 \%$, $1 - 3 \%$ and above 3%. This method yields satisfactory results when checked against the results of semiquantitative analyses of the samples. The content of copper can also be determined using a relatively rapid colorimetric method. Cuprometric mapping results in a map of metal-content isolines, according to which it is possible to distinguish the Cu-rich segments from Cu-poor segments and to determine places where the oxidation zone was leached and secondary mineralization can be expected.

Plumbometric mapping is only suitable for exploration of impregnation mineralization over large areas (Fig. 96). In the U.S.S.R. the colour reaction of lead with KI and HNO_3 was used for analysis of samples taken on the surface (limonite with supergene Pb minerals); the reaction of a sample containing $0-1\%$ Pb gave a faint but definitely yellow precipitate, in a sample with $1-2\%$ Pb the precipitate is easily perceptible and with $2-3\%$ it is of substantial volume. This method is particularly suitable for the delimitation of intensely mineralized parts within an extensive altered zone. It should generally be complemented by study of the alterations and jointing of the rocks.

Molybdenometric mapping can be employed on deposits of the Climax type and for molybdenum-bearing skarns. Niclometric mapping has so far been little developed.

In aurometric mapping of impregnation deposits, samples taken from the outcrops are crushed and the Au content is determined by panning. Mercurometric mapping is carried out with a portable emission spectrometer (for example HGG3, Scintrex), which enables faults (concentrations of $10^{-9}\%$ Hg), $Pb-Zn$ ($10^{-7}\%$) or Hg

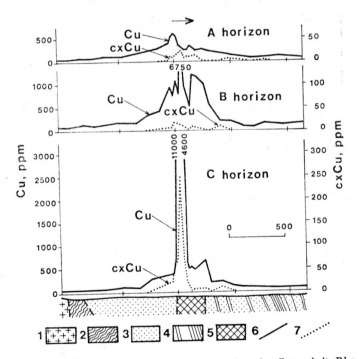

95. Metallometric (cuprometric) anomalies above a Cu deposit—Copperbelt, Rhodesia, Baluba district. The anomaly depends on the soil horizon from which samples were taken (according to Tooms and Webb, 1961). 1 — granite basement, 2 — crystalline basement (Archean), 3 — sandstone of Roan series, 4 — schists of Roan series (Proterozoic), 5 — mineralized schists, 6 — Cu content in ppm, determined by dithizone method, 7 — c×Cu, pp — Cu content in cold leach.

mineralization (10^{-5} %) to be determined. In Czechoslovakia mercurometric mapping with spectral evaluation of samples was used successfully in the Jeseníky Mts. The sensitivity of the spectral method was 7×10^{-6} % Hg and anomalies in the overburden above the deposit varied between 7×10^{-5} and 5×10^{-4} % Hg. Stannometric mapping using spectral analysis (Figs. 97, 98) or panning combined with study under a binocular magnifier gives good results on deposits of tin-bearing granites, greisens and skarns. Wolframometric mapping is facilitated by the possibility of determining the scheelite content using the luminescence method.

A new method of lithogeochemical prospecting and exploration is now being introduced in the USSR (Grigoryan 1973). The contents of metallic elementes

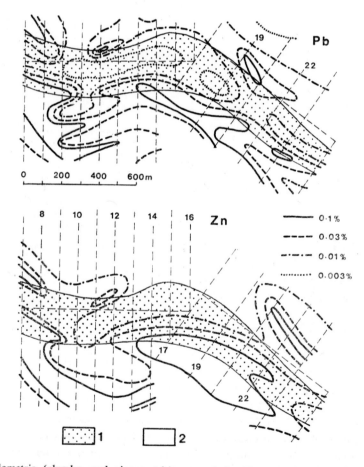

96. Metallometric (plumbo- and zincometric) map of the Zlaté Hory district, Czechoslovakia (after Gruntorád in Mašín—Válek, 1963). 1 — compact sericite quartzite, 2 — sericite-chlorite quartzite. Anomalies are over pyrite mineralization with accessory Pb and Cu.

in rocks around a hydrothermal body (primary aureole) are multiplied. The aureole of these metallic elements around the ore body constructed on the basis of the resulting values is more pronounced than the aureoles of individual elements. This method also proved very useful for the study of the lithogeochemical anomalies accompanying ore bodies near their wedging out to depth. Concealed deposits at depth were discovered in some cases when a broader multiplicative aureole of metallic elements was found in depth than above the known deposit. The deeper aureole belongs to the concealed deposit beneath the known one.

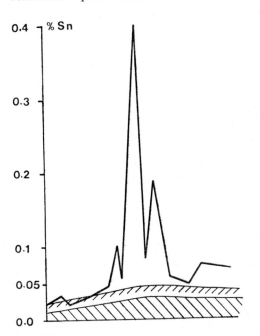

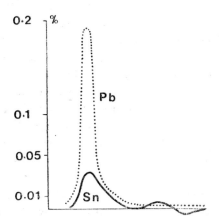

97. Curve showing the tin content in samples from eluvium and colluvium taken along a line perpendicular to the strike of vein (after Nesterov et al., 1938).

98. Ratio of tin and lead (the latter of low migration capacity) in colluvium above a polymetallic tin-bearing vein (after Nesterov et al., 1938).

STRUCTURAL RESEARCH OF ORE FIELDS AND ORE DEPOSITS

Structural mapping during the prospecting-exploratory stage should be related not only to the location and course of major faults and fault zones but also to the joint and fracture systems, bedding and schistosity, lineations, axes of small-scale folds and the orientation of the optical axes of some minerals. The plotting of small-scale tectonic elements (systems of joints and fractures, lineations) is only schematic and more or less subjective. A detailed, statistically precise record of the strikes and dips of joints and fractures is presented by a rosette diagram or a contour (point) diagram. The latter is obtained by plotting the poles of joint and fracture planes (i.e. the intersection points of normals to joints and fractures with the lower hemisphere) in a Lambert net in a pole position, and by contouring the zones with the same density of points. Field measurements of lineations (B axis) can also be plotted in this net. Preferred orientation of minerals determined by the Fedorov method is likewise represented by contour diagrams. Joint and fracture planes can be plotted in the Lambert net in the equatorial position as points of intersection with the lower hemisphere if the plane to be plotted passes through the centre of the sphere. The circles thus obtained are the nearer to the centre of the sphere, the steeper the planes. Poor clarity of the diagram is a drawback of this method. The diagrams of B axes, c axes of quartz, and the poles of joints and fractures reveal the character of deformation, suggest pre- and post-mineralization deformations and help to restore the history of the structures. Examples: limestones in the neighbourhood of faults are deformed into B tectonites (Kreiter, 1960, I. 271); the direction of movement on a slickensided surface is expressed by the orientation of c axes of quartz, which are parallel to this movement; recrystallization of crystalline schists and disturbance of the orientation of the c axes of quartz occurred near the contact with the Freiberg plutonic body (Hoffmann, personal communication); flexure folds showing gliding along foliation planes and puckering of incompetent interbeds differ from shear folds, which originate by non-affine planar deformation; distinguishing of undeformed and deformed intrusions (the optical axes, for example, of quartz and micas have sub-parallel orientation); metasomatic bodies developed along B axes on the Boliden deposit, Sweden (Ödman, 1942); magnetite deposits elongated parallel to the B axes in limestones, Vermont, U.S.A. (Bain, 1936); ore veins in Siegen (F.R.G.) are related to axial ramps (lines on which the plunge of all fold axes changes flexure-like); changes in the strike of fold axes, where mineralization is often preferentially localized (Kazakhstan); sulphide bodies of the polymetallic deposit Zlaté Hory (ČSSR) follow contacts between quartzites and chlorite-muscovite schists and fractures in quartzites (Constantinides, Pertold, 1974). Lenses rich in chalcopyrite are often embedded in the axial planes of folds in quartzites (Pertold, Constantinides 1974:.

DOCUMENTATION AND PROSPECTING-EXPLORATORY WORKS ON MINERAL DEPOSITS

In detailed geological mapping all outcrops and exposures of rocks and mineral deposits are documented. Great attention is paid to secondary aureoles or ore elements in the overburden, which is also examined with a view to future exploration (thickness, size of rock fragments, water content) and the construction of a mining and dressing plant (bearing capacity, suitability of rocks for backfilling or as a raw material for brick making). Outcrops are plotted on a detailed map, described, drawn at a scale of 1 : 10 or 1 : 100 and sampled. The spatial relations of rocks and mineral materials differing in age, strike and dip, jointing and the relationship of the deposit to country rocks are thoroughly studied. The floor of trial pits and one of their walls are drawn; in most cases it is not necessary to make a sketch of all four walls. Placers are examined by drill holes 15−20 m deep, or by test pits laid out in lines perpendicular to the valley. A placer may have the form of dispersed nests or lenses, or may fill an ancient river bed, the course of which does not correspond to the present river channel and the relics of which are buried by terrace gravels. The exploratory lines are spaced 800−1200 m apart (C_2 reserves) and 200−400 m (C_1 reserves); the drill holes or pits in the midst of the flood plain are 20 m apart and 40−80 m along the margins; this spacing is reduced to half in narrow valleys. Exploratory works are sampled in half-metre segments. In prospecting for gold, a heavy black mineral concentrate is obtained and a grey concentrate in prospecting for cassiterite.

Sedimentary Fe and Mn deposits are explored by a network of drill holes and the results of drilling are partly checked by digging trial pits.

Chromite deposits have the form of nests and lenses. The lenses are often examined by several oblique drill holes sunk from one point. Small, greatly disrupted deposits located near the surface are best explored by opening a small quarry to a depth of 3−5 m.

Contact metasomatic (skarn) deposits are explored by geophysical methods and by drilling combined with underground works. If the topomineral effects of the country rocks on hydrothermal mineralization have been established, great attention should be paid to the course of the active layer (e.g. graphitic shales) and its intersections with the vein, where it is enriched with useful minerals.

Silicate Ni ores are examined by a net of shallow drill holes and later by trial pits.

Transverse trenches and later underground works are driven on deposits of pegmatitic materials (feldspars, quartz, micas) and on quartz, barite and fluorite veins. To evaluate these deposits it is necessary to carry out mineralogical, petrographic and chemical analyses, qualitative spectral analysis (quartz, mica, barite, fluorite), firing tests (quartz, feldspar), fissibility test and determination of the dielectric constant (mica).

Crystalline magnesite deposits of Veitsch type are explored by drilling combined with underground works. Owing to the physical properties of talc, its deposits cannot be examined by drilling.

Rock salt deposits must first be mapped in detail; then they are explored by geophysical methods and by the least possible number of drill holes. A safety pillar, 20−200 m across, must be left around each drill hole to protect the salt from dissolution by water.

Magnetometric methods are applicable to the exploration of ferric bauxites and laterites.

On kaolin deposits, test pits with circular cross-sections and drill holes with a diameter of over 100 mm have proved to be satisfactory. The following technological tests are required for assessment of the quality of kaolins, clays and claystones:

sieve analysis, washing (kaolins), firing, absorption capacity, refractoriness, shrinkage on drying and firing, thermal analysis, whiteness (in kaolins), Andreasen analysis, water of plasticity and others. The exploration of building stone deposits requires the determination of defective components (sulphides, inclusions, schlieren, veins, concretions, shatter zones), of macro- and micro-tectonics, the character of jointing, the thickness of the overburden, the proportion of stones of suitable size and quality and the petrographical and physical properties (compression strength, impact strength, abrasion strength). Deposits of brick loams are explored chiefly by hand drilling and if necessary by test pits. Technological evaluation of brick loam is based on the determination of the amount of impurities (calcareous concretions, pyrite, gypsum, salts, buried beds of humus, gravel and sand, lignite, rock fragments, limonitic concretions and pelosiderites). The following laboratory tests are required: sieve analysis, determination of $CaCO_3$, disintegration in water, breakability, determination of the amount of water necessary to achieve a plastic paste, shrinkage and water loss by firing, adsorption capacity, volume weight and capillarity.

During prospecting-exploratory works on coal deposits, the thickness of seams and their structure (splitting, interseams), tectonic structure, grade and type of coal, the character of under- and overlying rocks and their water content are studied.

MAPS OF MINERAL RESOURCES AND RESERVES

On *regional maps of mineral resources and reserves,* individual deposits are plotted using various symbols on uncoloured geological maps at scales of 1 : 1,000,000 to 1 : 100,000. Endogenous deposits are usually depicted together with their extension. Regional maps of mineral deposits together with structural maps and a thorough knowledge of geochemical interrelations of deposits form a basis for the construction of a *metallogenic map* of a scale of 1 : 1,000,000 to 1 : 100,000. Metallogenic zones, metallogenic areas, provinces, districts, sub-districts and ore belts are shown on this map. On more detailed scales, *prognostic maps* are constructed on the basis of metallo-genic maps. Prognostic maps represent conclusions on the potential mineralization in definite segments; they show all structural features, known deposits, hydrothermally altered zones, haloes of metals, geophysical anomalies and other prospecting guides established during geological mapping, prospecting and exploration. On a special sheet (an overlay sheet on a map is most convenient) the area is divided into four categories:

1. prospects in the neighbourhood of industrial deposits, where new resources can be expected to be verified in the prospecting-exploratory stage;
2. prospects showing indications of mineralization associated with favourable structures and outcrops of rocks, which are known to accompany mineral deposits at other places; in this case detailed prospecting should be proposed;

3. segments as (2) but without direct indications of mineralization; prospecting works are designed;
4. segments prospected in detail and found unpromising.

The segments of individual categories are delimited and differentiated by various hachures. The potential mineralizations of the prospects are shown using symbols.

ECONOMIC ASSESSMENT OF THE PROSPECTING-EXPLORATORY RESULTS

The results of the prospecting-exploratory stage should decide whether the importance of the deposit studied justifies preliminary exploration or whether the deposit is subeconomic. In uncertain cases, the analogical method, scoring the characteristics of a deposit, or the method of technical-economic calculation can be used.

The analogical method is based on the results of exploration of a deposit which is genetically allied to the deposit under study. The thickness, the content of useful minerals, depth below the surface, size and other parameters are correlated.

Scoring of the parameters of a deposit (Kvasnikov, 1956 in Kreiter, 1960) expresses the subjective knowledge of the geologist on the deposit explored (Table 42). According to the scoring results, the deposits can be divided into five groups:

1. exceptionally important deposits (9 − 10 points),
2. very important deposits (7 − 8 points),
3. common industrial deposits (5 − 6 points),
4. deposits of uncertain value (3 − 4 points),
5. subeconomic deposits (0 − 2 points).

TABLE 42

Scoring of the parameters of deposits

Parameter	Points		
	2	1	0
1. size of the deposit	large	medium	small
2. quality of the material	high	normal	low
3. amount of the material per 1 areal or volume unit	large	medium	small
4. mining-technical conditions	very favourable	normal	unfavourable
5. economic conditions in the area	very favourable	normal	unfavourable

The method of technical-economic calculation (Pomerantsev, 1961) uses *the annual production of the future plant* (A in tons) for the determination of the importance of a deposit:

$A = k \sqrt{Z}$ (where Z = reserves in tons, k — a coefficient ranging from 100 to 300 depending on the amount of reserves), and the *necessary investment on exploitation* $W = 300A + 40 \times 10^6$ roubles. Further, the cost of the exploitation of 1 ton of ore $W + C/Z$ (C — floating capital) is calculated and compared with the economic parameters required in the respective country and with international costs, and only then is it decided whether the deposit is worth further exploration or should be abandoned. Both these formulas were valid in the economic situation of the Soviet Union in the fifties of this century for deposits of non-ferrous metals, particularly of copper. In other countries and for other types of deposits, the values of the coefficients will be different.

Prospecting-exploratory works provide the following *items for the feasibility study of the utilization of the deposit* (for example in developing countries):

1. a solid geological map of a scale of 1 : 5,000 to 1 : 25,000 (for sedimentary deposits) and a report on the geology of the area;
2. a map of heavy minerals concentrates;
3. metallometric maps with designated primary and secondary aureoles; results of hydrogeochemical research;
4. a map of geophysical anomalies;
5. a structural or facies-lithological map in which the outcropping deposits and the sectors with likely occurrence of concealed deposits are plotted; a report on the deposit;
6. a hydrogeological map; a report on mining conditions;
7. results of petrographical, mineralogical, chemical and technological studies;
8. a prognostic map of the area around the deposit;
9. calculation of C_2 (inferred) reserves or in very favourable cases of C_1 (probable) reserves;
10. calculation of the price of a ton of metal in the concentrate at the site of its usage (e.g. in a smelting plant) and at the nearest railway station or port, and
11. comparison of the calculated price with the valid world-market price (e.g. FOB in the shipping port).

GEOPHYSICAL METHODS OF PROSPECTING AND EXPLORATION OF METALLIC, NON-METALLIC AND COAL DEPOSITS

Geophysical prospecting methods have been steadily increasing in importance in the last few decades. They are based on the study of natural and artificial physical (gravity, magnetic, electric) fields, which correspond to the distribution of rocks with definite physical properties such as density, magnetic susceptibility and conductivity. Geophysical methods may directly locate mineral deposits when they possess anomalous physical properties (magnetization of skarn ores, conductivity of polymetallic ores) and are situated in a geologically and geophysically homogeneous environment. More frequently, in addition to indications produced by ore bodies, similar indications due, for example, to an insignificant magnetite admixture in rocks, to graphitization or pyritization appear. Geological interpretation is then facilitated by a simultaneous study of several physical parameters such as magnetization and conductivity. Some types of deposits are not reflected in physical fields, either because of unfavourable physical properties or due to a small concentration of the useful mineral.

Experience, however, has shown clearly that the application of geophysical methods also proves valuable when the mineral body does not produce any change in the physical field, since they help to solve many problems directly connected with prospecting. They are an important tool in geological mapping, in the establishment of structural-tectonic features, and in the study of the lithology, chemistry and differentiation of magmatic rocks, metamorphism and the size of intrusions. Geophysical methods in conjunction with geological and geochemical methods are an extremely important tool in both prospecting and exploration of mineral deposits.

The development of applied geophysics in recent years has also been promoted by progress in the construction of instruments and by automatic data processing. Thanks to scientific-technical progress precise devices (proton-resonance magnetometers, gamma-ray spectrometers, electromagnetic instruments) are available, which enable the survey to be carried out on the ground, in the air (airplane, satellite) and on water (ships). The immense amount of data thus obtained is then processed in automatic computers.

PHYSICAL PROPERTIES OF ROCKS AND MINERALS

The feasibility of using geophysical methods in geological mapping and prospecting for metallic and non-metallic deposits is based on differences in the physical properties of rocks. At present, technical equipment for ground and aerial measurements makes it possible to determine even very small changes in normal physical fields caused by the presence of geological bodies (e.g. basic rocks, ores) that differ in their physical properties from their environment. These changes in the normal physical fields are called geophysical anomalies. Their interpretation provides qualitative and quantitative data on the disturbing body (geological characteristics, depth, dimensions, dip and shape).

The magnitude of magnetic anomalies is related to the intensity of magnetization and the size and position of the disturbing bodies. The magnetization in rocks is of two kinds—induced and remanent magnetization. Induced magnetization ($I_i = T$) is directly proportional to the intensity of the earth's magnetic field, T, and to the magnetic susceptibility. Remanent magnetization, I_r, varies over a wide range (from a small fraction to many times the I_i value), depending on the physical conditions, particularly the temperature, existing when the rock was formed.

High magnetization of the rock is largely due to the presence of strongly magnetic minerals: magnetite, titanomagnetite, hematite and pyrrhotite. Approximate values of the magnetic susceptibilities of some minerals are listed in Table 43 and magnetic characteristics of selected rocks in Table 44. Apart from rare exceptions, sedimentary rocks are virtually non-magnetic. Increased susceptibility is caused by admixture of magnetite which is resistant to weathering. The magnetic susceptibility of igneous

TABLE 43

Magnetic susceptibility of minerals

Mineral	Magnetic susceptibility 10^{-6} CGS
magnetite	100,000—1,000,000
titanomagnetite	20,000—100,000
hematite	200—20,000
pyrrhotite	20,000—300,000
hornblende	50—400
augite	20—150
garnet	100—200
olivine	0
tourmaline	50—100
serpentine	250
biotite	100—200
phlogopite	200
quartz	0
limonite	100—150
pyrite	20—150
chalcopyrite	100
arsenopyrite	100—200
pyrolusite	500—600
sphalerite	250
galena	0
siderite	450

TABLE 44

Magnetic susceptibility of rocks

Rock	Practially non-magnetic 30×10^{-6} CGS	Very weakly magnetic $(30—100) \times \times 10^{-6}$	Weakly magnetic $(100—1000) \times \times 10^{-6}$	Magnetic $(1000—5000) \times \times 10^{-6}$	Strongly magnetic 5000×10^{-6}
limestone	●	–	–	–	–
dolomite	●	–	–	–	
sandstone	●	○	○	–	–
claystone	●	○	○	○	–
gypsum	●	–			–
granite	●	○	○	–	–
granodiorite	–	–	●	●	–
porphyry	●	–	–	–	–
syenite	●	○	○	○	–
gabbro	–	○	○	●	○
diabase	○	○	○	●	●
porphyrite	○	–	–	●	–
basalt		–	●	●	–
peridotite		–	○	●	●
serpentinite	–	–	○	●	●
amphibolite	●	○	○		–

● frequent occurrence
○ rare occurrence
– absence

rocks increases with increasing basicity. The magnetic properties of metamorphic rocks are diverse. As long as metamorphic processes were not connected with the supply or formation of magnetic minerals, metamorphosed rocks do not differ in their magnetic susceptibility from primary rocks.

The application of gravimetric methods is feasible where there are differences in rock density. A study of rock densities must precede the decision on the suitability of gravity survey, and interpretation of gravity maps. The range of densities of selected rocks and minerals is presented in Table 45. The range of variation of rock densities is appreciably narrower than that of magnetic susceptibilities. The density of igneous rocks increases with their basicity and the density of sedimentary rocks with their age and depth of deposition. It should be noted that the density of large bodies and the effects of tectonic movements on the rocks cannot be determined by laboratory measurements. The density of rocks decreases markedly with increasing porosity.

TABLE 45

Density of rocks and minerals

Rock	Density g/cm^3		Mean density g/cm^3
	from	to	
basalt	2.6	3.3	3.0
gabbro	2.7	3.4	3.0
peridotite	2.6	3.6	3.2
granite	2.4	3.0	2.7
gneiss	2.4	3.4	2.8
porphyry	2.3	2.8	2.6
clay	1.5	2.2	1.9
sand	1.4	2.0	1.7
clayey shale	2.1	2.8	2.4
sandstone	2.1	2.8	2.4
limestone	2.3	3.0	2.7
chalk	1.8	2.6	2.2
anhydrite	2.9	3.0	3.0
gypsum	2.2	2.4	2.3
rock salt	2.0	2.2	2.1
coal	1.2	1.7	—
hematite	5.1	5.2	5.1
corundum	3.9	4.0	3.9
magnetite	4.8	5.2	4.9
titanomagnetite	4.5	5.0	4.7
manganite	3.4	6.0	4.7
pyrite	4.9	5.0	4.9
chalcopyrite	4.1	4.3	4.2
pyrrhotite	4.3	4.8	4.6
chromite	3.2	4.4	3.7
scheelite	5.9	6.2	6.0
barite	4.4	4.7	4.5

The electric properties of rocks—resistivity, dielectric constant, electrochemical activity and polarization—vary considerably. Electric DC methods and low-frequency electromagnetic methods are based on resistivity, which is the principal electrical property.

Low resistivity is typical of most ore minerals—covellite, bornite, pyrrhotite, chalcopyrite, galena, magnetite, titanomagnetite, arsenopyrite, pyrite and marcasite. Graphite is a good conductor (admixture of it in rocks produces a considerable reduction in the resistivity) the same is sometimes valid for cassiterite, hematite and wolframite. Sphalerite, molybdenite, realgar and boulangerite have a high resistivity. The resistivity of ores is a function of their structure and not of the proportion of conductive minerals they contain. If the conductive component in the ore constitutes a continuous interconnected network, the resistivity of the ore is low. The resistivity is high if conductive minerals form isolated clusters surrounded by non-conductive minerals (e.g. Ni−Cu mineralization in basites). The suitability of electrical methods in prospecting for ore bodies can be assessed from the succession of minerals in thin sections. If conductive ore minerals were separated last, the applicability of electrical methods is warranted. Resistivities of several conductive ores are presented in Table 46.

TABLE 46

Resistivity of ores

Essential mineral	Resistivity Ωm	
	ore	pure mineral
pyrite	$10^{-4} - 10$	$5 \times 10^{-5} - 5 \times 10^{-2}$
chalcopyrite	$10^{-4} - 10^{-1}$	$10^{-4} - 7 \times 10^{-4}$
pyrrhotite	$10^{-5} - 10^{-3}$	$10^{-5} - 5 \times 10^{-5}$
arsenopyrite	$10^{-3} - 10^{-1}$	3×10^{-4}
galena	$10^{-2} - 3 \times 10^{2}$	$3 \times 10^{-5} - 3 \times 10^{-4}$
magnetite	$10^{-2} - 10$	10^{-4}

TABLE 47

Resistivity of rocks

Rock	Resistivity in Ωm
limestone	$2.1 \times 10^{5} - 4.2 \times 10^{5}$
dolomite	$3.5 \times 10^{2} - 5.0 \times 10^{3}$
sandstone	$3.5 \times 10^{4} - 1.4 \times 10^{5}$
claystone	$1 - 30$
gypsum	$1.0 \times 10^{4} - 1.0 \times 10^{6}$
clayey shale	$1.0 \times 10^{3} - 6.4 \times 10^{4}$
quartzite	4.7×10^{6}
gneiss	6.8×10^{4}
skarn	2.5×10^{2}
granite	$3.0 \times 10^{5} - 3.6 \times 10^{6}$
granite porphyry	$4.5 \times 10^{3} - 7.0 \times 10^{3}$
quartz porphyry	9.2×10^{5}
porphyrite	$10 - 5M10^{4}$
diorite	$2.0 \times 10^{4} - 2.0 \times 10^{6}$
gabbro	$5.0 \times 10^{2} - 1.0 \times 10^{5}$
diabase	$2.9 \times 10^{2} - 4.6 \times 10^{7}$
basalt	$1.6 \times 10^{3} - 2.3 \times 10^{4}$
peridotite	3.0×10^{3}

The resistivity of rocks made up of non-conductive or poorly conductive minerals is mainly determined by the porosity of the rock, the amount and degree of mineralization of pore water, the temperature and, to a lesser extent, by the resistivity of the rock-forming minerals. Resistivity of geological bodies depends on tectonic disturbance and on the amount of hygroscopic components produced by crushing and weathering. The range of resistivities of several rock types is given in Table 47.

The electrochemical activity and polarization of rocks are made use of in methods based on electrochemical principle, i.e. in the methods of spontaneous polarization and of induced polarization.

Detailed data on magnetic properties, densities and the electrical properties of rocks, together with data on the radioactivity of rocks, the velocity of seismic wave propagation and other parameters have been published, for example, in the Handbook of Physical Constants (Clark, 1966).

GEOPHYSICAL METHODS

Gravimetry studies the earth's gravity field; the gravitational acceleration is measured (in milligals, abbr. mgal, which is equal to 0.001 gal; gal = 1 cm s^{-2}). Gravity anomalies are obtained as the difference between the gravitational acceleration measured at a given point (this value expresses the effect of bodies whose density differs from that of their surroundings) and the normal gravitational acceleration. The latter is given by a formula and differs according to the geographical latitude. The Bouguer anomaly is a gravity anomaly calculated by allowing for the effect of masses between the topographical surface and the surface at sea level. The gravitational acceleration can be measured by gravimeters with a precision of 0.01 mgal. For calculation of anomalies, the altitude above sea level of the measured point must be determined with a precision of a few centimetres. A gravity survey is usually made on the ground, exceptionally from ships, airplanes or satellites. Gravity measurements from moving objects are not sufficiently accurate (ca. 5 mgal).

In prospecting for mineral deposits, gravity survey makes it possible to solve structural-tectonic problems, particularly mapping of intrusive bodies (heavy basic rocks and light granitoids), to delimit basins with light sedimentary filling, and so on. Very precise modern gravimeters are used to locate ore bodies deposited in an environment of homogeneous density and near the surface. The method is most effective in defining the extent of known ore bodies in mine workings. Measurements with special gravimeters in drill holes are conducted for precise determination of the density of rocks around the drill hole.

Gravity data are processed in automatic computers. In addition to Bouguer anomaly maps, derived maps, i.e. of regional anomalies, residual anomalies, second gravity derivatives, horizontal gradients and other values may be constructed.

Magnetic survey is based on the study of the earth's magnetic field; the parameters measured are the total field intensity, T, or its vertical, Z, or horizontal, H, component. The unit of magnetic field intensity used in geophysics is the gamma $\left(= 10^{-5} \text{ Oersted, i.e. } \frac{1}{4} 10^{-3} \frac{A}{m} \right)$. Magnetic anomalies produced by varying magnetization of rocks in the upper part of the earth's crust are calculated as the difference between the value measured at a given point and the normal intensity of the magnetic field (defined for a certain area and epoch). The magnetic field was formerly measured by a magnetic balance, the principal component of which is a permanent magnet. This device records both the vertical, Z, and horizontal, H components of the magnetic field. In recent years, proton-resonance magnetometers (total intensity, T) and fluxgate magnetometers (Z and H components) have been most widely used. Measurement with modern magnetometers is very simple and rapid (several seconds). Accurate measurements (± 1 gamma) can be conducted from both stationary and moving stations. Aeromagnetic survey operates far more effectively than ground survey. Magnetic field measurements are also carried out from ships during investigation of oceans and precise magnetometers belong among satellitic equipment. In recent years, airborne survey has been used for precise measurements of the vertical gradient of the earth's magnetic field.

Magnetic survey has proved suitable in all stages of prospecting for metallic and non-metallic deposits. Aeromagnetic survey is used for rapid and cheap mapping of basic rocks and rock complexes containing an admixture of ferromagnetic minerals — mainly magnetite and pyrrhotite — and for locating transverse tectonic lines in magnetically anomalous areas. On the basis of an interpretation of the depths of disturbing magnetic bodies beneath sediments, the thickness of the latter can be determined. Under favourable conditions, this method is effective in outlining major bodies of magnetic ores, such as iron ores and polymetallic ores with pyrrhotite. Magnetometry is often applied to measurements in drill holes, where it records intensely magnetized ore bodies in their neighbourhood.

Electrical methods are numerous; they examine natural or induced electric fields of direct or alternating current. The *resistivity method* (profiling and sounding) is based on the modified Ohm's law and it measures the apparent resistivity of rocks, expressed in Ωm. In *electrical profiling,* the depth of effective penetration remains constant and the position of the point of measurement is changed. In *electrical sounding,* the position of the point of measurement is constant and the depth of effective penetration increases with the increasing spacing of the electrodes. The *electromagnetic methods* (EM) utilize the laws of electromagnetic induction. A source of the alternating magnetic field (primary) on the earth's surface induces the current in a conductor (e.g. sulphidic ore body) below the surface, which sets up a secondary magnetic field. By measuring the resultant electromagnetic field, i.e. the sum of the primary and secondary fields, the presence or absence of prominent conductors in the study area may be determined. There are numerous kinds of electromagnetic methods; some of them are listed in the survey of geophysical methods in Table 48. The method

TABLE 48

Method	Parameter, characteristic physical property	Main causes of anomalies	Applications: direct investigation	indirect
GRAVITY ground, marine	gravity, milligal $(1\ \text{gal} = 1\ \text{cm/s}^2)$, density	deposits of heavy ores, differences in the distribution of densities	iron ore, chromite, pyrite, chalcopyrite	geological structural mapping, placer configuration
MAGNETIC ground, airborne, marine, logging	earth magnetic field: total intensity, vertical gradient, $1\gamma = 10^{-5}$ gauss, magnetic susceptibility	magnetic content of the material, contrast of magnetization	magnetite, pyrrhotite, titano-magnetite	iron ore, chromite, copper ore, kimberlites, geological structural mapping
RESISTIVITY ground, marine, logging	apparent resistivity, Ωm, resistivity, conductivity	conductive vein, ore body, sedimentary layer, resistive layer, limestone, volcanic intrusion, shear zone, fault, weathering	massive sulphides, quartz, calcite, special clays, rock salt	detailed tectonics, base metals, phosphates, uranium, potash, coal
INDUCED POLARIZATION ground, logging	time domain, chargeability, ms, polarizability, %, frequency domain, frequency effect, %, metal factor, ionic-electronic, over voltage	conductive mineralization, disseminated or massive	conductive sulphides, oxides	associated minerals, zinc, tin, gold, silver

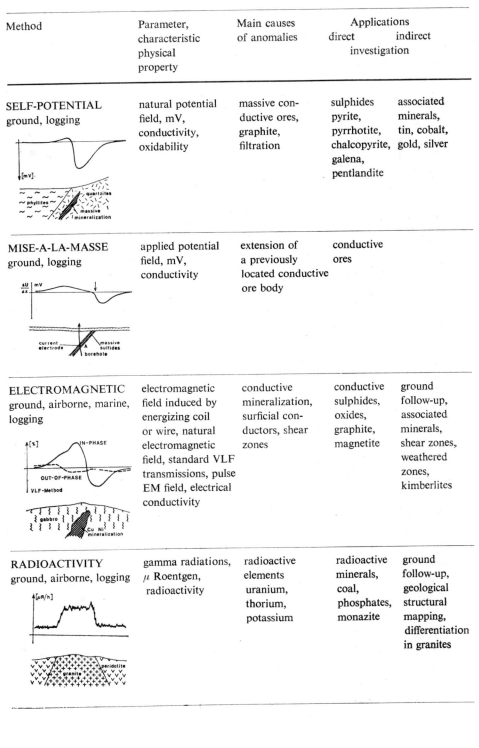

Method	Parameter, characteristic physical property	Main causes of anomalies	Applications direct investigation	indirect
SELF-POTENTIAL ground, logging	natural potential field, mV, conductivity, oxidability	massive conductive ores, graphite, filtration	sulphides pyrite, pyrrhotite, chalcopyrite, galena, pentlandite	associated minerals, tin, cobalt, gold, silver
MISE-A-LA-MASSE ground, logging	applied potential field, mV, conductivity	extension of a previously located conductive ore body	conductive ores	
ELECTROMAGNETIC ground, airborne, marine, logging	electromagnetic field induced by energizing coil or wire, natural electromagnetic field, standard VLF transmissions, pulse EM field, electrical conductivity	conductive mineralization, surficial conductors, shear zones	conductive sulphides, oxides, graphite, magnetite	ground follow-up, associated minerals, shear zones, weathered zones, kimberlites
RADIOACTIVITY ground, airborne, logging	gamma radiations, μ Roentgen, radioactivity	radioactive elements uranium, thorium, potassium	radioactive minerals, coal, phosphates, monazite	ground follow-up, geological structural mapping, differentiation in granites

Continued

Method	Parameter, characteristic physical property	Main causes of anomalies	Applications direct investigation	indirect
SEISMIC ground	refraction, reflection, travelling time of elastic waves, m/s, elastic wave velocity, dynamic modulus	contrast of velocity, marker at variable depth, fissured rocks	buried channels, faults, general tectonics, sand, gravel deposits, heavy minerals	tin, heavy minerals placers, coal, uranium

is also applied in airborne operation, usually combined with aerial magnetic and radioactivity methods.

The methods of spontaneous polarization (SP) and induced polarization (IP) are based on the electrochemical activity and polarization of rocks. The *spontaneous polarization* method measures the spontaneous electrical potential (in mV) resulting from the oxidation of sulphidic ores. An intensive spontaneous polarization field is produced only above an ore body containing conductive minerals with a favourable structure. Part of the ore body must lie in an oxidizing environment and part in a reducing medium. Unless this requirement is fulfilled, the method of *induced polarization* should be used. This technique is particularly suitable for the determination of sulphide impregnations; the polarization is given as a percentage. Other electrical methods measure the electromagnetic fields of long-wave transmitters, the natural alternating electromagnetic field of the earth and fields of other types.

Electrical methods are useful in the study of tectonic-structural conditions, such as tracing conducting formations and tectonic lines, or in determination of the thickness of sediments. They are, however, most effective in prospecting and exploration of mineral deposits. Conducting bodies can be searched for by ground and aerial electrical survey and their geological character can be inferred from the type of indications. Electrical methods are adaptable for measurements in boreholes and assist in finding ore bodies in their neighbourhood and linking up bodies established by drilling or underground works.

Radioactivity methods are based upon natural or artificial radioactivity. Instruments for the detection of natural radioactivity measure the overall gamma-ray intensity, the contents of uranium, thorium and potassium (gamma-ray surveying) and emanations of radon (Rn), thoron (Tn) and actinon (An) in soil air (emanometry). Field measurements of gamma-ray activity and gamma-ray surveying are suitable only where the thickness of the overburden does not exceed 1 to 2 metres. Since emanometry is effective to greater depths, it can also be employed where the thickness of the sedimentary cover is greater. Methods based on the measurement of artificial radioactivity enable the concentration of useful minerals in rocks, their density and porosity, their saturation with water and other properties to be determined. Radioactivity survey is carried out on the ground (gamma-ray surveying, emanometry), from the air (gamma-ray surveying), in drill holes (total gamma-ray activity), in underground workings and in ore-preparation plants (in uranium mines).

Airborne gamma-ray survey is used in prospecting for mineral deposits over large areas. On the radioactivity maps of areas free of sedimentary cover, active granitoid rocks are differentiated from inactive basic rocks, and major near-surface uranium deposits are detected. Because of its great depth of penetration emanometry is an extremely useful tool in ground prospecting for uranium deposits. Artificial radioactivity measurement is employed particularly in drill holes in determining the physical parameters of rocks and to establish the content of useful elements.

Seismic survey studies shock waves generated by an explosive charge or a vibrator. The principal parameter measured is the velocity of wave propagation (in km/sec). Seismic analysis provides a record of the depths of geological boundaries on which the waves are reflected (reflection method) or refracted (refraction method). Seismic survey is very expensive and is applied mainly in areas built up of sedimentary rocks and only rarely in crystalline areas. Measurements are carried out either on the earth's surface or from ships on lakes and shelf seas.

For geophysical well-log measurements, several methods are usually combined, depending on the type of drill hole and the problem studied. Probes with detectors attached to cables are lowered into holes to make electrical (resistivity log, self-potential log), radioactivity (natural and induced radiation), magnetic, thermal and other measurements.

Geophysical measurements in drill holes detail the lithology established by drilling, delimit precisely the mineralization zones, provide data on the physical properties of rocks necessary for the interpretation of the surface geophysical survey and determine the technical parameters of the drill hole (diameter, inclination, water flow, etc.). In recent years, methods allowing exploration of the area adjacent to a drill hole or between drill holes have been used, such as magnetic survey, induced polarization, the mise-a-la-masse method and electromagnetic methods.

Geophysical methods and their modifications are too numerous to be treated adequately in this book. The basic data on the most important methods, illustrated by typical examples, are given in Table 48.

APPLICATION OF GEOPHYSICAL METHODS

In field geophysical survey several principles must be borne in mind. Geophysical exploration should be connected closely with geological investigations and should be carried out comprehensively and in the proper sequence. The correlation of geophysical and geochemical results is of primary importance. Geophysics can locate geological bodies with very differing densities, conductivities or magnetic susceptibilities but the presence of particular elements can be corroborated only geochemically. The individual phases of a geophysical survey should be organized so that the regional survey is followed by prospecting and mapping and completed by prospecting and exploration. The full use of geophysical methods, i.e. simultaneous application of several methods studying different geophysical parameters, facilitates the geological interpretation of geophysical data. The set of methods should be sufficiently comprehensive to utilize all the possibilities offered by modern geophysics but not so extensive as to provide the same information in various forms.

Geophysical survey methods depend to a large extent on the geological knowledge of the area and its mining development, on the density of settlements and on climatic and other factors. In surveying large unexplored areas (e.g. in developing countries) different geophysical methods from those adopted in industrial countries, near mines and preparation plants, will be used. In the former instance, an integrated airborne survey, which can cover vast areas rapidly and at a low cost and discover unknown mineralization areas, is preferred. Ground and well-log geophysical methods will be used in the latter case, allowing exploration of a known mineral prospect at greater depth.

An *aerial geophysical survey is* usually combined with air photographs and, where necessary, with air photogrammetry. An example of a complete aerial record obtained by a modern Barringer apparatus is shown on Fig. 99. The INPUT system has six channels for various intervals after termination of the current pulse in a transmitting loop, which enables determination of the time constant of a transient phenomenon and thus qualitatively differentiates conductors — sources of anomalies. The record of the magnetic field is presented graphically and on a punched tape for automatic processing. The record of a gamma-ray spectrometer enables potential sources of radioactive anomalies to be identified. The height of flight above the ground is also registered. Programmes have been set up allowing computer construction of maps of ΔT isolines, maps of the intensity of gamma-ray radiation according to individual components (Th, U, K), and correlation schemes showing indications differentiated on the basis of their prospects.

An integrated geophysical air group is very efficient and it is necessary to employ a sufficient number of groups for ground follow-up of the anomalies determined by airborne survey. Using a scale of 1 : 25,000, the profiles are spaced 250 m apart so that 4 km profiles are needed for 1 km². An integrated air group can cover an area of up to 100,000 km² in one season. Seigel proposed the following procedure for ground assessment of air-determined anomalies (see the scheme in Fig. 100). After eliminating anomalies caused by artificial conductors, the remainder is divided into indications produced by a conductor deposited at greater depth and those connected with the overburden. The ensuing procedure can be seen from the figure.

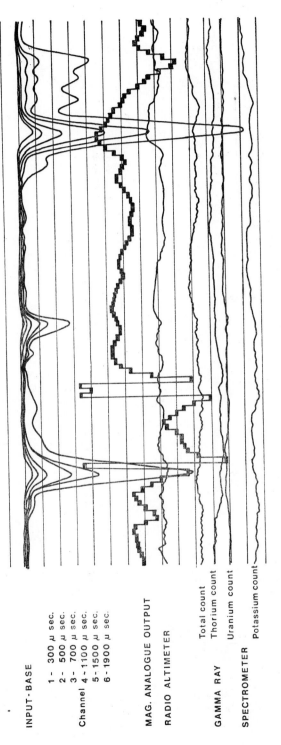

INPUT - BASE

1 - 300 μ sec.
2 - 500 μ sec.
3 - 700 μ sec.
Channel 4 - 1100 μ sec.
5 - 1500 μ sec.
6 - 1900 μ sec.

MAG. ANALOGUE OUTPUT

RADIO ALTIMETER

GAMMA RAY

Total count
Thorium count
Uranium count

SPECTROMETER

Potassium count

99. The Barringer INPUT airborne E-M exploration system.

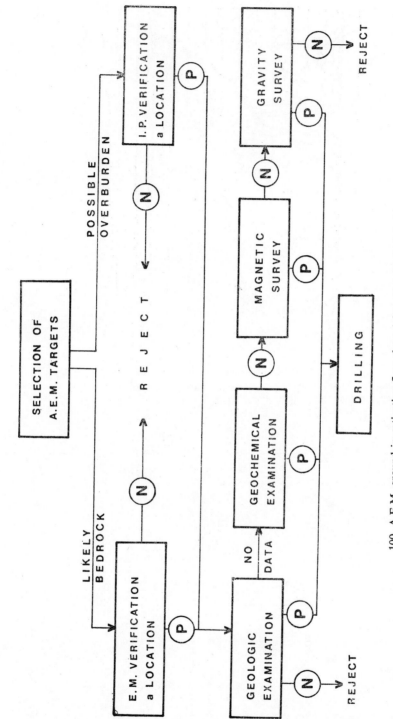

100. A.E.M. ground investigation flow chart (after Harold O. Seigel, 1972).

An airborne geophysical survey provides an immense number of indications. The paper by T. Verbeek et al. (1972) provides a useful example of the results obtainable by ground control of an aerial survey. The result of ground control of 772 anomalies established chiefly in the Abitibi area in Quebec is summarized in Table 49.

Geophysical survey methods for some types of metallic and non-metallic deposits are described briefly below.

TABLE 49

Investigation of electromagnetic conductors: Jan. 1968 to Sept. 1970

Airborne conductors selected for investigation[1]) 772
Airborne conductors defined on the ground[2]) 610
Conductors found during ground surveys 5
Total no. of conductors defined on the ground 615
Rejected on a geophysical basis[3]) 307
Conductors drilled . . . . 274 (398 drill holes totalling 152,000 feet)
Conductors trenched[4]) . . . . 34 (39 trenches)

Explanation of conductors investigated:

	Number	Per cent
Sub-economic[5]) massive sulphides	1	0.3
Sub-economic disseminated sulphides	3	1.1
Massive sulphides[6])	61	21.3
Graphite + sulphides	136	47.4
Graphite	39	13.6
Other conductors[7])	5	1.7
	287	100.0
Conductors not explained[8])	12	
Drilling abandoned[9])	9	
	308 = 50.2 % of total ground conductors	

[1]) Selection of airborne conductors for ground investigation is a complex process depending on the nature of the geophysical response, the apparent geological situation, and non-geological factors such as the availability of the ground and whether any previous work is known.

[2]) Approximately 220 airborne conductors were not located. However many of those located were resolved into more than one conductor on the ground.

[3]) Conductors were usually rejected because of: feeble ground EM response; lack of magnetic association; very strong magnetic association (suggesting iron formation); or evidence of previous drilling.

[4]) Manual removal of overburden and blasting into the bedrock.

[5]) Economic metal content and size approaching values required for a reliable mining prospect.

[6]) Massive sulphides contain more than about 30 % of sulphide minerals; disseminated sulphides less than this amount.

[7]) Usually schist zones.

[8]) No evidence of a conductive zone found in the bedrock.

[9]) Usually because of excessively thick overburden.

Iron deposits are generally amenable to geophysical prospecting, since they are formed of minerals displaying anomalous magnetization and density. The chief prospecting methods are airborne and ground magnetic survey, gravity survey and, occasionally, electrical methods. Figure 101 shows a section of an aeromagnetic map, where the ΔT maxima correspond to a detected deposit of metamorphogenic ores (with quartz) at Belozerka (Kursk area, U.S.S.R.). The applicability of a gravity survey to a deposit of the same type in the same area is illustrated in Fig. 102. A section

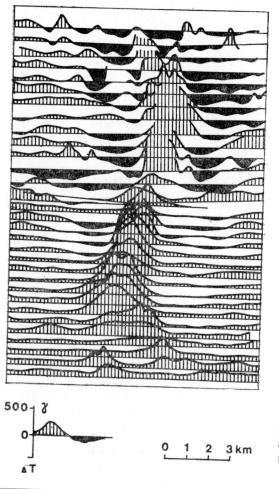

500 γ
0
ΔT

0 1 2 3 km

101. Magnetic field over Belozerka iron deposit, which was discovered by aeromagnetic survey (after Suslenikov in Krutikhovskaya et al., 1967).

102. The map showing Bouguer anomalies Δg and results of calculation of the lower boundary of iron ore deposit (after Krutikhovskaya, Kuzhelov, 1960). 1 — sandstones, slates, quartzites and dolomites of the upper series, $\Delta\varrho = 0.1\text{—}0.2$ g/cm^3; 2 — coarse laminated iron quartzites with slate intercalations, $\Delta\varrho = 0.50\text{—}0.55$ g/cm^3; 3 — iron quartzites and slates of middle and lower series, $\Delta\varrho = 0.35\text{—}0.70$ g/cm^3; 4 — granite and gneiss; 5 — faults; 6 — Bouguer anomalies Δg; 7 — calculated curve Δg; 8 — contour of magnetic anomaly in 1000 gamma isoanomaly; 9 — profiles.

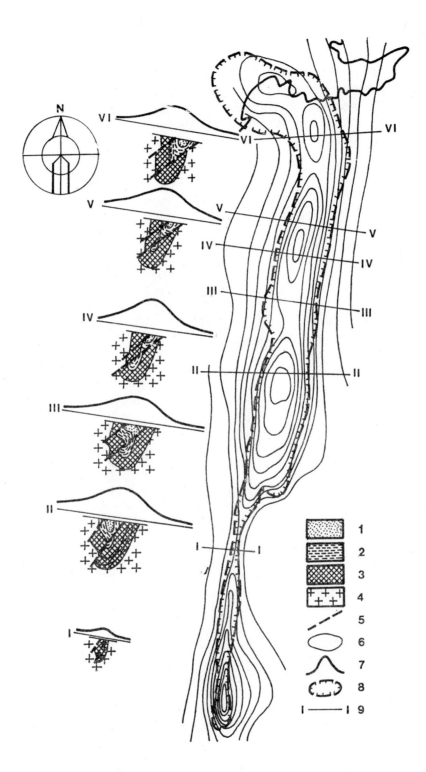

of the deposit and its lower boundary can be defined approximately from the Δg curves.

Skarn deposits are readily identifiable by geophysical methods. Magnetite bodies deposited in metamorphosed or carbonate rocks are revealed by prominent magnetic anomalies. The results of a magnetic survey in the area between Přísečnice and Měděnec in the Krušné hory Mts. (Bohemia) are cited here as an example (Fig. 103).

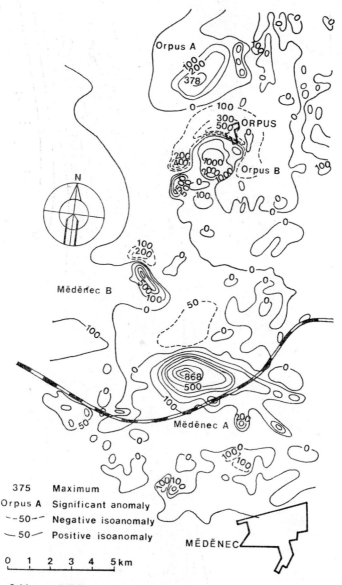

375	Maximum
Orpus A	Significant anomaly
--50--	Negative isoanomaly
—50—	Positive isoanomaly

0 1 2 3 4 5 km

103. Magnetic field over Měděnec iron deposit, Czechoslovakia, discovered by magnetic survey (after Pokorný)

Skarn bodies forming irregular lenses occur at the contact of orthogneisses with underlying crystalline schists, which dip at a moderate angle beneath the orthogneiss body. The entire body was surveyed. Magnetic anomalies that are bounded by a uniform gradient on all sides are of practical importance, since they are produced by flat lying lenses (Měděnec A, Orpus A). Uneven anomalies with a clear-cut minimum are caused by steeply dipping bodies of Tertiary basalts (Měděnec B, Orpus B).

Iron ore deposits of sedimentary and hydrothermal-metasomatic types are less suitable for geophysical prospecting. In this case, a gravity survey is the principal tool, but it can locate only sufficiently large ore bodies deposited at a small depth. Electrical methods are used where siderite mineralization is accompanied by sulphides, and have proved successful in determining the structural-tectonic conditions of a mineralization area.

Geophysical methods can also be employed in detailed exploration of iron ore deposits: magnetic survey in drill holes and gravity survey in mine workings.

Geophysical methods are effective in prospecting for nickel deposits, in their mapping and in resolving their structural tectonic pattern. Aeromagnetic and gravity surveys reliably delimit the massifs of basic and ultrabasic rocks (density $2.7-3.3$, high magnetization). With respect to the conductivity of sulphides, the IP method is the most important during prospecting, as it can determine which basic and ultrabasic rocks contain larger amounts of sulphides. In a detailed geophysical survey, the magnetic method, which enables the differentiation of basic rocks, resistivity profiling or inductive methods (which aid in studying tectonics) are employed as auxiliary techniques. In detailed exploration of a deposit, the IP measurement in drill holes, the mise-a-la-masse method and accurate gravity method (in mine workings) are useful.

The IP method furnished positive results at Staré Ransko (Českomoravská vysočina Highland, Bohemia). Under intricate geological conditions, measurements were carried out within a grid of 200×20 m over an area of 4.5 km^2, with AB intervals of $1000-1600$ m. All known occurrences of Ni$-$Cu mineralization are accompanied by intensive anomalies $\eta_z = 20-30$ %. Pentlandite, pyrrhotite and chalcopyrite are the most frequent sulphidic minerals; they occur in peridotites, troctolites and gabbros. The complicated situation can be seen from Fig. 104 showing IP profile measurements correlated with a geological section. All principal and minor ore bodies are reflected on the IP curve.

The results of measurements over a Ni$-$Cu deposit (Russian platform, U.S.S.R.) covered by a 120 m thick overburden are shown in Fig. 105 to illustrate the penetration depth of the IP method. Sulphidic mineralization in peridotite produced an intensive polarization anomaly. The great depth of penetration of the pulse EM method is apparent from Fig. 106 (Karelia, U.S.S.R.). A flat lying chalcopyrite-pentlandite lenticular body deposited at a depth of 150 m is the source of a marked anomaly in the pulse EM method.

Residual nickel deposits occur above massifs of basic and ultrabasic rocks. These massifs can be delimited precisely by magnetic and gravity surveys and the depth of unweathered basement can be determined by vertical electrical sounding.

Geophysical methods are generally successful in locating copper, lead and zinc deposits, particularly those of hydrothermal and skarn types. Since the bodies consist of conductive sulphides with a favourable structure, they can produce intensive anomalies in SP, resistivity and electromagnetic surveys. Porphyry copper ores, cupriferous sandstones and impregnated ores are less suitable for geophysical prospecting, since the ore minerals form isolated impregnations which cannot be determined using DC electrical measurements. The IP method is most suitable in prospecting for this type of deposits.

In the initial prospecting stage integrated airborne geophysics is most useful. During regional investigation gravity and magnetic surveys should be employed for mapping promising prospects. Both these methods aid in delimiting major geological structures and in locating major tectonic lines. Exceptionally, vertical electrical sounding is applied to trace the graphitized and pyritized rocks at greater depths.

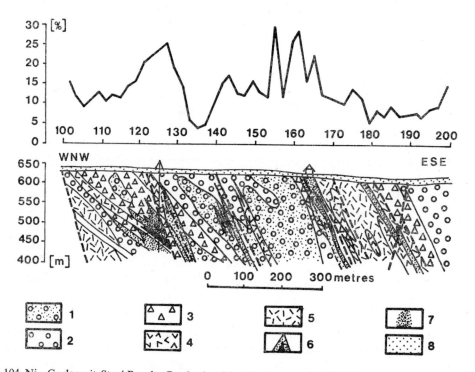

104. Ni—Cu deposit, Staré Ransko Czechoslovakia (after Gruntorád, Jůzek, Kněz, 1967). Correlation of the apparent polarization curve with a geological section. 1 — peridotite; 2 — peridotite with 10—20 % plagioclase; 3 — troctolite; 4 — olivine gabbro; 5 — pyroxene gabbro; 6 — Cu—Ni deposit with massive ore; 7 — Cu—Ni impregnations; 8 — clay and debris.

The prospecting itself is carried out to advantage using the SP method, resistivity or electromagnetic profiling, magnetic survey, and the IP and pulse EM methods along selected profiles. The self-potential (SP) method is a useful tool in prospecting for ore bodies located at a small depth below the surface and the IP method can be used to locate deeper lying bodies. Magnetic surveys have proved useful in the search for ore bodies rich in pyrrhotite. The IP, SP, magnetic and the mise-a-la-masse methods are applicable to detailed exploration in drill holes and gravity survey in mine workings.

The possibilities of geophysical methods in prospecting for polymetallic hydro-thermal deposits will be shown using an example from the Jeseníky Mts. (Czecho-

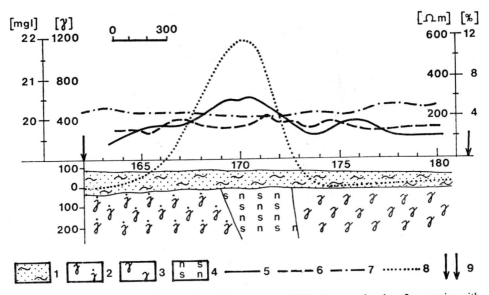

105. Ni—Cu deposit, Russian platform (after Komarov, 1972). 1 — overburden; 2 — gneiss with pyrrhotite impregnations; 3 — gneiss; 4 — peridotite; 5 — IP curve; 6 — resistivity curve; 7 — Δg curve; 8 — Δt curve; 9 — current electrodes.

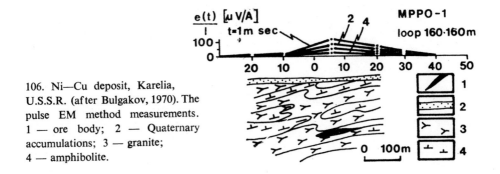

106. Ni—Cu deposit, Karelia, U.S.S.R. (after Bulgakov, 1970). The pulse EM method measurements. 1 — ore body; 2 — Quaternary accumulations; 3 — granite; 4 — amphibolite.

slovakia). A structural-tectonic pattern constructed from the results of regional
gravity and magnetic surveys was correlated with the results of geochemical analysis
(Fig. 107). Geochemical anomalies not only follow bands of Devonian rocks containing polymetallic deposits but also occur near geophysically recorded structures in

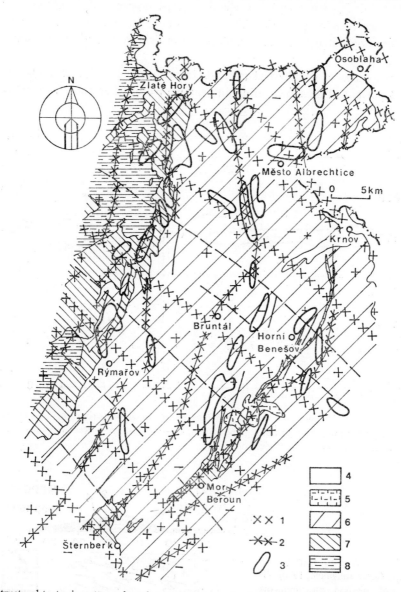

107. Structural-tectonic patterns based on regional geophysical measurements in the Jeseníky area.
1 — transverse tectonic lines; 2 — strike faults; 3 — lithogeochemical anomalies; 4 — Tertiary
formation; 5 — young volcanics; 6 — Carboniferous formation; 7 — Devonian formation;
8 — schists.

Culm (Lower Carboniferous) sediments. On the basis of geophysical data (Fig. 108) geological sections were constructed along regional profiles surveyed by electrical (VES SP) methods. This procedure makes it possible to delimit sectors where productive Devonian rocks occur at depths attainable by present technical means and those where prospecting for polymetallic deposits buried by Culm sediments can be taken into consideration. The IP method is particularly suitable for the assessment of concealed ore bodies. The correlation of isolines of apparent polarization and apparent resistivity with a geological section is plotted on Fig. 109. The target assessed by drilling is expressed by higher polarization and strikingly reduced resistivity. Geophysical anomalies have been verified by increased contents of mercury in soil air samples.

A detailed geophysical survey was carried out along the principal body of the Zlaté hory deposit.

Measurements with SP and magnetic methods and resitivity profiling were conducted over a grid of $50-100 \times 10-20$ m; part of the area was covered by ground metallometry. The whole area shows intensive SP anomalies, whose values range from 250 to 500 mV and exceptionally amount to 800 mV. Ore beds are easily traceable by resistivity profiling and by magnetic survey if pyrrhotite is present. The results of all the geophysical methods used are summarized in a correlation scheme in

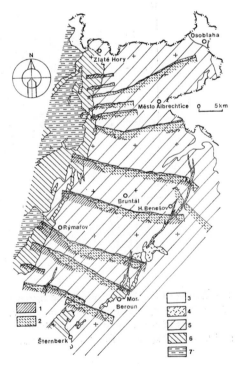

108. Geological sections of the Jeseníky area constructed from results of vertical electric sounding, gravimetry and magnetometry.
Geological sections: 1 — Devonian formation; 2 — Carboniferous formation; geological map: 3 — Tertiary formation; 4 — young volcanics; 5 — Carboniferous formation; 6 — Devonian formation; 7 — schists.

208

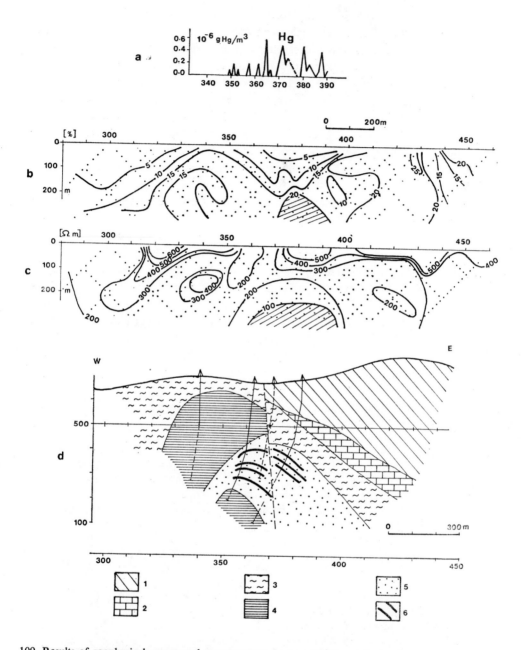

109. Results of geophysical survey and mercurometry in a profile above the southern continuation of the Zlaté Hory ore structure, Jeseníky area. a — mercurometry; b — isolines of polarization; c — isolines of resistivity; d — geological section. 1 — Culm slates; 2 — limestones; 3 — graphitic schists; 4 — chloritic schists; 5 — quartzites; 6 — mineralization.

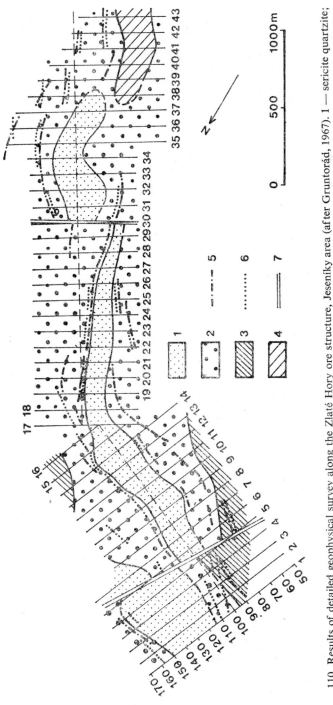

110. Results of detailed geophysical survey along the Zlaté Hory ore structure, Jeseníky area (after Gruntorád, 1967). 1 — sericite quartzite; 2 — sericite-chlorite quartzite; 3 — greenschists; 4 — graphitic phyllites; 5 — axis of conductivity; 6 — axis of magnetic anomalies; 7 — faults.

Fig. 110. It can be seen that conductivity indications agree with the axes of magnetic anomalies. A typical geophysical profile is illustrated in Fig. 111. Curves of resistivity profiling show three well-marked conductivity indications, which are accompanied by maxima on the ΔZ curve; the whole deposit is reflected by a broad SP low. The

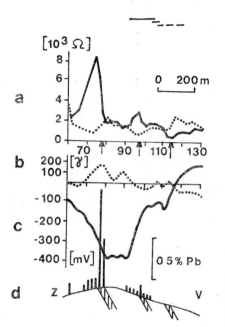

111. Geophysical profile above the Zlaté Hory ore structure, Jeseníky area. a — resistivity profiling; b — magnetometry; c — spontaneous polarization; d — metallometry.

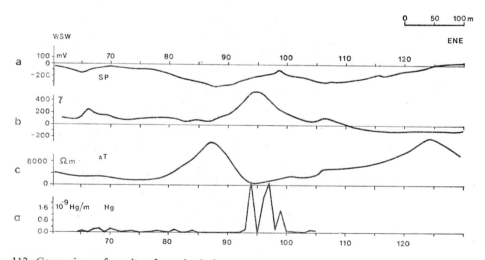

112. Comparison of results of geophysical survey and mercurometry above the Zlaté Hory ore structure, Jeseníky area. a — spontaneous polarization; b — magnetometry; c — resistivity profiling; d — mercurometry.

anomaly at point 80, accompanied by an intense geochemical anomaly, is particularly promising. Trial measurements with a Scintrex mercurometer confirmed that mercurometry can be used for verification of geophysical indications over a postulated polymetallic deposit (Fig. 112).

Polymetallic hydrothermal veins are economically important when their Ag content is appreciable. Polymetallic veins show a high conductivity and polarization; they are amenable to geophysical prospecting if they are deposited in a sufficiently homogeneous environment. Resistivity or electromagnetic profiling can trace the structural-tectonic zones to which the vein deposits are related. The self-potential and magnetic methods are also useful. The isoline map in Fig. 113 showing part of the Havlíčkův Brod ore district (Czechoslovakia) illustrates a successful application of the SP method. Ore veins in homogeneous migmatites are manifested by anomalies over a length of 300–400 m. An anomaly of −300 mV corresponds to a vein composed mainly of pyrite and chalcopyrite, while values of −100 mV correspond to galena-sphalerite mineralization. A large SP anomaly is produced by pyrite and pyrrhotite

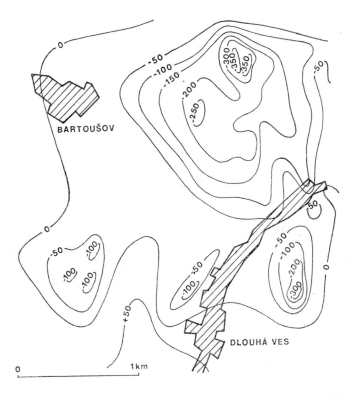

113. The map of isolines of spontaneous potential field in the Havlíčkův Brod ore district, Bohemia (after Gruntorád, Skopec, 1963).

impregnations in migmatites. A geophysical profile of the same area is shown in Fig. 114. Indications obtained by electrical methods are in good agreement with geochemical results and jointly locate a polymetallic vein.

In detailed exploration of polymetallic deposits, magnetic survey, the IP, SP, electromagnetic and the mise-a-la-masse methods can be carried out in drill holes (Fig. 115). Measurements in drill holes that have not encountered an ore body yielded reliable indications of mineralization buried by a 50 m thick sedimentary cover that was not detected by the ground survey.

The largest amount of data on the physical properties of rocks and ores is gathered

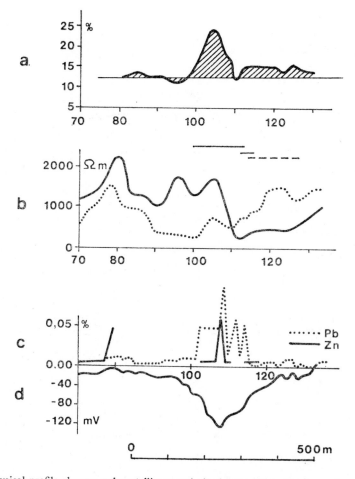

114. Geophysical profile above a polymetallic ore vein in the Havlíčkův Brod ore district, Bohemia (after Gruntorád, Jůzek, Kněz, 1967). a — induced polarization method; b — resistivity profiling; c — metallometry; d — spontaneous polarization method.

during detailed exploration or exploitation of the deposit. The interrelationship between physical fields and the geological structure can be inferred by correlating the detailed geological sections and the detailed geophysical records. This is of special importance where an analogous geological-geophysical situation is to be expected in other, as yet unexplored, areas.

The methods of ore logging provide reliable data on the physical properties of rocks and ores (Fig. 116). The best results are obtained by the electrode potential method, which accurately records the individual ore layers. In recent years, activation-analytical methods have also been used, providing a continuous record of the content of the element throughout the drill hole.

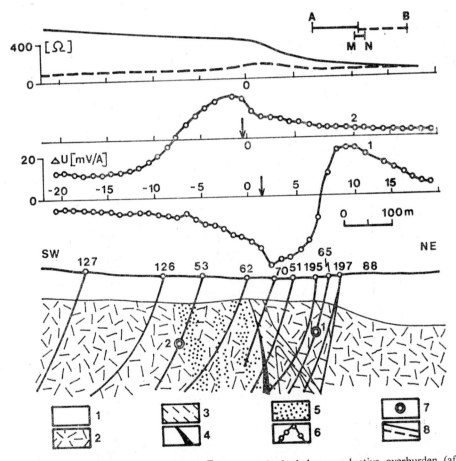

115. Mise-a-la-masse method, localization of an ore body below conductive overburden (after Volosiuk, Safronov, 1971). 1 — overburden; 2 — effusive rocks; 3 — graphitic schists; 4 — ore body; 5 — pyrite impregnations; 6 — curves of potential gradient; 7 — current electrode; 8 — resistivity profiling.

Ore deposits of some types are less suitable for the application of geophysical methods, which can be used as auxiliary techniques in the study of structural problems and in geological mapping.

Tungsten, molybdenum and tin deposits, for example, are usually connected both spatially and genetically with acid granitoids. In prospecting for these deposits, a precise gravity survey proves suitable for determining the morphology of concealed granite bodies. The suitability of other geophysical methods depends on the type of deposit. Skarn deposits can be localized by magnetic survey, whereas hydrothermal veins enriched in sulphides can be recorded by electrical methods.

Antimony ores are detectable by electrical methods following the tectonic zones to which the antimony veins are related, that is, by resistivity and electromagnetic profiling. Antimonite is non-conductive, but is often accompanied by sulphides. Larger amounts of sulphides can respond to the IP or SP method.

Uranium deposits occurring at very small depths can be located directly by radioactivity methods, particularly by gamma-ray surveying (ground and airborne) and emanometry. Figure 117 shows the location of an ore body by gamma-ray

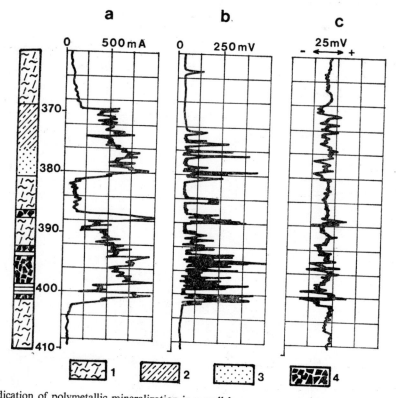

116. Indication of polymetallic mineralization in a well log. a — current logging; b — potential of electrodes; c — gradient of spontaneous polarization. 1 — chlorite-sericite schists; 2 — graphitic schists; 3 — impregnations of sulphides; 4 — rich mineralization.

profiling in southern Bohemia. Since radioactive emanation is dampened by a very thin overburden, uranium is very often not reflected in radioactivity logs and geophysical methods are used mainly in the study of the structural pattern especially in prospecting for sedimentary deposits. In crystalline complexes, resistivity profiling and electromagnetic methods are often employed, since uranium mineralization is frequently connected with steeply dipping tectonic zones. These zones have an appreciable strike length and increased conductivity and are the source of definite anomalies. Using emanometry, sectors promising for uranium mineralization can be delimited along zones determined by electrical surveying.

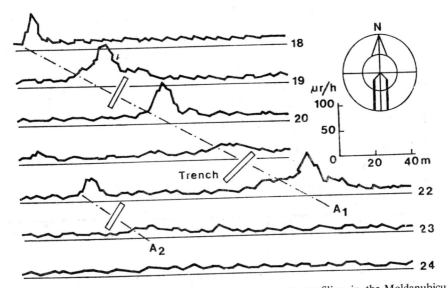

117. Localization of uraniferous mineralization by gamma-ray profiling in the Moldanubicum, southern Bohemia (after Matolín).

Deposits of pegmatite, quartz, barite and fluorite most frequently occur as veins. The feasibility of their direct localization depends on the contrast between the physical properties of the veins and of the wall rocks. The above-mentioned types of vein deposits are distinguished by a high resistivity, differing little from granitoid rocks. Electrical prospecting is therefore useful only if the deposit is located in sedimentary or metamorphosed rocks possessing a lower resistivity. An example of discovery of quartz veins in phyllites (Dětkovice, Moravia) is shown in Fig. 118. In the central gradient configuration they produced prominent apparent resistivity maxima. Pegmatite veins situated in basic rocks (Poběžovice, SW Bohemia) were localized on the basis of magnetic lows (Fig. 119). The piezoelectrical method can be employed for direct location of quartz or pegmatite veins deposited at a small depth (Fig. 120).

Graphite deposits are very suitable for electrical prospecting, because of the high conductivity of graphite. Intensive SP anomalies are present over these deposits and definite indications are also obtainable by resistivity or electromagnetic profiling.

Geophysical methods are a useful tool in exploration for building and ceramic materials. Electrical methods and precise gravity survey can determine blocks of compact, undisturbed and impervious rock near a quarry. The disturbed zones are

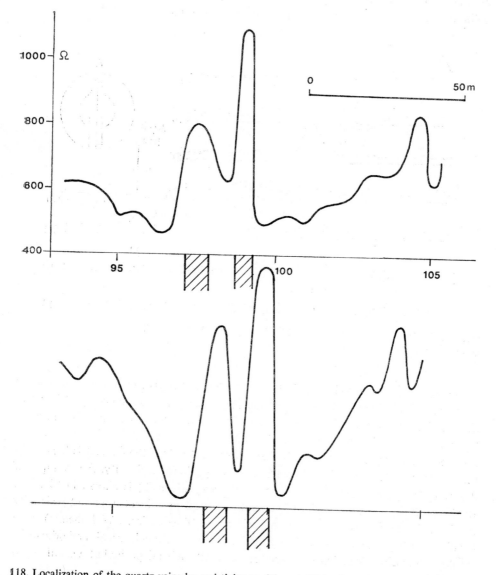

118. Localization of the quartz veins by resistivity profiling at Dětkovice, Moravia (after Karous).

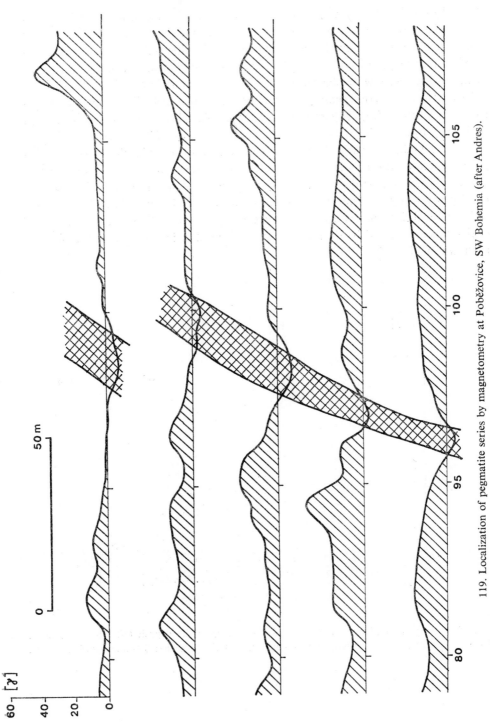

119. Localization of pegmatite series by magnetometry at Poběžovice, SW Bohemia (after Andres).

218

reflected by reduced resistivity and by a gravity low. Karstified limestone portions are associated with particularly pronounced electrical indications. Gravel and sandy gravel are characterized by substantially higher resistivities relative to their environment, so that they can be reliably followed by resistivity profiling. In contrast, brick loams having very low resistance, produce distinct minima in resistivity profiling.

In coal prospecting, geophysical methods are used principally to solve structural-tectonic problems. A gravity survey is usually employed for the delimitation of coal basins and to determine the approximate depth of the crystalline basement. Vertical electrical sounding and seismic methods can be used to study localities where geological boundaries dip at a small angle. Structural maps can be constructed for key horizons, i.e. horizons discriminated by marked differences in electrical resistance, density or velocity of seismic wave propagation. Logging is a useful tool in the exploration of coal deposits, as it allows very accurate determination of the thickness of coal seams established by drilling. An example of electric logs from a coal deposit is shown in Fig. 121.

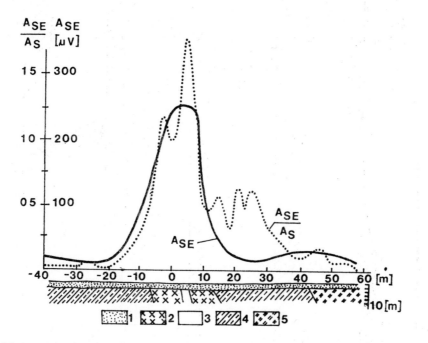

120. Measurements by piezoelectrical method above a pegmatite deposit (after Sokolov, 1966). 1 — sedimentary cover; 2 — pegmatite; 3 — quartz; 4 — biotite gneiss; 5 — quartz-sericite gneiss.

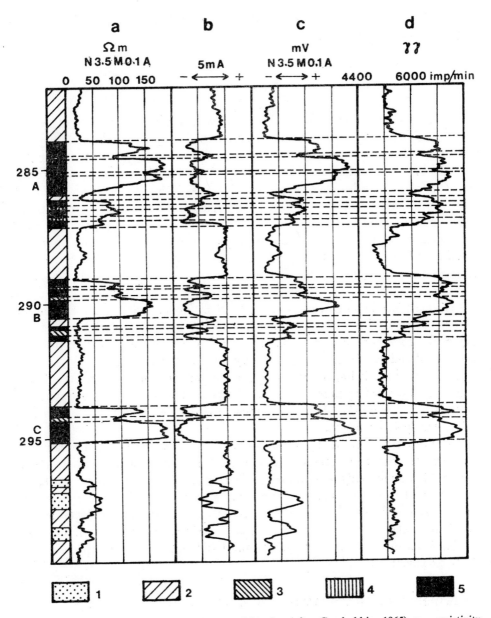

121. Delimitation of coal seams with the use of logging (after Grechukhin, 1965). a — resistivity logging; b — current logging; c — IP method; d — gamma-gamma logging. 1 — sandstone; 2 — argillite; 3 — coal claystone; 4 — coal argillite; 5 — coal.

PROSPECTING AND EXPLORATION OF OIL
AND GAS DEPOSITS

Oil and natural gas differ from solid mineral materials in that they are liquids, which enables their easy migration in the earth's crust. Owing to their genesis and physical properties, oil and gas accumulations occur in pervious rocks, invariably in the presence of ground-water; they fill the elevations covered with impervious rocks, which are called oil traps. Prospecting for oil and gas is thus, in fact, a search for oil traps which, in turn, requires a detailed investigation of the geological structure of the reservoir rocks. Exploration is aimed at determining whether or not the oil or gas accumulations are of economic value, the shape and extent of the accumulation, the pressure and temperature of the fluids and their chemical and physical properties; determination of the perviousness of reservoir rocks and of hydraulic conditions in the deposits and in their vicinity, and assessment of the quality and reserves of oil and gas are other important parts of exploration. Prospecting and exploration for oil and gas are closely allied to the methods of hydrogeological investigation. Experience from procedures applied to the detection of solid mineral deposits can be used for oil and gas deposits only to a very limited extent.

The great economic importance of oil and gas and their enormous reserves in the earth's crust has led to independent development of prospecting and exploration methods since the middle of the last century, following a different trend from that of hydrogeological research.

Today the science of oil and gas deposits and their prospecting and exploration is a separate branch of economic geology using specific working methods and is discussed in an unusually large volume of literature. Within the scope of this handbook the reader can become acquainted with only the basic principles of the search for and exploration of these deposits; those interested in detailed problems are referred to special textbooks, of which Geology of Petroleum by A. I. Levorsen (W. H. Freeman and Co., San Francisco and London, 2nd ed., 1967) can be particularly recommended.

1. THE GENESIS OF OIL AND NATURAL GAS AND OF THEIR DEPOSITS

The scientifically controlled search for and exploration of deposits must necessarily be based on sound knowledge of their origin. The problem will therefore be dealt with in more detail, particularly since the genesis of oil and natural gas has no analogy in the formation of other mineral materials.

Today the organic theory is universally accepted to explain origin of oil and gas deposits, which is postulated to occur by the following processes:

1. Origin of dispersed globules of oil and/or gas from organic matter in the "*source rocks*". These may be terrigenous pelites or carbonate pelites which were generally deposited in marine euhaline and brackish basins of neritic or shallow bathyal zones. They are distinguished by high contents of organic matter with 0.5 — 3 weight per cent (rarely more) of total carbon; the proportion of humic carbon usually makes up 10 — 20 % and that of bituminous carbon 3 — 6 % of the total organic carbon.

In the opinion of one group of geochemists, only primary hydrocarbonaceous substances of extinct organisms, making up at the most 0.3 — 0.5 % of the total buried organic matter, are converted to oil. Many geochemists, however, believe that fats, organic acids and some albumins can be converted into oil. Even cellulose can be altered into hydrocarbon gases, especially methane. The alteration of organic matter into hydrocarbons is assumed to occur partly by biochemical processes, that is, by *bituminization*, in unconsolidated sediment with a high water content in a reducing environment.

2. The dispersed oil and/or gas globules are squeezed from the source rocks together with superfluous synsedimentary water, mainly in the initial stages of diagenesis, either to the surface of sediments from which they are dissipated into the atmosphere or hydrosphere, or into overlying and intercalated porous and pervious rocks. These are called carrier beds (*collector rocks*). The transfer of liquids from pelitic source rocks into carrier beds is designated as „*primary migration*".

3. The expelled fluids percolate through the carrier beds, and separate depending on their specific weights. The lightest accumulate in the top layers of the carrier beds, usually in the axial elevations of anticlines or at an impervious fault. This process is termed "*secondary migration*". If an impermeable cap rock prevented the oil or gas from escaping upwards in the course of geological eras, a permanent accumulation — a deposit — originates. The elevated part of a carrier bed sealed by an impermeable cover, from which the fluid can migrate only downwards, is an *oil or gas trap*. If the trap is not filled with hydrocarbons, it is designated as unproductive, The part of a carrier bed filled with oil and/or gas is called a "*reservoir rock*". From what has been said, it follows that oil and gas accumulations are hidden (*concealed*, "*blind*") *deposits*, which do not crop out on the earth's surface. Where the oil-bearing series are overlain by younger uncomformable beds or (rarely) by nappes, buried deposits are formed. These relations are of primary importance in the choice of prospecting and exploration methods and working procedures.

2. PROSPECTING AND EXPLORATION

In prospecting for and exploration of oil and gas deposits a definite succession of working stages must be preserved. This does not imply, of course, that under favourable conditions some stage cannot be omitted. The complete sequence of works comprises the following stages:

1. mapping of sedimentary areas, assessment of potential oil and gas-bearing strata and the selection of suitable areas;

2. the search for oil or gas traps;

3. test drilling to determine whether and in which horizon the traps are productive. If they prove to be barren, further exploration is stopped. Otherwise,

4. contouring and exploratory drilling is conducted in order to

a) delimitate the extent and shape of the deposit, and

b) determine the thickness of the reservoir rock and its physical properties, in particular the effective porosity, permeability, lithology, pore space, and so on;

c) determine the regime of the deposit, that is, the pressure and temperature of the fluids, their chemical composition and physical properties, saturation of the reservoir rock with gas, oil and water, and the forces that drive the liquids into the well;

5. calculation of oil reserves based on the parameters derived from exploratory drilling;

6. elaboration of the project of deposit development with regard to requirements on mining capacity, exploitation time, economic effect, etc.

Stages 1−3 are usually termed prospecting works and stage 4 exploration of oil and gas deposits.

Mapping

Mapping is carried out to study the stratigraphy, lithology and, usually, the tectonic structure of the region and to ascertain the presence of source, carrier and reservoir rocks. Areas where geological conditions are not yet known satisfactorily are mapped at scales of 1 : 200,000 to 1 : 100,000 (rarely at smaller scales of up to 1 : 500,000). In areas that have been investigated in greater detail, either available maps are used, complemented by geochemical analyses of organic carbon, or mapping at scales of 1 : 75,000 to 1 : 20,000 is carried out. Mapping methods are described in many textbooks and manuals as, for example, in F. H. Lahee: Field Geology, McGraw-Hill, New York−Toronto−London (1961), Chapters 14−20.

In prospecting for oil and gas deposits, mapping is aimed mainly at the identification of source rocks and of suitable impervious cover rocks. Pelitic series with at least 0.6 wt. % of organic carbon and with a maximum of 80 % relict carbon are conventionally regarded as oil-producing rocks. Mapping the marginal parts of basins and their elevation areas alone, where these sequences crop out on the surface, generally proves sufficient. In depressions where the thickness of the sedimentary filling is up to several thousand metres, geological mapping can only provide information on the geology of the youngest (uppermost) sequences. The complexes at a greater depth which are most important for oil accumulation must be explored by drilling.

To obtain an overall picture of the regional distribution of facies, the variation of source, carrier, reservoir and cover rocks, the range of hiatuses etc., stratigraphic (key) holes are drilled after reconnaissance mapping has been completed. These drill holes should be cored throughout their length and sunk either to the crystalline basement or, when the thickness of the sedimentary filling exceeds 4000 m, to the maximum possible depth. Depending on the complexity of the structure, one stratigraphic (key) drill hole is planned per 3,000 to 10,000 km². A key drill hole is also used for the purposes given under 4a—4c.

Geophysical measurements using magnetic and gravity surveys are usually performed simultaneously with reconnaissance geological mapping for the delimitation of volcanic regions and depression and elevation areas. The key boreholes are usually placed in depressions which do not show volcanic features and positive magnetic anomalies so that drilling is carried out through as complete a succession of strata as possible.

At the same time direct and indirect indications of oil- and gas-bearing beds are noted, such as emanations of hydrocarbon gas and oil, mud volcanoes and saline waters with increased iodine content. The final report on the mapping should provide, in addition to general information, the following:

a) the stratigraphic levels at which the source rocks occur, their regional distribution, thickness, facies variation (including the variation in carbon content);

b) ditto for carrier and reservoir rocks,

c) ditto for impervious cover rocks,

d) the basic tectonic features and delimitation of structural elevations and, if possible, of oil traps,

e) list and evaluation of direct and indirect oil and gas indications,

f) recommendation of areas suitable for detailed prospecting.

3. PROSPECTING FOR OIL TRAPS

Prospecting for oil traps is one of the most responsible and most difficult tasks in oil geology. It is very difficult to find and delimit an oil trap occurring at a depth of several hundred or thousand metres, from the surface or subsurface geological structure of the area and from the results of geophysical measurements. The type of oil trap, its depth, structural complexity and other features control the exploration procedures and working methods, particularly the most costly exploration method, that is, drilling.

Principal types of oil traps

The oil traps can be divided into two groups on the basis of their morphology and genesis. Traps caused by fold or fault tectonics are placed in the first group (it is not important whether, for example, a domal structure originated by folding or as an "envelope structure" of morphological elevation in the substrate). The second group comprises traps whose origin is due to other than tectonic factors, such as changes in the lithology of the carrier and reservoir beds, or sealing of a reservoir bed by impermeable transgressive beds. In many earlier papers, traps of the first

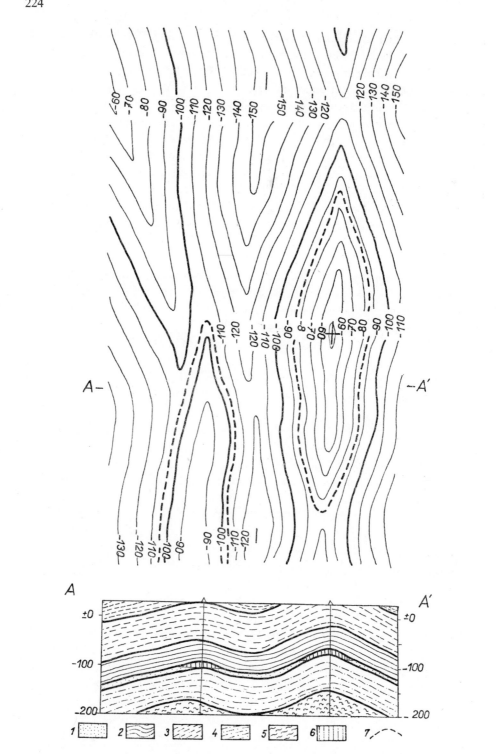

type were called "structural traps" and those of the second type "non-structural, lithological or stratigraphic traps". Since the term "structure" is not interpreted uniformly by all geological schools (it often denotes not only geolocial structure but also material composition), the use of terms "structural" and "non-structural" is not to be recommended, although this division proved to be most comprehensible and most useful in practice.

I. The first group of oil traps ("structural" traps) comprises the following types:

A. traps in anticlinal bends and morphological elevations,

B. traps in fault structures.

These basic types are often combined and each of them is divisible into several subtypes.

A. Oil traps in anticlinal bends

a) *Traps in linear anticlines* whose length is several times their width. Their height varies widely, from a fraction to many times the width. This subtype in upright or oblique folds is characteristic of folded foredeeps, for example North Caucasian or Mesopotamian foredeep; when present in overturned folds, it is distinctive of the flysch fold mountains (often combined with thrust faults). If axial elevations and depressions occur at the apex of the fold, a morphologically uniform anticline can contain several traps confined to axial elevations. Virgation of the folds results in changes in the spatial distribution of traps (segregation, coulisse arrangement, etc.) (Fig. 122).

In depicting the structure of large areas and individual traps and delimitation of oil- and gas-bearing beds, isopachous structural maps are used almost exclusively. Geological sections and block diagrams are employed only as auxiliary graphic techniques.

b) *Traps in domal (cupola) elevations of plateaus and unfolded or slightly folded foredeeps of fold mountains.* These have a characteristic oval or moderately elongated shape (length: width ratio = 1 : 1 − 3), large extent (length of axes up to tens of kilometres) and moderately plunging limbs (at angles of tens of seconds to a few degrees). In the apical parts of the domes there are frequently local depressions so that the contours of oil- or gas-bearing beds are diversified and irregular (Fig. 123).

c) *Traps in monocline flexure bends* are characteristic of plateau regions. In the moderately dipping limb (at angles of tens of seconds to several degrees) they reflect the character of the domal elevation; the steep (flexure) limb often plunges at angles of up to 90°. They resemble subtype (a) in their linear course (Fig. 124).

◄————————————————————————————————

122. Oil traps in upright anticlines. Isopachous map (above) and section (below). 1 — permeable beds, 2 — clay, 3 — calcareous clay, 4 — sandy clay, 5 — limestone, 6 — oil or gas accumulation, 7 — outline of deposit (only in the map).

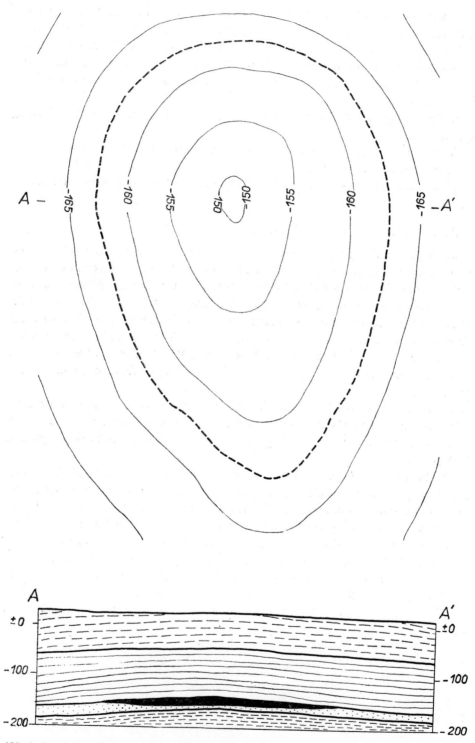

123. A domal trap (for explanation see Fig. 122).

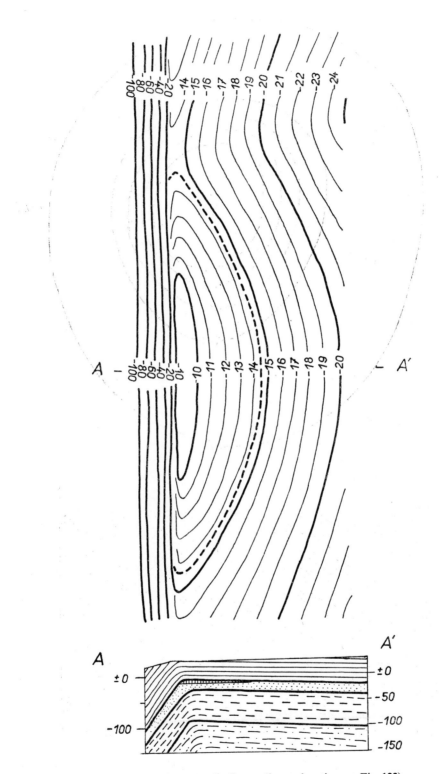

124. A trap in the apex of a flexure (for explanation see Fig. 122).

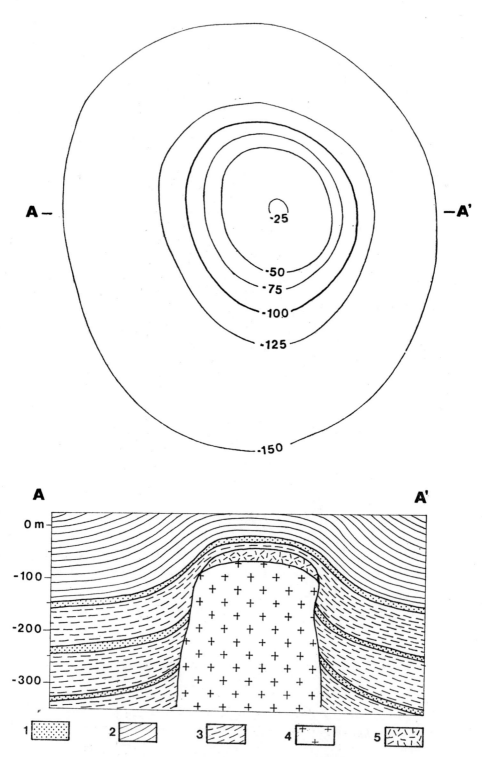

d) *Cupola-like elevations above salt plugs.* These are oval or elongated, following the shape of the salt body; their length rarely exceeds 1 km. The limbs plunge periclinally at various angles, depending on the intensity of squeezing-up the salt plug. They increase with depth until the beds break at the contact with the salt body (Fig. 125).

e) *"Envelope" structures* more or less copying morphological forms of the basement relief. They usually originate as a result of differential settlement of clayey sediments which are deposited on a diversified substratum. The unevennesses become fainter nearer the surface, finally disappearing altogether (Fig. 126).

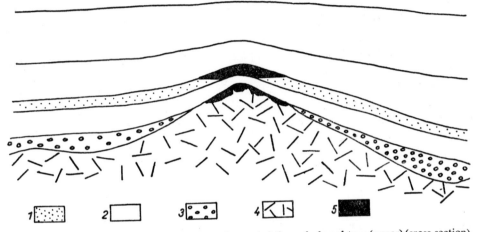

126. An oil trap in buried elevation (lower) and a copied-through domal trap (upper) (cross section). 1 — permeable beds, 2 — impermeable beds, 3 — fossil debris, 4 — buried elevation, 5 — oil.

B. Near-fault oil traps develop where a reservoir rock contacts sealing beds along an impervious fault. As a rule they occur only in unconsolidated or poorly consolidated sediments; in firm sediments the fault has a sealing function when the cover rocks are of great thickness. Traps of this type are smaller than fold traps.

a) *Oil traps adjacent to normal faults* exhibit very varied morphology and the greatest variation with depth. They are distinguished by syntectonic sedimentation, which is responsible for the differences in the thickness of beds in individual blocks. They are generally subdivided into

1. half-domes,
2. synthetic traps with drag structures as an extreme kind,

125. A domal trap above a salt plug. Isopachous map (above) and section (below). On the section, traps in the second and third sand horizons (from above) are sealed by the salt plug forced up tectonically. 1 — sand, 2 — clay, 3 — calcareous clay, 4 — salt, 5 — cap rock.

3. antithetic traps,

4. corner traps occurring where two normal faults meet.

1. Half-domes usually develop only to a depth of several hundred metres below the ground surface where faults abruptly change their strike or dip (Fig. 127). Such half-dome structures can pass downwards into

2. synthetic fault structures. The trap forms in transversal elevations of the reservoir (i.e. more or less perpendicular to the fault); its extent is defined by the position of the nearest transverse depression (Fig. 128). Drag structures with steeply dipping beds may originate in the deepest parts of the basin when the beds in the downfaulted block are dragged out (Fig. 128, bottom).

3. If the monoclinal beds dip in the opposite direction to the fault plane, an antithetic fault trap originates in transverse elevations of the reservoir complex (Fig. 129).

4. If two normal faults unite in the up-dip direction of strata, a corner trap forms in the wedge-shaped space (Fig. 130).

Typical oil traps adjoining normal faults (not combined with other tectonic elements) are characteristic of intramontane depressions in fold mountains, where grabens and horsts alternate (Intra-Alpine basin, Danube plain). They also occur as subsidiary forms complicating the linear-fold traps of foredeeps and domal traps of plateau regions.

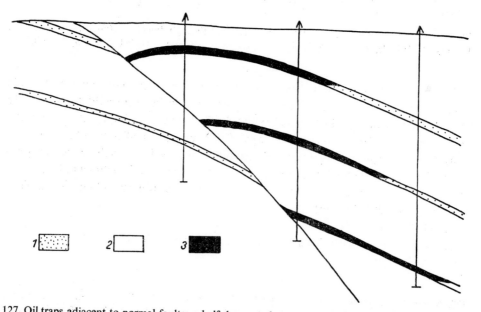

127. Oil traps adjacent to normal faults: a half-dome at the top and in the middle, a synthetic trap at the bottom (cross-section). 1 — permeable beds, 2 — impermeable beds, 3 — oil.

b) *Oil traps developed at reverse faults* are unknown although their existence is theoretically feasible. Their absence is obviously due to the intense fracturing of rocks which accompanies reverse faults, frequently over large areas, causing the perviousness of the cover rocks. Fold overthrusts, which are numerous in flysch areas (esp. in the Carpathians), give rise to structures which are actually anticlinal traps complicated by a fault (Fig. 131).

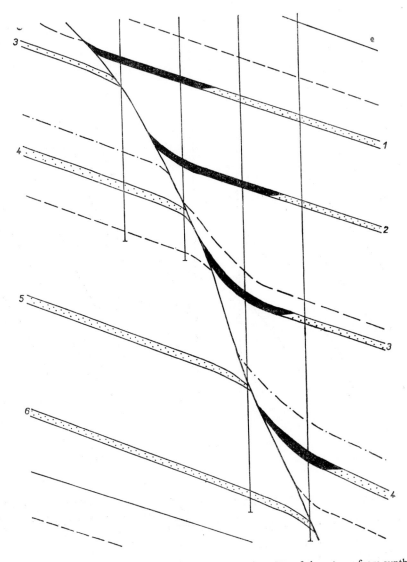

128. Oil traps adjoining normal faults (cross-section): evolution of drag traps from synthetic traps (for explanation see Fig. 127).

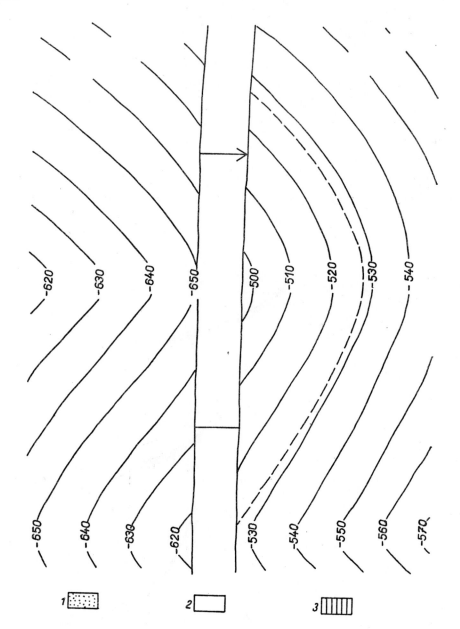

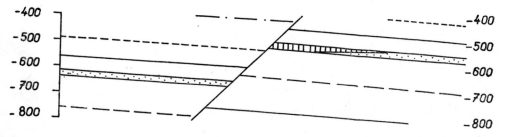

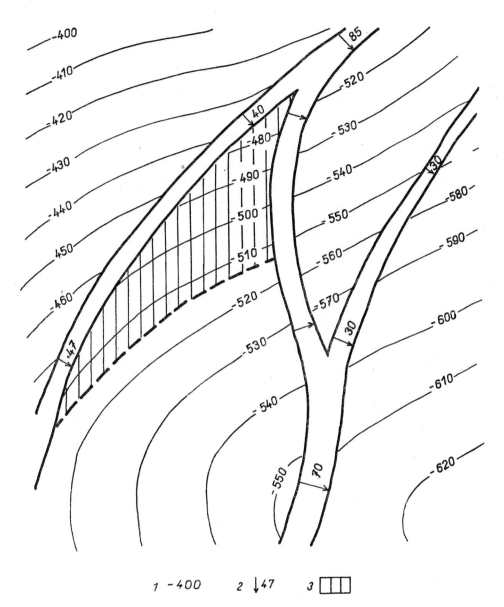

130. Corner trap adjoining a normal fault (isopachous map). 1 — contour of permeable bed, 2 — amount of throw, 3 — corner trap.

129. Antithetic trap adjacent to normal fault (isopachous map above, and cross-section below). 1 — permeable beds, 2 — impermeable beds, 3 — gas.

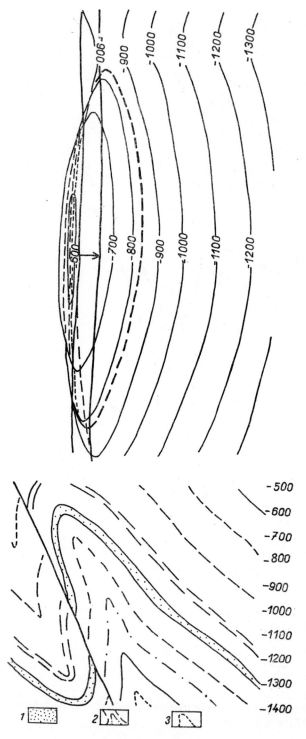

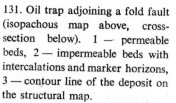

131. Oil trap adjoining a fold fault (isopachous map above, cross-section below). 1 — permeable beds, 2 — impermeable beds with intercalations and marker horizons, 3 — contour line of the deposit on the structural map.

c) *Oil traps sealed by planes of nappe overthrusts* (Fig. 132) are extremely rare. They are confined to the interior parts of foredeeps over which flysch nappes were translated.

II. Lithological traps. This heterogeneous group includes traps in reservoir rocks having primary permeability, which thin out up-dip as, for example, beach sands (Fig. 133), off-shore sand bars, transitional parts between pure permeable sands

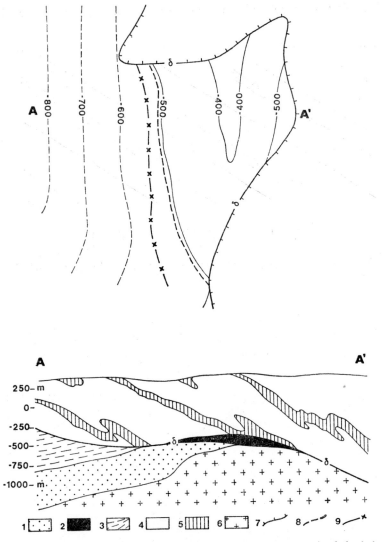

132. Oil trap sealed by an overthrust nappe (isopachous map above, cross-section below). 1 — permeable beds, 2 — oil, 3—5 — impermeable beds (3 — in autochthonous position, 4, 5 — in the nappe), 6 — granite, 7, 8 — boundaries of the deposit (7 — controlled by the overthrust plane, 8 — controlled by the course of permeable bed), 9 — boundary of permeable bed beneath the overthrust plane.

and impermeable clayey sands, and traps in rocks whose secondary permeability is due to fracturing, dolomitization and karstification of limestones, etc. Coral reefs covered by impermeable beds form a special type of lithological traps. Being highly cellular and often of large extent, such traps are among the most productive sources of oil.

The search for oil traps of the first type and in coral reefs can justifiably be based on the sedimentary facies concept. Since the construction of a facies map needs a considerable amount of geological data — only available in the advanced stage of

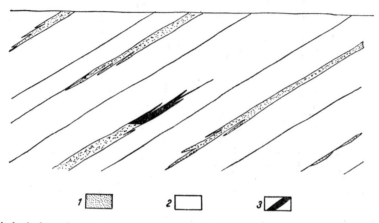

133. Lithological trap in a wedging-out sand horizon (cross-section). 1 — permeable beds, 2 — impermeable beds, 3 — oil.

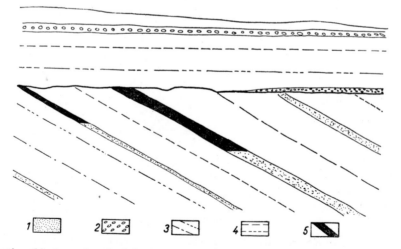

134. Stratigraphic traps (on the left and in the middle) sealed by a transgressive impermeable complex; no trap developed on the right since permeable beds are at the base of the transgressive complex (cross-section). 1, 2 — permeable beds, 3, 4 — impermeable beds, 5 — oil.

subsurface exploration — prospecting for primary lithological traps invariably follows the search for structural traps. There are no regular indications to locate lithological traps in rocks with secondary permeability and their discovery is mostly fortuitous.

III. Stratigraphic traps originate in permeable rocks on which younger impermeable rocks lie unconformably. The shape of the trap is predetermined by the dip of the reservoir rock with respect to the dip of the plane of unconformity. If the permeable rock dips in the direction of the plane of unconformity but at a larger angle, a trap develops in transversal depressions (Fig. 134). If the dip is antithetic, the trap occurs in transverse elevations.

A special case of stratigraphic traps are "buried hills", i.e. morphological elevations of ancient relief which are covered by transgressive impermeable beds. If the whole elevation is formed of permeable rocks, the structure is similar to a dome (Fig. 126 bottom). If only one horizon is permeable, the trap originates in the uppermost part of the collector below the surface of transgression.

From the above survey it is evident that in rock complexes free of apparent unconformities the presence and position of deep oil traps can be inferred from the surface geological structure only if they are of I Aa—d subtypes and, in favourable circumstances, also of I Ae and I Ba—b subtypes. The existence of other types of oil traps cannot be deduced from the surface structure.

The trap subtypes enumerated above can be sought by *direct geological methods*, i.e. mapping (usually complemented by drill holes denoted as *"structural" holes*). Prospecting for traps of the remaining subtypes and for all traps in complexes covered unconformably by younger series must be carried out by *indirect methods*, either *geophysical measurements* or, if the area was investigated by a dense network of drill holes, by facies analyses of individual sequences.

Direct prospecting methods

The cycle of direct prospecting works usually begins with detailed geological mapping at a scale of 1 : 50,000 to 1 : 20,000; mapping on more detailed scales is rare. In the areas where the beds dip at angles greater than $10-15°$, the mapping may reveal surface structures corresponding to oil traps at depth. Simultaneously, direct and indirect shows of oil and gas deposits are noted, such as emanation of gas or oil seepage, salty water with increased iodine content and dissolved hydrocarbon gas or naphthenic acid. In places where oil traps are presumed to exist, geochemical analyses of soil air for hydrocarbon gas content or bacteriological research can be carried out during or at the end of this stage (see p. 152).

Detailed geological mapping does not generally provide sufficient information for the geological structure to be established or for the location of oil traps. Boreholes are therefore sunk; in some countries they are termed "structural holes" as their primary objective is to determine the geological structure of the area.

The drilling should be planned so as to furnish the maximum amount of information within the shortest time and at the lowest possible cost. This demand is fulfilled

238

most adequately by drilling small-diameter holes using truck-mounted drilling rigs allowing continuous coring and satisfactory drilling speed. In holes deeper than 300 m water, oil and gas eruptions must be prevented.

The depth and spacing of structural holes is controlled by the dip of the beds and the intricacy of the geological structure. In principle, the holes should be so distributed that a key horizon (marker) encountered in one drill hole near the land surface is encountered in the neighbouring hole at two-thirds or three-quarters of its depth (see Fig. 135). This agrees with the formulas

$$h_v = \frac{2}{3}(h_2 - h_1),$$

$$l = (h_2 - h_1) \cotan \alpha,$$

where h_v = depth of the hole, h_1 = depth of the key horizon below the surface in the first hole (V_1), h_2 = depth of the key horizon below the surface in the second hole (V_2), l = distance between V_1 and V_2 drill holes, α = apparent dip of beds between V_1 and V_2 holes.

The surplus depth (of 1/4 to 1/3) relative to the formula is left as a safety factor if the presumed dip increases or the key horizon disappears so that correlation with another auxiliary horizon is necessary.

The depth of structural holes is determined by the type of geological structure. In prospecting for oil traps of domal and flexure types on platforms it ranges from 1200 to 1800 m or even more; the drill holes are spaced several kilometres apart. In folded

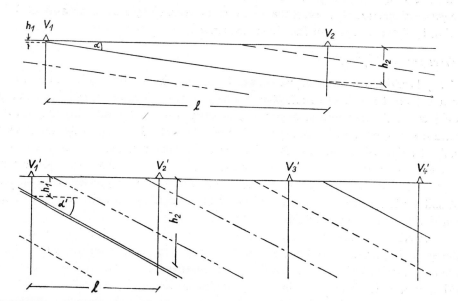

135. Relation between the depth and spacing of deep boreholes at different dips of beds (cross-sections).

piedmont areas structural holes are usually 600 – 1200 m deep; the distance between the holes is up to 1 km at right angles to the strike of beds and several kilometres parallel to it. In strongly folded piedmont and intermontane depressions and heavily faulted intermontane depressions structural holes 300 – 500 m deep laid out in a triangular grid are used; the spacing of drill holes varies from several hundred to one thousand metres. Where the geological structure suggests the presence of an oil trap or where the geological pattern is unclear, additional drill holes are sunk to make the grid denser.

Geological information gathered from the cores of structural drill holes and geophysical measurements (see below) should be processed simultaneously with drilling so that necessary changes can be made if actual conditions are at variance with the assumptions (premature interruption of drilling or, conversely, drilling to a greater depth than was designed, wall coring, etc.).

On the basis of boring data the following operations are carried out:

a) The construction of a lithostratigraphic, occasionally also biostratigraphic boring log, either in written or graphic form (lithology shown by hatching and symbols, sediments in colour).

b) Correlation of drill holes, choice of principal and auxiliary correlation horizons. The principles of lithostratigraphic correlation are described, for example, in M. S. Bishop: Subsurface Mapping, John Wiley and Sons, 1960. It should be noted that in pelitic-psammitic series, which make up most oil-bearing areas, the characteristic coloration or tint of sediments and their spottedness or banding provide a more stable correlation than the lithology. Lithologically, a prominent, easily discernible correlation horizon is generally absent from these series. Correlation is then based on characteristic rock sequences (the "packets" of Soviet authors) varying in thickness from a few to several dozen metres. Correlation is carried out to determine the difference in the altitudes of a particular correlation horizon in individual drill holes; its absolute altitude in a given hole is calculated from these relative elevations.

c) Construction of an isopachous map (structure contour map) of the principal correlation horizon (marker). This is not difficult when the horizon is actually encountered during drilling, but it may plunge deep beneath the bottom of drill holes if these are of constant, relatively small depth (e.g. 300 m) and drilling proceeds along the dip of beds. In this case, an approximate structural scheme of the principal correlation horizon is compiled by computing the thickness of the complex between it and an auxiliary higher horizon using an "ideal profile". An ideal profile is obtained by superimposing the individual bore logs and by generalizing the profile thus constructed (Fig. 136). This procedure, however, does not take account of the facies variation and changes in the thickness of sediments. The depths established for the principal correlation horizon on the isopachous map are usually smaller than in reality, since the thickness of sediments in the centre of a basin is normally greater than that determined by boring on its margin. A correction is possible where a key drill hole is available.

Individual bore logs   »Ideal profile«

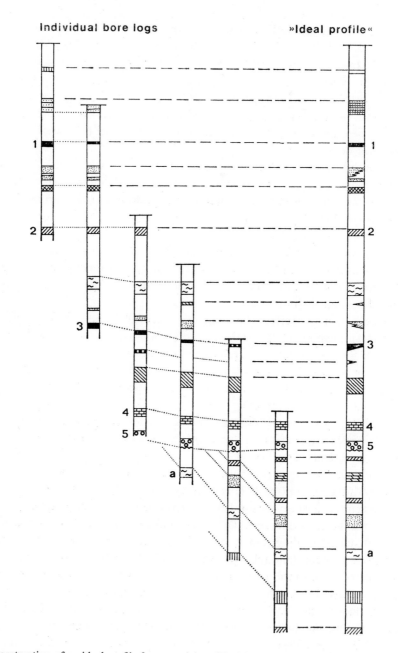

136. Construction of an ideal profile from partial profiles of deep boreholes. 1—5 marker horizons in the transgressive formation; a — marker in the underlying unconformable formation.

d) The location of oil traps is inferred from the isopachous map of the correlation horizon, where the traps are indicated by the closed contours with the highest relative altitude. The geologist should determine whether or not the position of the trap changes with depth, i.e. with the stratigraphic levels of the reservoir rock, as is the case with traps in inclined and overturned anticlines and fault-bounded traps. The horizontal shift of the trap with depth can be derived from geological sections (Figs. 127—129, 139).

e) Location of a pione er test drill hole and the design thereof. The geological data for the design of the drill hole (such as the depth of the producing horizon, lithological and stratigraphical characteristics of the drilled formations, the dip of beds, zones of abnormal pressure, etc.) are obtained from the geological cross-sections.

Indirect prospecting methods

Until recently, indirect prospecting methods, particularly geophysical methods, were used only when the oil-bearing formations were overlain by a younger formation with angular unconformity, or when the thickness of the formations changed frequently and randomly. At present, however, there is a tendency for economic reasons and because of the speed and comparative simplicity of surveying to replace drilling exploration.

Geophysical methods are based on the fact that the structures sought differ from their surroundings in some of their physical properties, such as mass-unit volume, magnetic susceptibility, resistivity, velocity of wave propagation and radioactivity. According to the differing properties, gravity, magnetic, seismic, electric and radio-activity methods are used in both regional and detailed prospecting.

Geophysical methods are also used for drill-hole measurements; well-logging comprises a set of surface geophysical methods adapted to drill-hole conditions.

The aim of regional investigation is to determine distinct and extensive structural elements, particularly the zones of elevation and depression of sedimentary basins, to demarcate the distinct tectonic lines and to determine the presence of volcanic series in the sedimentary strata. During the search for and demarcation of extensive structures it is possible to carry out geophysical prospecting in a wide-spaced grid; this is both faster and cheaper.

It is only fairly recently that a new geophysical technique in seismic prospecting has been developed making it possible to detect oil- and gas-bearing strata directly. This method, called "bright spot", will be considered later.

A geophysical survey should either provide the details of the geological structure of a larger structural body or to identify smaller or less distinct structural elements. This requires measurement in a relatively closely-spaced grid and the use of sensitive instruments. These measurements are time-consuming and expensive.

Geophysical prospecting is conducted either in a system of points (gravimetric and magnetic methods, vertical electric sounding) or along profiles (electric profiling, seismic prospection). Most geophysical methods can also be carried out in air or at sea (mainly offshore); such measurements are usually made along profile lines. The results of such measurements are plotted on isanomal or structural maps, i.e. iso-pachous maps of a definite physical horizon (either a key or fictitious horizon) or in vertical sections.

A gravity survey is generally used for the regional interpretation of geological structures of sedimentary basins. It can be carried out on land, sea-shelf or air. Using a gravity meter, differences in gravitational acceleration between given points are measured. Not only differences in the density of the basin being explored but also the structure and mass distribution in the underlying formation or deeper parts of the earth's crust contribute to the measured values of gravitational acceleration. The values obtained are plotted on Bouguer's isanomal maps for geological inter-pretation. To eliminate inhomogeneities in the density of deeper complexes, these maps are converted to maps of residual anomalies or second-order derivatives of gravity. With the use of highly sensitive (centesimal) gravity meters this method makes possible the demarcation of even less distinct geological structures and faults. As the surveying network has to be very fine, this method is used only for detailed prospecting, in areas of around 10^2 km^2 at the most.

Gravimetric maps show, though often only approximately, the positions of zones of elevation and depression (positive and negative closed isanomals), faults (large horizontal isanomal gradient, relative to the prevailing density), structural platforms (strongly decreased isanomal density) and very distinct transverse undulation of beds ("noses" of isanomals normal to their regional course) (Fig. 137). They thus provide first-rate data for the selection of areas suitable for detailed prospecting by direct and indirect methods.

A magnetic survey is undertaken in prospecting for oil-bearing areas as a comple-mentary method mainly for the delimitation of magmatic series with increased magnetic susceptibility, i.e. mostly of intermediate and basic volcanic bodies. Acid volcanics usually cannot be identified owing to their low susceptibility. Magnetic measurement can be carried out by ground or aerial survey. In a ground magnetic survey either the vertical component ΔZ or the total vector ΔT is determined, and in aeromagnetic profiling, the total vector ΔT is measured.

In geological interpretation of magnetic anomalies it is often difficult to determine the stratigraphic position of volcanic bodies, that is, whether they are emplaced in the formation underlying the oil-bearing series, in the series itself or in the overlying formation. Generally, sedimentary areas with more intense manifestations of vol-

137. Gravity anomaly in Bouguer anomalies over various geological structures (isanomalous maps). a — anticline disturbed by a cross fault, lower block relatively sank; b — salt plug, influence of the salt body dies out near —2 mgl isanomal; c — antithetic half-dome adjacent to normal fault; d — synthetic half-dome adjacent to normal fault, the course of the fault indicated by closely spaced isanomals.

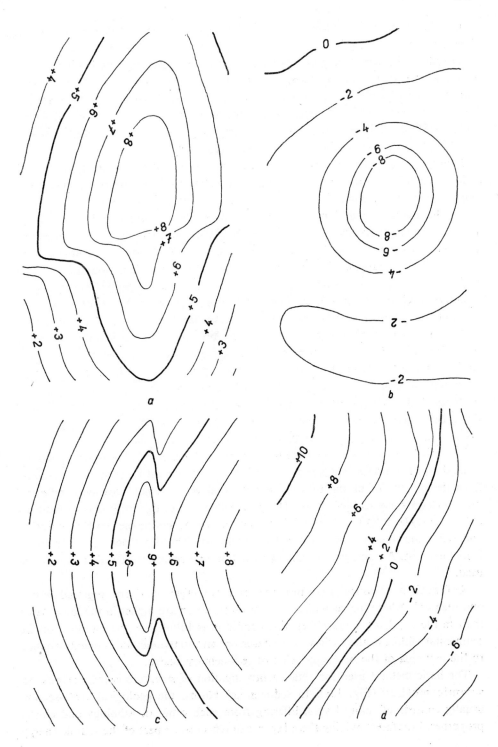

a

b

c

d

canism in or above the oil-bearing rocks are less promising as major oil and gas deposits.

Together with aerial magnetic reconnaissance of sedimentary basins, an aerial radioactivity survey is carried out to determine the intensity of gamma-ray radiation above the individual lithological types of rocks. In contrast to the magnetic method, this method demarcates acid volcanic complexes if they are exposed on the earth's surface. Radioactivity survey on the ground is occasionally used in prospecting for oil and gas in conjunction with geochemical methods.

Electrical survey is applied in both regional and detailed prospecting. The telluric current method is used in regional prospecting and vertical electrical prospecting and/or electromagnetic methods are used in detailed prospecting.

Of the methods mentioned above, vertical electrical sounding is particularly useful in providing information on the depth of the boundary plane from deviations in the electric conductivity. By other methods only the isolines of the measured electric parameters are obtained, which give a rough idea of the shape and limitation of geological structures and possibly also of the strike and dip of beds. The geological structure can be interpreted only when the electric properties of the whole succession of strata are known at least approximately. In sedimentary basins with clastic-pelitic filling, electrical survey usually provides only data on the depth of the basin filling, and in basins with carbonate and evaporite rocks on the depth of the upper surface of the top layer. In well-defined structures with moderately dipping beds containing intercalations of rocks with a substantially higher resistivity, it is possible to decide on the position of a test borehole directly from the results of this measurement. Otherwise, the results of electrical measurement are verified either by structural drilling, seismic survey or by the two methods combined.

Seismic surveying is also used in both regional and detailed prospecting works, in land and sea-level modifications. The depth of individual horizons is determined from the velocity of propagation of longitudinal waves which have been reflected or refracted by physical interfaces. The great advantage of this method is its remarkable accuracy ($\pm 5 \%$) and the possibility it provides of observing the depth of reflecting and refracting interfaces directly and continuously. The drawback of this method is its high cost. A number of modified refraction and reflection seismic methods are used.

Refraction seismic prospecting is more often used for regional works and then only where the velocities of wave propagation are higher in the lower layer than in the overlying ones. Often this condition is fulfilled at the interface of the sedimentary filling of the basin with its basement and so the method is mainly applied to the mapping of the basement relief of sedimentary basins.

The basic method used for detailed investigation of structure is the reflection seismic method (Fig. 138). In sedimentary basins, the geological cross-section usually consists of many layers differing from each other in lithology and elastic properties. Interfaces dividing these layers can reflect some part of the seismic energy back to the earth's surface. By constructing sections from seismic records we obtain the individual reflection patterns more or less linked together. A continuously reflecting physical interface (the real key horizon) then coincides with the lithostrati- graphic boundary. In uniform eomplexes, especially of a sandy-clayey character, the reflecting interfaces are usually discontinuous and are often produced by local lithology alone, that is by a change in the degree of diagenesis, by lenticular inter- calations of sandstone, marl, etc., or by complicated interference patterns of seismic waves coming from several interfaces. The reflecting patterns are then not linked together and it is necessary to draw a parallel line with the nearest reflecting patterns in the seismic cross-section and to construct a fictitious horizon.

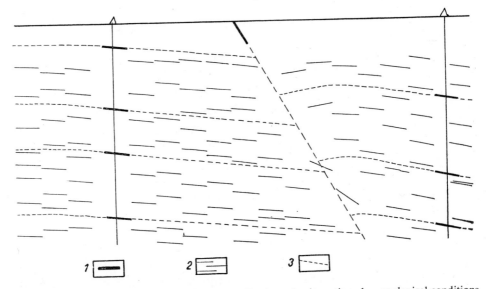

138. Example of geological interpretation of a reflection seismic section. 1 — geological conditions determined by bore holes and by mapping near the fault, 2 — seismic reflection interfaces, 3 — interpreted course of stratigraphic boundaries established by drilling.

The main technique used in seismic survey is the Common Depth Point (CDP) method, now used extensively in geophysical oil prospecting. Seismic waves are recorded on magnetic tapes and these records are later transformed into digital records, which are processed by digital computers. The processed seismic records are then transformed so that each corresponds to a set of seismic waves reflected from one point on the reflecting interface; they are then stacked together. The use of computers enables improvement in the quality and effectiveness of the survey by suppressing many kinds of seismic noise. Results are plotted in the form of time cross-sections or depth cross-sections, where reflecting interfaces, fold and fault tectonics and other geological features are visible.

The most advanced modification of the CDP method is the "bright spot" technique using the intensity, polarity and dip of seismic reflections to determine whether a reflecting interface corresponds to the top of an oil- or gas-bearing layer. This approach was made possible by the introduction of the method of relative amplitude preservation into seismic data processing system.

Another modification of these basic seismic methods widely used in prospecting work is the correlation method of refraction waves, which has proved useful in prospecting for stratigraphic traps. For detailed prospecting of a refracting interface (e.g. the top or flanks of a reef, salt dome or intrusive igneous body) the drilling refraction method was developed. The geophone below the interface is placed in a deep hole drilled to the interface under examination and the propagation of waves spreading radially around the borehole is measured at various distances. In this way it is possible to extrapolate the course of a physical boundary to a certain distance from the drill hole.

In zones disrupted by faults, the construction of isopachous maps from seismic cross-sections is very difficult. Faults are often not particularly distinctive so that their position in seismic cross-sections cannot be determined reliably. A fictitious horizon in individual tectonic blocks need not always correspond to the same stratigraphic horizon. Correct correlation is possible only if at least one deep hole situated in a seismic cross-section occurs in every tectonic block. The methods of geologic interpretation of seismic cross-sections and the construction of isopachous maps are discussed in the paper by Dlabač and Adam (1959).

Along with ground survey, geophysical methods are also applied in drill hole measurements. Logging of oil and gas wells is an indispensable part of the prospecting process, since it provides precise and reliable information on the lithology of formations, the existence of permeable zones and layers, the porosity, clay content, water saturation, oil and gas saturation, residual oil saturation in the flushed zone, content of mobile hydrocarbons, recoverable hydrocarbon volume, oil-water contact, technical control of the well, dip and strike of beds, and the existence of faults and fracture zones. Complete evaluation of logging data in the area permits a reliable correlation of layers between wells, furnishes an idea of the geological structure and the thickness and shape of oil-bearing horizons and represents an objective basis for estimation of producible hydrocarbon reserves. The logging service also carries out a very important technical control in the wells, e.g. control of cementation quality behind the tubing and thus control of the sealing of individual producing horizons. It helps to clear up the causes of different technical complications during well completion, carries out perforations and other tasks in opening producing horizons, and controls of hydrocarbon production from individual producing horizons and of the state of producing wells.

The complex of logging methods in each area depends on the lithological character and the age of formations and on the type of prevailing permeability (caused by intergranular or fracture porosity). A modern logging programme in sedimentary formations includes the following logging methods: induction electrical log, dual laterolog, gamma-ray log, formation density log, borehole compensated sonic log, compensated neutron log, high resolution dipmeter, sidewall sampling and wire line formation testing. The logging equipment is furnished with both analog and digital recording systems so that quantitative evaluation of logs can be processed by a computer. The first step in computer evaluation consists of thorough statistical analysis of logging data leading to the determination of matrix physical parameters (matrix resistivity, density and neutron porosity), the clay content and the ground-water resistivity. A system of cross plots is most efficient for this purpose. The results of a computer-processed interpretation (Fig. 139) presented by Schlumberger Ltd. (1974) furnish running information in analog form on the average matrix density and under appropriate conditions on the permeability, water saturation, hydrocarbon saturation, hydrocarbon volume and weight, porosity, fluid analysis by volume (residual, mobile hydrocarbons, water) and formation analysis by volume (clay content and matrix content).

Computer-processed interpretation of high resolution dipmeter data furnishes a reasonable picture of the dip and strike of beds throughout the well profile, of the boundary between different stratigraphic units, of the existence of transgression and of the presence of faults.

After completion of the well, a special logging programme ensures efficient control of actual conditions. For this purpose, casing collar log and cement bond log will be carried out in order to estimate the cementation quality.

In producing wells, logging methods furnish information on the depth intervals of water, oil and gas inflows, on the water-oil ratio at different depth intervals in producing wells and on the vertical flow (based on data from a high resolution thermometer, gradiomanometer, density meter, and flowmeter) and on the positions of tubing defects or diameter diminution (through tubing caliper).

From the above it follows that the geophysical methods mentioned are applicable and very useful in prospecting for oil and gas. The type and spacing of geophysical measurements, as well as the localization of survey points and profiles, are determined by the geologist in consultation with a geophysicist, taking into consideration the geological structure of the area. In non-folded regions with moderate dip of beds isometric surveying grids are generally used. In folded regions, measurements along profiles are made. In this case the basic and closer network of profiles (cross profiles) is drawn at right angles to the general fold strike and the subsidiary network of subsidiary network of profiles is wide-spaced (connecting profiles). With seismic and electric methods, the connecting profiles join the cross profiles in the vertical direction. The application of geophysical methods and the mode of grid spacing in regional prospecting is shown in Table 50.

Of the above-mentioned methods it is necessary to select only those that can give relevant and geologically interpretable results under the given geological conditions. A rational working system demands that exploration works are carried out in a definite

248

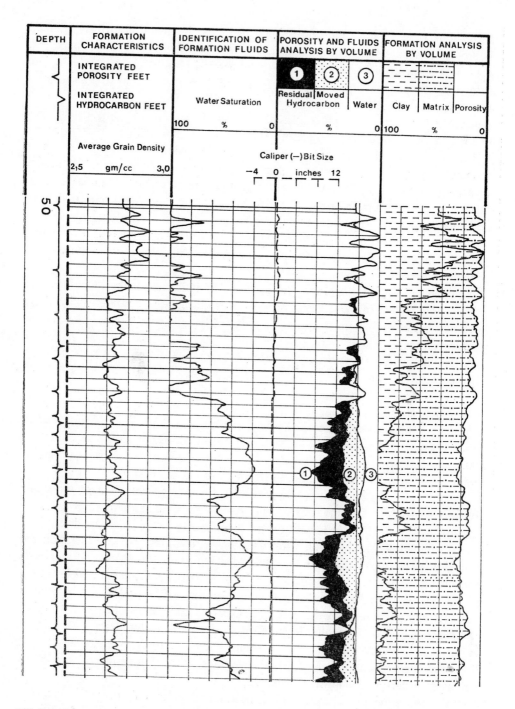

139. Results of the computer processed interpretation of logging data from an oil well (after Schlumberger Ltd., 1974, Well Evaluation Conference, Nigeria).

TABLE 50

Geophysical methods used in regional exploration

Method	Platform areas type of grid	point (profile) spacing	Folded areas type of grid	point (profile) spacing
1. gravimetric	triangular	3—5 km	triangular	1—2 km
2. aeromagnetic	parallel profiles	2—4 km	parallel profiles	1—2 km
3. ground magnetic	triangular	3—5 km	triangular	1—2 km
4. airborne radioactivity survey	parallel profiles only exceptionally	2—4 km	parallel profiles	1—2 km
5. resistivity profiling	parallel profiles	2—4 km	parallel profiles	1—2 km
6. resistivity sounding	triangular or parallel profiles	3—5 km 2—4 km	profiles normal to the structure	1—2 km step 0.5—1 km
7. seismic reflection	profiles in square or rectangular grid	3—5 km	profiles in rectangular grid, denser at right angles to the structure	1—2 km × 2—4 km
8. seismic refraction	profiles in square or rectangular grid	every second to fifth reflection profile is measured	profiles in rectangular grid, denser at right angles to the structure	every second to fourth reflection profile measured

sequence. An orientative geophysical prospecting scheme, from regional to detailed survey, may be expressed as follows (the numbers correspond to the methods in Table 50):

$$1(+2 \text{ or } 3 + 4) - 8(+6 \text{ or } 5) - 7,$$

(in parantheses are given less suitable methods or methods applicable only under special conditions).

The planning of a geophysical survey, as well as the geological interpretation of its results, must be based on sound knowledge of geophysics, geology and economics and also on experience. Planners and interpreters, i.e. geophysicists in close co-operation with geologists, or specially trained geologists can save an enormous volume of drilling work and thus contribute to expeditious and cheaper execution of prospecting operations.

PIONEER BOREHOLES

The purpose of pioneer boreholes is to assess whether the trap discovered (or a set of traps situated at different stratigraphic levels below one another) contains oil or gas or whether it is unproductive, i.e. filled with water or dry. In addition they should determine the reservoir properties, i.e. the petrography, porosity and permeability of the rocks, chemical and physical properties of fluids filling their pore space, the chemistry of the oil, gas or water, the reservoir pressure and temperature, the pore saturation, the content of dissolved gas, the viscosity of the reservoir fluids, etc. The geologist examines these characteristics in close co-operation with a drilling technician and laboratory workers. At this stage of subsurface exploration the geologist is charged 1. with the design of drilling from the geological point of view, 2. with control of the drilling procedure and 3. with interpretation of the data obtained.

The geological part of the pioneer borehole design must define:
a) the precise location of the borehole (and permissible deviations from the starting point);
b) whether the hole should be vertical or what departure from the vertical is presumed;
c) total depth of the hole (potential depth extension);
d) stratigraphic and lithological log (column) of the hole; strata which may cause the failure of the borehole by swelling, caving-in, pressure of water, oil or gas (danger of eruption) or loss of drilling fluid;
e) if necessary for geological reasons, the setting depth of the casing and its cementing;
f) kind, number and depth of coring (bottom coring, coring interval, run length, required core recovery; side-wall coring; cuttings from sludge barrel or removed by drilling mud) and laboratory testing of samples;
g) geophysical measurements (electrical and radioactivity logging, measurements of inclination, temperature conditions, etc.);
h) required pumping tests, i.e. the total number of tests, intervals of tapping of individual horizons, type and length of pumping tests, sampling of reservoir fluids, laboratory tests of reservoir fluids;
i) remarks on other geological features which may affect the drilling procedure.

On the basis of this geological information and requirements, technicians prepare the technical design, i.e. they choose an adequate drilling rig and equipment, determine the casing-bit-size programme, cementing operations, technology of pumping tests, and estimate the cost.

The data which the geologist can establish for the pioneer borehole are, understandably, hypothetical and rather schematic. Consequently, he should control the drilling procedure continuously so that the assumptions and requirements may be modified whenever necessary. In addition, the geologist acts as a technical supervisor, whose main duties are
a) the control of hole logging: depths attained (including occasional checking of the

length of the drill rods), quality of the drilling mud and its loss, gas content of the mud, water inflow, pinching or caving-in of the hole;

b) the taking of core samples. He should personally supervise the lifting of the drill string and removal of the core from the core barrel when the core is of major importance;

c) to be present at the geophysical survey, to control the length of logging cables;

d) co-operation in the interpretation of geophysical measurements;

e) determination of the precise depth to which the casing shoe should be sunk depending on the geological conditions established in the well, and of intervals for cementing of casing strings; in producing wells also supervision of the placing of a filter;

f) determination of the number and depth intervals of pumping tests according to the geological conditions established in the well, prescription of the length of pumping tests, sampling of fluids, and evaluation of the pumping tests;

g) in case of failure in the course of drilling, co-operation in remedial measures;

h) when the true geological log of the well differs markedly from the assumptions, preparation of new geological data for a new casing-bit-size programme;

i) recommending the well for gas or oil production or, if the hole is dry or water producing, prescribing its liquidation or its use for some other purpose.

The activity of the geologist during drilling for oil and gas is so specialized in some respects that a new branch of geology has developed. We shall briefly discuss the most important aspects of this specialization, on the assumption that the student has sufficient knowledge of geophysics (electrical and radioactivity logging) and boring.

a) Location of a pioneer test hole. In the last decade, pioneer test holes have almost invariably been located so as to strike the highest point of the trap. Since this implies that the well will tap the gas cap, great care must be taken to prevent eruption.

b) If the traps are enclosed in an upright fold (or in envelope structures) the vertical well may encounter several oil or gas pools below one another (Figs. 122–126). The position of traps in inclined or overturned folds or near faults changes with depth (Figs. 127–131); thorough exploration necessitates a number of vertical drill holes or one inclined hole.

c) The pioneer test borehole should be extended to the barren rocks underlying the oil- or gas-bearing series. Generally only the crystalline basement is regarded as unproductive. Of sedimentary series, only those that were intensively folded and exposed to denudation for long geological periods (as suggested by the lack of cover rocks) are assumed to be unproductive.

d) The lithological and stratigraphic well log (columnar section) compiled at the project stage is hypothetical and can differ markedly from reality (thinning out of complexes ascertained by exploration at the basin margins, setting on of unknown series, changes in facies and thickness, etc.). The parts of the borehole which may impede drilling can only be identified by analogy with holes bored in identical rocks at other places in the basin or in other basins, or deduced from the properties of rocks cropping out at the periphery of the basin.

The pressure of fluids in permeable rocks and thus the danger of eruption or loss of drilling mud can also be inferred from the geological conditions.

e) Geological conditions that may demand driving of the casing to other than the technically optimum depth include: a thick series of water-bearing, slaking or caving rocks or horizons with low pressures of reservoir fluids underlying high-pressure horizons. The casing is usually lowered so that the shoe sits in impermeable rocks. The space between the wall of the hole and the casing is always sealed by cementation in modern operations. The conductor casing and surface casing are cemented from the shoe up to the surface as a safety precaution in case of eruption. The height of the cement filling behind the protective (intermediate) casing varies. If the well encounters several aquifers and water circulation between them is undesirable, the casing is cemented up to the shoe of the next upper casing segment. If the aquifers are absent or if it is not necessary to isolate them, a cement column reaching 100—150 m above the casing shoe is usually sufficient. Since the geologist in planning the well cannot predict the actual conditions accurately, he should always ask for cementation throughout the length of the hole.

Geological factors also affect the installation of the last (producing) segment of casing. If the deepest rock complex, on which a pumping test is to be performed, is lithologically uniform and only contains either oil or gas, a filter is set in to cover the whole tested horizon. Cementation is then effected through a window close above the upper end of the filter. If the horizon tested consists of firm rocks with a low probability of caving-in, no filter is necessary; the shoe of the full casing is placed above the horizon tested and cementation is effected from the shoe. The higher permeable horizons that should be tested are isolated from the rest by cementation and perforated by a casing gun. Examples of various constructions of pioneer boreholes are shown in Fig. 140.

f) If the test boreholes for oil and gas are very deep and heavy drilling rigs are to be used, coring is time-consuming and expensive. The cores are therefore recovered from the bottom discontinuously; for example, a core is taken at 20, 25 or 50-m intervals. A reasonable mean must be chosen between geological needs and technical possibilities. For information, samples of cuttings are taken from the drilling mud, usually at 5-m intervals. Side-wall cores are taken using a hydraulic or a percussion

──►

140. Several constructions of deep exploration or production wells. a — medium depths (1,000 to 2,500 m), cementation of 6 5/8″ casing string from the shoe; b — medium depths (1,000—2,500 m), cementation of 5″ casing string through the window; c — great depths (2,500—4,000 m), cementation of surface casing string 6 5/8″ and of 4 1/2″ oil casing string from the shoe. In all cases conductor casing string is cemented from the shoe up to the surface. Oil deposits in black, gas deposits — triangles.

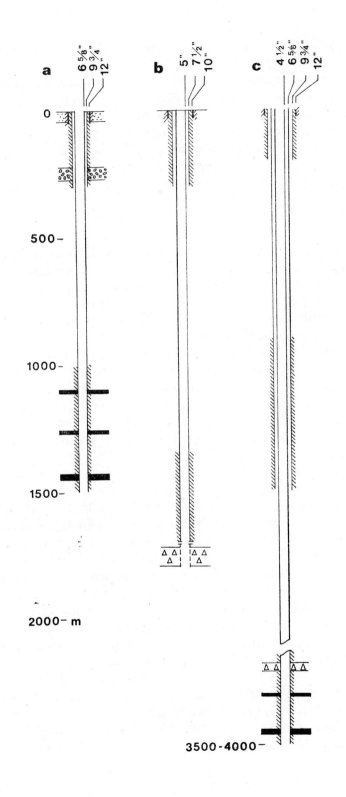

type of side-wall coring device before the casing is lowered into the hole in order to obtain a more precise stratigraphic and lithological log.

The core samples are examined for their stratigraphy (according to macro- and microfauna and flora) and lithology. The effective porosity, permeability, oil saturation and other parameters are also determined in permeable rocks. Samples of permeable rocks must be sealed with paraffin immediately after being extracted from the core barrel. Recently, a core sampler was developed which enables the sample to be delivered to the laboratory under natural pressure conditions, so that the gas saturation and the true content and composition of dissolved gases can be examined.

g) In boreholes, lateral electrical logging is carried out for the determination of rock porosity and resistivity of fluids. In carbonate rocks, where electrical logging is ineffective, various radioactive and accoustic methods are used to determine the porosity and saturation of the rocks and the type of fluids present. The geologist determines the depth at which the geophysical survey will be carried out, which methods will be applied and the intervals between individual measurements. Thermal logging for determination of cement backfilling, taking of side-wall cores, inclinometry and other special procedures, such as seismic logging and seismic survey in drill holes, should also be planned at this stage.

h) The programme of pumping tests, which is only broadly outlined at the pioneer well stage, is made more precise after finishing the hole. The number and the length of the pumping tests and their technology, the kind and amount of laboratory tests and the mode of sampling of fluids are determined. The geologist should insist on taking samples using a tester and on measuring the pressure and temperature simultaneously.

If the result of a pumping test on some of the horizons is positive, that is if there is an economic oil or gas supply in the well, pumping tests are usually not performed on the overlying horizons. The well is "conserved", that is filled up with drilling mud. Exploitation is generally begun only after the next exploration stage has been finished. If all horizons designated for pumping tests prove to be negative, that is water-bearing or dry, the well is liquidated. The geologist decides which horizons should be protected from interconnection or from the penetration of surface water. According to the circumstances, the manner of sealing the well is chosen and the parts of the well from which the uncemented casing may be withdrawn are determined.

If the first pioneer well is barren, a second, third or even fourth test borehole is sunk. These should, in addition to the objectives mentioned above, assess whether the original interpretation of the deep geological structure was correct and whether the reservoir is dry only because it developed after the primary migration of oil or gas. The subsequent test holes are planned, executed and tested according to the same principles as the first, but they can be prepared more precisely and in greater detail, being based on the results of the first hole. This relates especially to the coring programme, which is oriented to defining the character of the reservoir rocks.

EXPLORATORY DRILLING

Exploratory drilling is aimed at delimitation of the deposit (or deposits) discovered by the pioneer hole, determination of changes in the thickness, porosity, permeability and saturation of the reservoir rocks, measurement of pressures and temperatures in individual wells, and determination of the deposit regimen and of the data necessary for the calculation of oil and/or gas reserves on the basis of pumping tests or trial exploitation.

Exploratory drilling differs from that of test drilling mainly in that it relates to a certain group of wells. The data to be obtained from the wells are practically the same as in pioneer drilling. Information from the pioneer hole makes it possible to anticipate the lithology and stratigraphy of the drill holes, difficulties in boring etc., with relatively high precision. Experience gained during pioneer boring facilitates the choice of the most adequate drilling method. Since the geological conditions and physical properties of the reservoir rocks must be known throughout their thickness, continuous coring within a certain length is often planned. The number of cores for stratigraphic purposes is reduced to a minimum. The number and depths of pumping tests are determined more precisely. All principles given for pioneer drilling are valid for planning and testing the exploratory drill holes.

The location of exploratory holes is decisive for economic subsurface exploration and the development of the oil field. It is advisable to delimit the deposit as soon as possible, to determine its extent and shape and thus assess the probable reserves, and subsequently to establish the physical properties of the reservoir rock, which control the spacing of productive wells. The particular distribution of exploratory drill holes depends on type and size of the geological trap, which were determined by the structural or geophysical survey. The network of wells is adapted to the shape of the oil trap; the method is the same as used in structural boring, except that the spacing of the wells is different.

The most important task in the initial stage of exploratory drilling is delineation of the oil-water boundary. Since it is most frequently horizontal, the boundary of the deposit can be readily extrapolated as an isopachous line of the relevant reservoir.

Because of the high cost of deep drill holes, relatively small distances between the first exploratory holes are chosen, i.e. 2 to 5 times the presumed spacing of producing wells. Drilling proceeds from the apex of the oil trap down-dip to the oil/edge-water boundary. Hence, the wells are spaced so as to investigate the largest possible area of the deposit by the smallest number of drill holes.

Each new well either confirms or corrects the assumptions of the planner. The programme of exploratory drilling may therefore have to be modified during the work, especially the location of new wells. As, theoretically, all exploratory holes should strike the productive part of the reservoir, the programme must be changed if any of the planned wells encounters only the water-bearing part. (This does not hold for wells that are intended to increase the reservoir energy for easier oil extraction.)

It is quite a complicated exercise to determine the number of drilling rigs to be used simultaneously in exploratory drilling. Since it may happen that only a small part of the reservoir is productive, one rig is usually put into operation in the initial exploration stage. More drilling rigs are only installed after two or three exploratory wells have proved productive. In this way, the risk of unnecessary expenditure, which is particularly high in the virgin areas of developing countries, incurred by the transfer of drilling rigs is eliminated, but the time between the discovery of the deposit and its development is prolonged. Several drilling rigs are therefore often put into operation as soon as the pioneer well yields positive results.

The principles described above are applicable to oil traps of large extent, covering more than about 10 km². In minor traps or large traps with complicated tectonic structure, where the occurrence of negative wells is very probable, a more economical policy is chosen. In order that the exploration stage may also involve the development of the deposit, exploratory holes are spaced at a maximum of twice the presumed distance between producing wells. This procedure is employed with most oil and gas deposits in Czechoslovakia.

The exploratory drilling programme is complicated when the reservoir contains several overlying productive horizons. Depending on the state of drilling and oil production engineering, one of the following alternatives will be selected in such cases:

a) If equipment for simultaneous exploitation of two (or three) horizons is available, two (or three) of the deepest horizons are explored at the same time. When the results obtained are unsatisfactory and, according to geophysical measurements, the higher horizons appear to be more promising, the explored horizons are shut in and exploration is carried out at a higher level. One drawback to this practice is that the casing may be damaged during exploitation and oil extraction from the lower horizons is then impossible.

b) All exploratory wells are drilled to the lowest horizon and the deposits are exploited from the bottom upwards. This method is used where the oil reserves of individual horizons are relatively small.

c) If the reserves of the horizons (or groups of horizons) are satisfactory, drilling and development are carried out separately for each level.

As mentioned above, the main objective of exploratory drilling is to determine the physical and hydraulic parameters in the reservoir bed and the pressure and temperature conditions in the deposit. Description of these works is beyond the scope of this book. The exploratory drilling stage is completed by calculation of the reserves and preparation of the development project.

After completion of the exploratory drilling the amount of oil reserves can be computed by a volumetric method, using the formula

$$V_{\text{geol}} = S \times m \times \varphi \times \mu \times \tau \times \alpha,$$

where V_{geol} — volume of oil in place in m³, S — area of the productive part of the reservoir in m², m — mean thickness of the reservoir rock (weighted mean) in m, φ — coefficient of effective porosity, μ — coefficient of pore saturation, τ — coefficient of thermal expansion for the difference in the reservoir and atmospheric (15.3 °C) temperatures, α — oil compressibility coefficient for the difference in the reservoir and atmospheric pressures.

The determination of these parameters is complicated and the results may even be unreliable. Economically valuable reserves are obtained by multiplying the oil in situ by the coefficient of recoverability, which varies between 0.1–0.2 (gravitational regime), 0.2–0.4 (the regime of dissolved gas), 0.4–0.6 (the regime of uneffective gas cap), 0.5–0.7 (the regime of effective gas cap) and 0.6–0.8, exceptionally 0.9 (the regime of effective hydraulic pressure).

It can be seen that only recoverable reserves of oil are considered, in contrast to ore deposits where the whole reserve of a given deposit is taken into consideration.

The amount of gas reserves can also be computed after completion of the exploratory drilling by the volumetric method according to the formula:

$$V_{geol} = S \times m \times \varphi \times \mu' \times f \times \tau' \times P_{res},$$

where S, m, φ are the same as in the previous formula, μ' — coefficient of pore saturation with gas, f — coefficient of deviation from the compressibility of an ideal gas, τ' — coefficient of thermal expansion of the gas, P_{res} — mean reservoir pressure in MPa.

Economic gas reserves are computed by multiplying the amount of gas in place by the coefficient of recoverability and subtracting the amount of gas which it is uneconomical to exploit because of the low residual pressure. This varies between 3 and 10 and even more MPa, depending on the operation pressure in the gas conduits.

The classification of oil and gas reserves differs from that of solid mineral materials; only the general principles for differentiating economic and subeconomic reserves are the same. Wells of low yield and extremely difficult exploitation (e.g. strong sanding up at small yield) are indications of subeconomic deposits.

Class C_2 includes reserves presumed to be present in new traps, the existence of which has been verified by structural drill holes or geophysical measurement but where oil or gas has not been proved by pioneer test well. The existence of promising reservoir and source rocks in the corresponding series is known from other places in the same sedimentary basin. Oil reserves in as yet unexplored tectonic blocks of the known deposits are grouped in this class.

Class C_1 involves traps in which at least one well has yielded a commercial amount of oil or gas, and areas adjacent to oil fields with reserves of a higher class. The characteristics of reservoir rocks can be determined from the logging records or by analogy with known deposits in the same basin at the same stratigraphic position.

Class B: The presence of oil or gas in commercial quantities is assessed by a pumping test in at least two wells and is indicated by logging results in others. Properties of the horizon are broadly known from the cores and logging results, and the chemistry and physical properties of the oil or gas have been established. If the reservoir shows a varied facies development or the tectonic setting of the trap is complicated, these reserves are also placed in class C_1.

Class A includes oil reserves assessed by pumping tests on producing wells and those in areas which have been fully delimited by exploratory drill holes and in which the productive horizon was examined by pumping tests. Physical properties of reservoir rocks, the yield of wells, reservoir pressure and the regimen of the horizons are known with satisfactory precision.

For the development of the deposit it is necessary that reserves of class A cover at least $10-20\%$ (according to the size of the deposit), A + B at least $30-40\%$ and C_1 $60-70\%$ of the total reserves. The development project is prepared by experts in oil geology and ground-water hydraulics. At present, electric modelling or computers are generally used for these computations.

DRILLING OF PRODUCING WELLS

Producing wells are planned in a similar way to exploratory drill holes. The assembled geological data make it possible to base the project on precise information regarding the lithology, stratigraphy and tectonics of strata that the drilling will encounter, to design accurately the casing and cementation of the well, and to anticipate complications in drilling. Producing wells are only cored in exceptional cases, for example, where the reservoir rock shows a strong facies variation or where no cores were recovered from nearby drill holes. The programme of pumping tests is detailed and precise.

Geological work during drilling and pumping tests on producing wells is the same as during exploratory drilling.

PART II

EXPLORATION OF MINERAL DEPOSITS

PRELIMINARY AND DETAILED EXPLORATION

The exploration of mineral deposits is conducted in two stages, preliminary and detailed. *Preliminary exploration* is carried out on those deposits which, according to prospecting results, proved to be most promising. Brief characteristics of the two stages and methods applied are presented in Table 51. The most important task of preliminary exploration is to assess the economic value of the deposit with reasonable accuracy. Such exploration must also provide sufficient information so that an adequate method of detailed exploration may be selected. On the basis of a technical-economic evaluation the decision is taken on whether detailed exploration is advisable or not. If a deposit proves to be workable but its exploitation is postponed for 10 – 15 years, it is considered as a state reserve. *Detailed exploration* is started on economically important deposits which are intended for immediate development as part of a programme for exploiting mineral resources and in agreement with local industry. The aim is to appraise the reserves and assemble all data necessary for the construction of a mining plant. Apart from certain details, the working methods for the two exploration stages are practically the same, but as the stages differ in their objectives, the degree of preciseness to which the deposit is explored also differs.

Deposits of mineral materials are heterogeneous owing to their different genesis and geological environment in which they originated. Despite this diversity, several basic principles can be defined that should be observed throughout geological exploration (Kreiter, 1961):

1. *The principle of integrated exploration* includes the following requirements: a) Complete delimitation of the ore district, coal basin or deposit. The exploratory works should intersect the whole deposit and the whole productive rock complex. b) Overall evaluation of the principal and secondary economic components. c) Interdepartmental evaluation of the expensive exploratory works in accordance with modern geological knowledge.

2. *The principle of gradual accumulation of data* on the deposit; in other words, of exploration step by step. If this principle is obeyed, unpromising deposits can be eliminated in time and the risk of negative results in further exploration and in exploitation is lessened.

3. *The principle of uniform degree of exploration* refers to the need for uniform

TABLE 51

Basic data for planning and required results of prospecting and exploration stages

Prospecting or exploration	A. Reconnaissance prospecting	B. Detailed prospecting	C. Prospecting — exploratory stage
1. geological	map of 1 : 200,000 to 1 : 100,000 scale (determination of prospecting criteria)	map of 1:50,000 to 1:25,000, aerial mapping, assessment of tectonics, stratigraphy + hydrogeology	map of 1 : 10,000 to 1 : 5,000 scale, aerial mapping, detailed analysis of geology, tectonics, metallogenesis, history of investigation and mining
2. petrographic	determination of principal rock types	determination of rock type variability	alterations in the wider area around deposits
3. mineralogical	location of prospects, compilation of registration deposit map	distinguishing of mineral occurrences from deposits; construction of metallo-genetic + prognostic maps	differentiation of industrial + non-industrial dep.; determination of deposit type, and of analogy with known deposits
4. geochemical		metallometry, heavy minerals, stream sediments method	see (B); clay from the alluvium-slope contact, hydrochemistry
5. geophysical	airborne radiometry and magnetometry	radioactivity, magnetic, aero-radioactivity, aeromagnetic, seismic + gravity surveys	mapping: electric methods; location of dep., gravity, seismic, radioactivity surveys, electrical methods
6. technological		sampling at one point	sampling at several points, laboratory study of samples, or pilot-plant examination

D. Preliminary exploration	E. Detailed exploration	F. Exploration during operation
solid geological map of a large scale, showing detailed geological structure of the deposit. Geological documentation of all exploratory works. Delimitation of most promising parts of large deposits for mine opening	geological documentation of exploratory works, geological mapping of underground workings at scales of 1:1000 to 1 : 200. Study of the geological structure of single deposit bodies	geological mapping and documentation of underground workings that are important for the recognition of detailed geological structure of mining blocks. Study of geological conditions of excavability and ore degradation
complete petrographic study of rocks of the deposit and its surroundings, alterations connected with mineralization, and of rocks lithologically suitable for location of a deposit	refining of data on the petrographic character of rocks, when needed	
mineralogical study is based on all positive exploratory works. Study of paragenesis of the oxidation and primary zones; of structures and textures; grain-size distribution. An overall characterization of useful minerals of the raw material	definitive differentiation of genetic and technological types	study of changes in paragenetic conditions of the deposit. Cooperation in solving problems of ore dressing
study of primary aureoles, completion of the study of secondary aureoles. Systematic sampling of the deposit. Geochemical research of the raw material—distribution of useful and deleterious components	detailed division of useful and deleterious components in natural and technological types of the raw material. Potential utilization of trace elements, full utilization of the raw material	cooperation in solving problems of dressing and treatment of the raw material; prevention of pollution impact on the environment
conclusion of geophysical survey of the whole deposit, delineation of the most promising structures. Logging in drill holes, inclinometry, geophysical determination of the quality of the raw material	geophysical survey in drill holes. Geophysical determination of the quality of the raw material	verification of minor deposit bodies (e.g. by wave method) whenever necessary. Geophysical determination of the quality of radioactive raw materials
study of technological dressing and treatment of natural and technological types of the raw material. Application of the data obtained during treatment of similar deposits	termination of technological investigation. Choice of the most adequate dressing and treatment scheme	cooperation in changing the treatment technology when an unexpected change in technological properties of the raw material takes place

Table 51 (continued)

Prospecting or exploration	A. Reconnaissance prospecting	B. Detailed prospecting	C. Prospecting — exploratory stage
7. hydro-geological and eng. geological	—	—	—
8. categories of reserves	speculative resources	hypothetical resources	possible ore, inferred reserves, C_2 category

exploration of the economically important parts of the deposit. However, this does not imply the same density of exploratory works throughout. On the contrary, deposits of varying quality and morphology are explored equally only where the exploratory works are distributed unevenly in the directions of the minimum and maximum variations of the deposit.

4. *The principle of minimizing labour and material losses* should be respected, particularly during the later exploration stages, when subsurface exploration works are extensive. The higher the labour productivity, the more economic will be the exploration. The principle that time losses should be minimized should not interfere with other factors; it only becomes the decisive criterion in special circumstances as, for example, when there is a shortage of mineral materials for the maintenance of an extensive industry.

METHODS OF SUBSURFACE EXPLORATION

During both preliminary and detailed exploration underground *mining works and drilling* are used. Underground workings are expensive and technically demanding; they are used where drilling cannot provide a sufficiently precise picture of the complicated structure, morphology and quality of the deposit. On such deposits most exploratory systems utilize the advantages of underground mining works and

D. Preliminary exploration	E. Detailed exploration	F. Exploration during operation
general study of mining-technical and hydrogeological conditions of the deposit on the basis of the geological structure of the deposit and its surroundings. Possibility of damage on surface constructions—landslides, diverging of water courses	detailed study of physical and mechanical properties of rocks and raw materials which control the development of the deposit and choice of extraction method. Design of drainage of the deposit	observation of rock pressure, shock disturbances in mines, effects of undermining; solution of dumping problems
approximately one third of C_2 category is transferred into C_1 category (indicated reserves). Technical-economic study determining conditions under which the deposit is commercially profitable	achievement of the optimum ratio of reserves $(C_2:C_1:B:A)$. The highest B and A categories $(=$ measured reserves) ensured in amounts sufficient for several years' production. In large deposits, total reserves in C_2 category	transfer of reserves into higher categories, establishment of increments or shortage of reserves from the results of exploration during operation

drilling. Underground workings are divided into 1. *opening* and 2. *development works.*

In exploration, the same opening works are used as in exploitation: vertical and inclined shafts, cross-cuts and galleries. Development works following the horizons are *drift galleries,* which on deposits of great thickness are combined with cross-cuts or drill holes in order to assess the thickness, and with *raises or winzes.* The technique and methods of these costly works are the subject of various branches of mining engineering.

The possible use of underground mining works during future exploitation should be condisered with great care. Essentially, four alternatives are possible: 1. The exploratory working can be used for exploitation without any reconstruction and no additional opening. 2. The exploratory working has an auxiliary function during exploitation; it is used as a second exit, for upcast work or for transport of personnel, material and fill. 3. Exploratory work requiring reconstruction, most frequently enlargement of the cross-section. 4. The exploratory work is of no use for exploitation.

The possibility of using exploratory works for exploitation can be expressed by a coefficient calculated from the ratio of the cost of usable exploratory works to that of all works. The profiles of exploratory works are divided into *exploratory profiles* that are economical but cannot be used for exploitation without reconstruction, especially on large deposits, and *producing profiles* suitable for exploitation. All debatable questions concerning the usability of exploratory works should be settled with the mining organization during exploration.

Drilling methods are widely adopted at present in geological exploration. Effective modern drilling rigs make it possible to deal successfully with problems that have so far been insoluble, as, for example, the deciphering of deep geological structures or the search for hidden deposits. When intelligently applied, exploratory drill holes lower the cost and time of exploration.

According to the purpose for which they are intended, drill holes are designated as "key", "structural", "reconnaissance", "exploratory" or "hydrogeological". Deep *key holes* contribute to the recognition of the regional geology of the area. *Structural holes* should verify the detailed stratigraphy and structure of the given area. *Reconnaissance holes* are mainly applied to assess the presence of the deposit inferred from geological, geochemical and geophysical data. *Exploratory drill holes,* laid out on an exploratory grid, are used in preliminary and detailed exploration. *Hydrogeological wells* are sunk to establish all data on the ground water in the drill holes.

Core drilling is at present most widely used in exploration of solid mineral materials. The advantage of this method is that it yields detailed geological information and often enables reliable sampling, provided core recovery is satisfactory. Most rigs have a spindle head revolving in a vertical plane which allows drilling of inclined holes. The direction of a drill hole can be modified using deflection wedges. Branched drill holes allow much shorter holes to be used. The principal index of efficiency of drilling is the core recovery, which must be satisfactory especially in key, structural and reconnaissance drill holes (75—80 % on an average). In exploratory drill holes, sufficient core recovery is required from the productive part of the deposit; it may be smaller from overlying rocks whose character is known. A higher drilling effectiveness can be achieved in several ways:

1. *Reduction of the diameter of the core boreholes;* this is practicable with high-speed rigs with diamond bits. Drilling with diamond bits in rocks of medium and high hardness greatly increases the length of runs and the quality of the core. The development of the wire-line method, in which the core barrel with the core is withdrawn through the drill rod set using a high-speed winch, is an improvement on this system. The rods remain in the hole until the bit is exchanged. The wire-line method can be used in medium-hard to hard rocks at depths greater than 100 metres. The rate of exploratory drilling can also be increased by small-diameter holes drilled by diamond bits from an underground work. The holes are arranged fanwise in a vertical or horizontal plane and provide detailed exploratory sections at a very low cost.

2. *Rotary non-core drilling* is widely used in oil and gas exploration. Under certain conditions it can be used in exploration for solid minerals in rocks of low and medium hardness. Roller and fishtail bits are applied in hard rocks. Rotary non-core drilling is most often employed for holes laid out on a grid to establish the precise position of the deposit, which is then drilled through by core boring. If the thickness and quality of the deposit can be determined by logging, core boring is omitted. In this

case, cored and non-cored holes alternate in the exploratory grid in an adequate ratio. Core drilling facilitates the correlation of beds and detailed interpretation of logging. In relation to core drilling, rotary non-core drilling enables an output 30 – 190 per cent higher, chiefly because the lengths of individual runs are substantially greater.

3. *Percussion drilling* — mainly cable drilling — is used to advantage on deposits whose projections onto the horizontal plane are sufficiently large, since only vertical holes can be drilled. It is conducted with larger diameters than core drilling, and is therefore suitable for deposits with low contents of useful minerals, such as placers and disseminated stockworks. A disadvantage of the method is that the structure of samples cannot be studied in detail and logging of the rocks drilled through is impossible. This drilling method is widely used in the exploration of deposits worked in opencast quarries.

4. *Auger boring and vibration drilling.* At present highly effective rigs are available for both these procedures. They can be used with advantage in the exploration of near-surface deposits of ores of low hardness. Vibration drilling is suitable for sinking shallow mapping holes.

In describing the drill holes, the geologist classifies the rocks according to their drillability; he therefore needs to know the classification schemes accepted for different drilling methods.

EXPLORATORY SYSTEMS

The term *"exploratory system"* denotes the distribution of exploratory works that permits the most effective verification and computation of commercial reserves of a mineral deposit.

Exploratory systems have not yet been defined uniformly. Kreiter (1961) and V. I. Smirnov (1957) distinguish three principal systems: drilling, underground working and a combination of these two. This classification has the disadvantage that it is based on the technical means employed. Shekhtman (1963) distinguished three exploratory systems according to the attitude of the deposits: 1. a system for exploration of horizontal deposits, consisting principally of parallel or transecting vertical profiles; 2. that for moderately inclined deposits in which vertical profiles are mainly used; 3. that for steeply inclined deposits, with horizontal profiles as the principal exploration tool. This classification based on geological principles is satisfactory but does not permit more detailed subdivision.

In the text below we shall use the classification established by Biryukov (1962) on the basis of the exploration of 1321 mineral deposits in the Soviet Union.

The elements whose spatial arrangement constitutes an exploratory system are as follows:

Exploratory section through the whole deposit, i.e. an intersection of the exploratory

TABLE 52

Exploratory systems used in solid mineral deposits

Usage of exploratory systems in relation to morphological types of deposits (in %)

Group of exploratory systems	Types of explor. systems	Names of groups and types of exploratory systems	Usage of explor. systems in %	Strata-bound deposits	Large deposits (thick layers and massifs)	Veins and lenses	Stocks, pipes, branched layers	Very small irregular deposits
1.		vertical sections obtained by drilling (Fig. 141)						
		1. shallow vertical drill holes	11.4	85.6	3.8	10.6		
		2. deep vertical drill holes	9.6	31.6	56.4	9.0	3.0	
		3. deep vertical drill holes combined with shallow drill holes	1.4	37.5	62.5			
		4. inclined and branched drill holes	6.7	9.4	48.9	36.8	4.9	
		5. vertical and inclined borings combined	3.9	29.6	51.9	21.2		
2.		vertical sections obtained by underground works (Fig. 142)						
		6. exploratory trenches	1.1	100.0				
		7. test pits	3.3	100.0				
		8. test pits combined with cross-cuts, raises and winzes	1.1	25.0	17.9	8.9	48.2	

3.	vertical sections obtained by drilling and underground works, Fig. 143						
	9. shallow drill holes combined with test pits	12.3	87.2	12.8			
	10. test pits combined with deep drill holes		2.7	39.2	50.0	10.8	
	11. test pits combined with cross-cuts and drill holes		8.8	18.1	35.3	28.5	18.1
	12. galleries and inclines combined with drill holes		2.2	27.9	23.3	17.4	18.1
4.	horizontal sections of drills used only exceptionally						
5.	horizontal sections of mine workings, Fig. 144						
	13. drifts, galleries		1.7	5.1	75.0	25.0	
	14. drifts and cross-cuts		3.7	34.0	60.9		
6.	horizontal sections of mine and drill works, Fig. 145						
	15. drifts combined with fanwise arranged horizontal drill holes		0.4	52.6	47.4		
	16. drifts and cross-cuts combined with horizontal drill holes		0.4	31.6	68.4		
7.	vertical and horizontal sections of drill holes	used only sporadically					

Table 52 (continued)

Group of exploratory systems	Types of explor. systems	Names of groups and types of exploratory systems	Usage of explor. systems	Strata-bound deposits	Large deposits (thick layers and massifs)	Veins and lenses	Stocks, pipes, branched layers	Very small irregular deposits
8.		vertical and horizontal sections of mine workings, Fig. 146						
		17. drifts with raises or winzes	0.5			45.8	54.2	
		18. drifts with crosscuts, raises and/or winzes	0.5		18.2	81.8		
		19. horizontal drifts in a square or rectangular grid	0.2				100.0	
9.		vertical and horizontal sections of drill and mine workings, Figs. 147, 148						
		20. drifts and inclined or vertical drill holes	4.9	2.0	6.6	73.5	17.9	
		21. drifts and vertical, inclined or horizontal drill holes	0.7		32.0	68.0		
		22. drifts with crosscuts and vertical or inclined drill holes	13.2	5.1	15.2	33.5	46.2	
		23. drifts with crosscuts and drill holes of various orientation	5.1	3.5	21.3	28.7	46.5	
		24. drifts with raises and inclined drill holes	1.2			12.3	87.7	
		25. drifts with crosscuts, raises or winzes and inclined or vertical drill holes	1.4	7.9	11.4	20.5	60.2	
		26. drifts with crosscuts, raises or inclines and drill holes of various orientation	1.1	7.6	15.1	16.7	60.6	
		27. horizontal drifts in a square or rectangular grid, combined with drill holes	0.7		7.5	5.9	19.1	

work with the deposit is the basic parameter. It should enable documentation, sampling and, if necessary, geophysical measurement to be made. Technically, the exploratory sections may be drill holes, test pits and cross-cuts, which will form an exploratory grid over the area of a deposit.

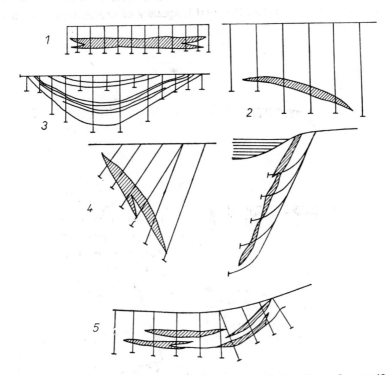

141. Scheme showing exploratory systems of group 1. 1 — vertical sections of a stratified deposit at a small depth; 2 — vertical section of a lens-shaped deposit at a large depth; 3 — vertical section through a thick stratified productive formation; 4 — vertical sections through steep lens-shaped deposits; 5 — vertical section through an ore zone whose deposits dip at various angles. For explanation see Fig. 143.

An *exploratory profile* (vertical or horizontal section) across a deposit is another important element of the exploratory system. Projected onto a vertical or horizontal plane, or a plane parallel to the average strike and dip of the deposit, it appears as an exploratory line. The profiles may be continuous when they consist of continuous exploratory workings or discontinuous, formed by exploratory sections, between which the geological conditions are interpolated. Continuous profiles are far more accurate than discontinuous ones.

A set of exploratory sections and lines, which is usually plotted in an adequately chosen projection, makes up an *exploratory grid*.

270

Exploratory profiles are the fundamental form of exploration for deposits. In most exploratory systems, the works are distributed so as to enable the most precise construction of profiles. Biryukov justifiably based his classification on the character of profiles and the type of exploratory works. By combining these parameters, Biryukov defined nine basic groups of exploratory systems consisting of 27 types (see Table 52). The Table shows the total frequency of exploratory systems in per-

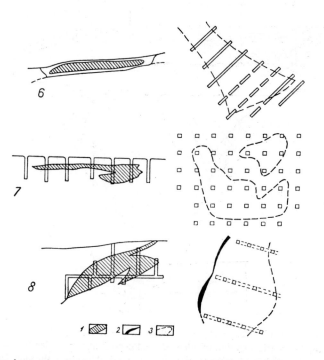

142. Scheme showing exploratory systems of group 2. 6 — plan of trenches and section of a trench through an eluvial-colluvial placer; 7 — plan of test pits and vertical section through a weathering deposit; 8 — plan and vertical section along a line of test pits with cross-cuts on a lens-shaped deposit at a small depth. 1 — ore bodies, 2 — outcrops of deposits on the surface, 3 — boundaries of deposits at depth.

centage and their frequency on deposits of various morphological types. Very small and irregular deposits are not included since their exploration and exploitation are sometimes carried out simultaneously. This classification is unnecessarily detailed but, on the other hand, it greatly facilitates the choice of an adequate exploratory system. The application of individual exploratory systems is demonstrated by typical examples in Figs. 141 — 148; it largely depends on the type of deposits.

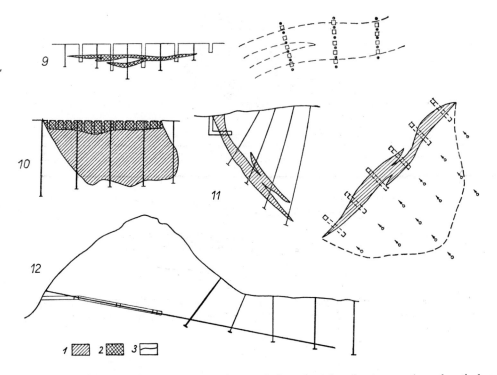

143. Scheme showing exploratory systems of group 3. 9 — plan of exploratory works and vertical section through a discontinuous placer; 10 — vertical section through a stockwork; 11 — horizontal projection and vertical section of a lens-shaped deposit; 12 — cross-section through an ore seam. 1 — primary sulphidic ore, 2 — secondary ore, 3 — stratified deposit.

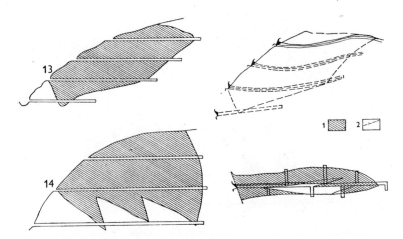

144. Scheme showing exploratory systems of group 5. 13 — horizontal projection and vertical section showing the distribution of drifts on a narrow vein; 14 — geological map of a level and vertical section of a lens-shaped deposit. 1 — ore bodies, 2 — depth boundaries of deposits.

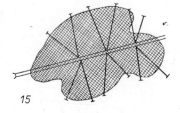

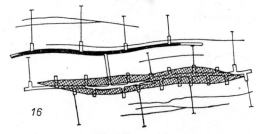

145. Scheme showing exploratory systems of group 6. 15 — geological map of a level on a large ore pipe; 16 — geological map of a level on a deposit formed of veins and lens-shaped ore bodies. 1 — large ore bodies, 2 — narrow ore veins.

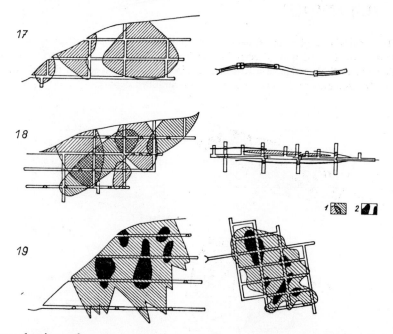

146. Scheme showing exploratory systems of group 8. 17 — longitudinal section and geological map of a drift on a vein with ore shoots; 18 — longitudinal section and geological map of drifts on a deposit formed of a system of parallel impersistent veins; 19 — geological map of a level and vertical section of an isometric stockwork. 1 — ore bodies, 2 — pay streaks.

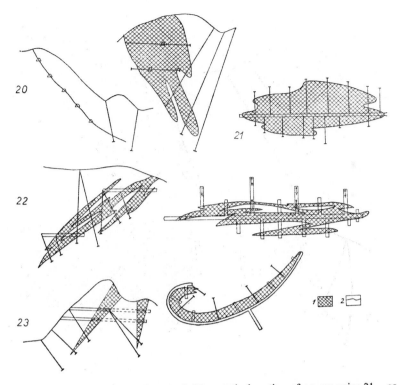

147. Scheme of exploratory systems of group 9. 20 — vertical section of an ore vein; 21 — geological map of a horizon and vertical section of a large lens-shaped deposit; 22 — geological map and vertical section of an irregular deposit; 23 — vertical section through a horseshoe-shaped deposit. 1 — ore bodies in sections, 2 — ore veins.

EXPLORATORY GRIDS

The principal criterion used for the definition of exploratory systems is the character of horizontal and vertical profiles. Their distribution in the projection of the deposit onto a suitable plane or in an exploratory grid is equally important. It should be realized that the required dimensions and shape of the grid are established by exploratory works within the body of the deposit. In the text below a brief description of exploratory grids is given, arranged according to decreasing geometric regularity.

1. *Exploratory grid of regular geometric shape consisting of discontinuous profiles of drill-holes and test pits.*

a) *Square grid*: The basic figure is a square. Two systems of equally precise profiles can be constructed from the parameters of this grid. It is suitable for the examination of horizontal deposits, pipes and stockworks of approximately isometric shape and for initial development of a deposit, when the directions of maximum morphological and qualitative variations are not yet known.

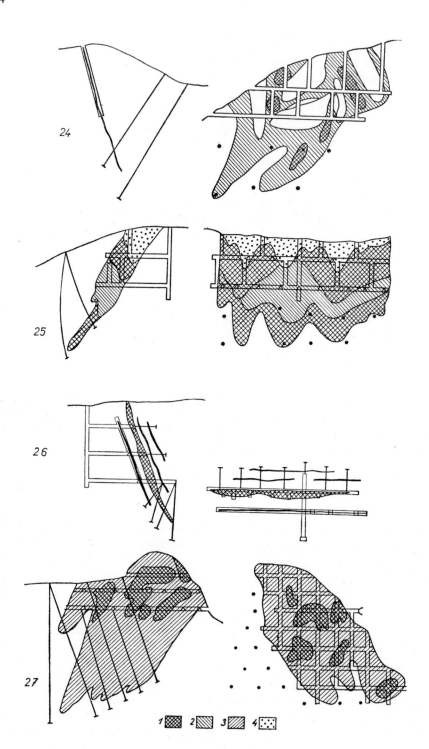

24

25

26

27

1 2 3 4

b) *Rectangular grid*: The basic figure is a quadrangle. Two systems of profiles with different densities of exploratory workings can be constructed. The system with the greater density of workings is to be oriented in the direction of the greater qualitative and morphological variation of the deposit.

c) *Rhomboid grid*: The basic figure is a rhomboid. Using this grid it is possible to construct exploratory profiles along the sides and diagonals of the rhomboid. The maximum density of exploratory works following the shorter diagonal should be oriented parallel to the maximum variation of the deposit. The rhomboid grid provides the best possibility to consider changes in the structure of the deposit that were discovered later. A special case of the rhomboid grid is a triangular net with an equilateral triangle as the basic figure. It is suitable for the exploration of isometric deposits and has three directions of exploratory works of the same density. All these types of grids can be used in exploration by drill holes or test pits.

2. *Exploratory grids of regular geometric shapes consisting of two systems of continuous or discontinuous profiles*. The basic figure is a square or a rectangle whose four sides are defined by underground works following the strike or dip of the deposit. The profiles are continuous when the thickness of the deposit is smaller than the width of the gallery. If the thickness of the deposit can be determined only by using cross-cuts, the profiles are discontinuous. This type of exploratory grid is represented by the systems in Fig. 148 (24, 25).

3. *Exploratory grids consisting of one system of profiles*. A network of this type can be constituted by drill holes, test pits or trenches located along lines oriented parallel to the maximum morphological and qualitative variation of the deposit. In projection, the profiles appear as exploratory lines. The spacing of exploratory sections on the lines is usually constant and appreciably smaller than the spacing of the lines. The systems in Fig. 143 (9, 11) are examples of exploratory grid of this type.

4. *Irregular exploratory grids*. Exploratory works which are not distributed on a regular geometric grid or on a system of other lines, are used for the delimitation of small composite deposits when exploration is conducted on the ancient miner's principle of "holding to the ore". However, it may also be effective on large deposits of complicated structure. In laying out exploratory works along short lines it is desirable to place them so that the course of the main structure can be determined as precisely as possible. A schematic example of this grid is shown in Fig. 149, depicting part of a large ore deposit in a folded sedimentary complex, which was examined by preliminary and detailed exploration. The preliminary exploratory works were

◄——

148. Scheme showing exploratory systems of group 9. 24 — vertical and longitudinal sections of an ore vein having a very complicated internal structure; 25 — vertical and longitudinal sections of a complicated lens-shaped deposit; 26 — geological map of a level and vertical section of a thick zone with veins and lens-shaped ore bodies; 27 — geological map of a horizon and vertical section of a structurally complicated stockwork. 1 — pay streaks, 2 — poor segments, 3 — impregnations, 4 — oxidized zones of the deposit (Figs. 141—148 after Biryukov, 1962).

arranged in two lines at right angles to each other. In the advanced stage of preliminary exploration, the short lines had to be reoriented perpendicular to the axes of fold structures. During preliminary and detailed exploration the short lines were drawn so as to determine the eastern limit of the deposit, anticlinal axis, the limit of the barren part of the deposit at its western border, detailed geological section at the site of the projected shaft, position of Z_1 isohypse at which level the main haulage gallery was planned, and the positions of two more isohypses. Although the course of structural elements was interpolated, it is accurate enough, since the spacing of exploratory works on short lines is small. It roughly equals the width of a working block and the spacing of the lines is about the length of the block. A substantially larger number of exploratory sections would be necessary if the structure of the deposit were examined with the same precision using a regular grid.

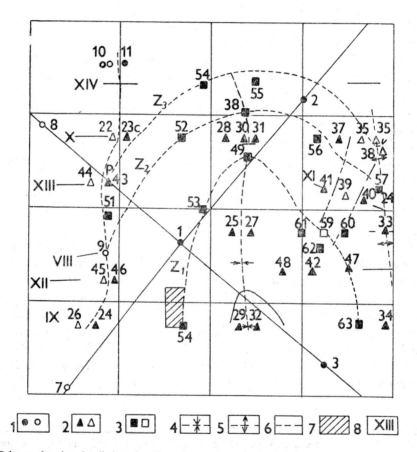

149. Scheme showing detailed exploration of a moderately folded deposit. 1 — exploratory works of the 1st order, 2 — of the second order, 3 — of the third order, 4 — axis of a syncline, 5 — axis of an anticline, 6 — boundaries of the deposit and of barren "windows", 7 — dimensions of a mining block, 8 — numbers of exploratory lines (after Shekhtman, 1963).

THE USE OF MINING WORKS AND DRILLING IN EXPLORATORY SYSTEMS

The reliability and accuracy of geological exploration and the cost per unit of proved reserves depend on the adequate choice of mining works (both open and underground) and drill holes or a combination thereof. Apart from exclusively underground or drilling exploration, the following three combination of these procedures are most frequently used in an appropriate time sequence:

1. Mining works and exploratory drill holes. The outcropping parts of the deposit are examined by shallow shafts or trenches and the remaining parts by drilling. This procedure is most applicable to large deposits with simple structure and composition which crop out on the surface. (Fig 143, 10, 12).

2. Exploratory drill holes and mining works. This method is suitable for the exploration of hidden deposits. The potential ore-bearing area is explored by drill holes and the deposit is then examined by mining methods. Exploration of large sedimentary deposits can be carried out only by boring.

3. Mining works, exploratory drill holes and underground works. This combination is generally used on deposits of irregular shape which crop out on the surface. The outcrop is examined by shallow shafts or trenches and deeper parts by drilling. Detailed exploration of the deposit is conducted by underground working combined if necessary, with subsurface drilling. Exploratory procedures in this succession can be used, for example, for the system shown in Fig. 147.

The use of drilling or mining works is controlled by a whole set of geological, mining-technical, geographical and economic conditions.

Geological conditions. The following factors suit the use of drilling methods: deposits of simple form and internal structure; continuous deposits without numerous, randomly distributed barren "windows"; horizontal or moderately dipping deposits; water-bearing rocks; incoherent, strongly compressible and swelling rocks. In deposits and rocks with the opposite characteristics, underground exploratory works are recommended.

Geographical conditions. Drilling methods are advantageous in remote, poorly accessible areas, in a terrain with flat relief and with sufficient water supply and under climatic conditions that are not extremely adverse, especially in winter.

Other conditions. Drilling is preferable whenever the time factor is important. Exploration by drilling can be performed 2—3 times faster and 2—3 times more cheaply than by mining methods. On the other hand, the results thus obtained are less reliable and precise and the interpretation of intricate geological structure is difficult. A serious drawback lies also in the fact that drill holes cannot be used for operating purposes on deposits of solid raw materials. Many practical examples show that drilling exploration cannot ensure reliable data on a deposit and its reserves. Chumakov (1965) reports on a deposit of non-ferrous metals where 400 drill holes 450—700 m deep were insufficient to determine the structure of the deposit unequivocally and to calculate the reserves. Eventually, underground exploration had to be undertaken for this purpose.

This example shows the importance of a suitable choice of exploration procedures in the geological assessment of deposits. The geological-economic and mining-technical comparisons of several exploratory variants are useful in solving this problem. The question has been studied on a broader scale by several authors (Čilík 1968, Bartalský 1966). Chumakov (1965) analysed the results of exploration on deposits of non-ferrous and precious metals. He determined the coefficient of effectiveness of underground and drilling works which would give the ratio of necessary lengths (in metres) of galleries and drill holes for verifying the ore reserves, assuming that the two methods are equally detailed. Table 53 shows that the coefficients of effectiveness (K_{dv}) vary over a wide range. On complex deposits of non-ferrous and precious metals the length of drill holes should be 8−20 times greater than that of underground workings to obtain equally detailed results. Considering the cost of drilling and underground works and the value of the K_{dv} coefficient, Chumakov came to the conclusion that on deposits of non-ferrous and precious metals drilling can be used for reconnaissance and detailed exploration only on large uncomplicated deposits (cf. Table 53) suitable for surface working and where exploration need not be extended to any depth. In all other cases exploration on deposits of this type should be carried out by mining techniques combined with subsurface drilling.

TABLE 53

Ratio of equivalent volumes of underground and drilling exploration on deposits of non-ferrous and precious metals

Group of deposits	Characteristics of deposits	Coefficient of effectiveness of underground and drilling exploration K_{dv}
I	deposits of non-ferrous, rare and precious metals; large size, simple shape, uniform distribution of useful mineral	1 : 2
II	deposits of non-ferrous, rare and precious metals; variable morphology, uneven distribution of useful mineral	1 : (3.5—5.4)
III	deposits of rare, non-ferrous and precious metals; very variable morphology, very uneven distribution of useful mineral	1 : (8—20)

LOCATION OF EXPLORATORY WORKS

The position of each exploratory work in the exploratory system is dictated by the general geological setting and, consequently, must be selected with due respect to it. Shekhtman (1963) distinguished four alternative geological positions:

1. Favourable geological position: the exploratory work is located at a site where the discovery of a deposit can be inferred from geological, geochemical or geophysical indications. For a favourable geological position reconnaissance drill holes are the rule.

2. Boundary geological position: the exploratory work is sited at the boundary of the deposit or in a barren window in order to delimitate it. Exploratory works in this situation usually fix the parts of the deposit with the lowest parameters, such as the thickness or content of useful minerals. Most of these works will give negative results. Most workings in the boundary position are performed during reconnaissance and at the beginning of detailed exploration. At this stage the limits of the deposits are defined and the unit cost of proved reserves will be at its highest.

3. Intra-deposit position: most exploratory works in the advanced stage of detailed exploration occur in this position. They are undertaken with the aim of examining the deposit along a sufficient number of exploratory sections so that the average parameters can be computed and the morphology and tectonics of the deposit determined with the necessary precision. They should also enable the structural and qualitative boundaries within the deposit to be established satisfactorily. If these exploratory procedures are carried out in the proper order, they are usually positive and the unit costs of exploration of proved reserves decrease at this stage.

4. Random position: the intra-deposit exploratory works are laid out regardless of the geological structure of the deposit. Their location is given by their position in the geometric grid or on an exploratory line.

The necessary care is not always taken to select the optimum position of the exploratory works in a regular network.

Locating shallow exploratory works (trenches, pits) is very simple. They should verify the outcrops of deposits, whose approximate position has been determined by mapping or from geochemical and geophysical anomalies.

The laying-out of exploratory drill holes is more difficult. Apart from shallow mapping drill holes, a plan of operation is worked out for each drill hole. The plan contains the following items: locality, number of the drill hole, its inclination (vertical, inclined), x, y, z co-ordinates, projected depth (predicted minimum and maximum depths); required core recovery from the deposit and the rock; geological setting at the drilling site; permitted deflection of the hole from the designed axis; tentative stratigraphic column; rock classes in the drill hole profile according to drillability; description of the physical properties of the rocks; tectonics, aquifers and gas-bearing horizons; demands on logging, inclinometry and pumping tests. On the basis of geological conditions a technician prepares the technical part of the project, such as the type of casing, drilling method, kind of washing, etc. It is advisable to complement the description of the geology by a tentative section drawn across the drilling site, in which the drill hole axis and the deposit to be explored are plotted.

From the section, the principal data on the projected drill holes will be determined: depth, course of the hole axis, the site of drilling which, after plotting in the map, will be determined by x, y, z co-ordinates.

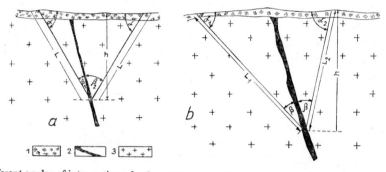

150. Different angles of intersection of a deposit by two drill holes of the same length and inclination, carried out from the hanging wall and footwall of a deposit. α — inclination of drill holes, β — angles of intersections, L — length of drill holes, h — depth of intersections; 1 — waste mantle, 2 — deposit, 3 — wall rock (after Kallistov in Volarovich, 1956).

Exploration of horizontal and moderately inclined deposits is almost invariably carried out by vertical drill holes. Great stress is laid on keeping the axes vertical. Inclined holes drilled from the hanging wall are most suitable for the exploration of steeply dipping deposits (Fig. 150). They cut the deposit at a larger angle than those drilled from footwall and the risk of axis deflection on striking the deposit or a steep fault is smaller. This angle should be at least $25 - 30°$. If the same angle of intersection is chosen for drilling from hanging wall and footwall, the length of the hole is shorter for the former (L_2). The hole is drilled from the footwall only when the configuration of the terrain does not permit location of the drill hole from hanging wall.

In laying out the axis of the drill hole, the position of the intersection of the drill hole with the deposit is first determined. This position is either defined by the assumed geological situation or derived from the required section density in the profile. The axis of the drill hole is then extended onto the surface and this point is plotted on the section line in the map.

Since an exploratory drill hole is used to verify tentative geological situation, the minimum and maximum depths of the hole should always be determined in such a project; these values are derived from the presumed variation of stratigraphic conditions. The maximum depth of an inclined drill hole on steeply plunging endogenous deposits is often fixed on the assumption that the deposits may even dip at $90°$. The drill hole is therefore designed up to a vertical line drawn from the lowest

Known point of the deposit. The minimum dip that can be expected in the part of
the deposit under examination is estimated analogously from the geological structure of the deposit (Fig. 151). Within economically valid limits some margin is
permissible for maintaining the strike and dip of the hole axis. The minimum and
maximum depths at which the deposit can be encountered by drilling are given
in relation to these conditions (Fig. 151).

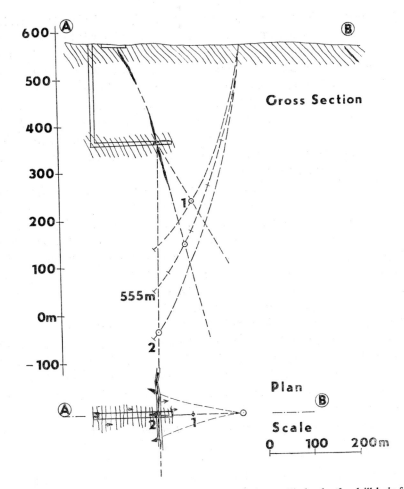

151. Determination of the presumed minimum (1) and maximum (2) depth of a drill hole from the
potential dispersion of the deposit and borehole inclination.

In sedimentary and schistose rocks the axis of the drill hole tends to assume a direction at right angles to the bedding or schistosity. If the drill hole is laid out irrespective
of factors that cause curving, its course must be adapted by lowering deflection wedges,
which complicates the work and increases the cost. Planning such drill holes is very

difficult and has not yet been developed reliably. It is based on empirical data on the degree of azimuthal and zenithal curvature (given in degrees per 100 m of the drill hole). The records are obtained from holes drilled in the same rocks using the same drilling procedure.

DELIMITATION OF THE DEPOSIT

The delimitation of a deposit is one of the main tasks of geological exploration. During preliminary exploration the entire deposit should be delimited; only in the case of exceptionally large deposits is exploration concentrated on that part which, from its geological position and results of prospecting, appears to be the most promising. During detailed exploration the limits of the deposit should be defined with greater accuracy, barren parts within it should be noted, the various types of mineral material determined and sectors of diverse engineering-geological and mining-technical characteristics should be distinguished. The boundaries that divide the deposit into blocks of the same type are based on geologico-structural, qualitative and mining-technical characteristics.

1. Geological boundaries. Sedimentary and endogenous deposits of tabular and lenticular shapes are terminated by the "boundary of zero thickness". The boundary of isometric deposits, stockworks, stocks and metasomatic deposits has the form of a closed topographic surface. Its complex course is depicted by a system of vertical or horizontal profiles and by projections onto a vertical or horizontal plane. Limitation of the deposit is due to various causes, which must be recognized if exploration works are to be executed adequately. Common modes of termination of deposits of various genetic types are the following:

a) Magmatic deposits: gradual transition into non-mineralized rock; wedging out of the parent rock; change in the structure of the magmatic massif; post-ore faulting.

b) Hydrothermal deposits: termination of the ore-bearing structure or a change in its character; physico-chemical conditions controlling the course of the ore-bearing interval; disappearance of rocks favourable to mineralization; barriers preventing hydrothermal solutions from circulation; post-ore faulting.

c) Residual deposits: thining out of the parent rock; termination of shatter zones which cause intensive weathering; erosion of the weathering mantle; fault boundaries.

d) Sedimentary deposits: wedging out along the periphery of the sedimentary basin; syngenetic or postgenetic erosion; facies changes; fault boundaries.

2. Qualitative intra-deposit boundaries delimit blocks of homogeneous technological properties of the ore material, or those parts of the deposit whose contents of useful and harmful components are within the determined limits, etc.

3. Mining-technical boundaries divide the deposit into zones with different exploitation conditions. The following boundary types are encountered most frequently: boundary of the maximum ore/overburden ratio, boundary of the maximum

depth of opencast and underground working, boundary delineating parts of different hydrogeological conditions and boundary of safety pillars.

Delineation of the deposit in the exploratory grid

For the delineation of a deposit in the exploratory grid records of the deposit boundary obtained in individual exploration sections and profiles are used. The tracing of boundary lines and determination of their origin require an adequate lay-out of exploratory works and the use of suitable construction methods. Since the boundary lines are rarely followed continuously by underground works, extrapolation and interpolation are widely used for their construction. In exploration we can differentiate between limited extrapolation, i.e. extrapolation of the deposit parameters in a sector between a positive exploratory work and a negative one, which represents the limit of the possible range for extrapolation, and unlimited extrapolation, when there are no negative exploratory works beyond the marginal positive works. The methods used for the construction of the key points for delineating the wedging-out of the deposit can also be applied to the construction of intra-deposit boundary lines.

1. The delineation of the deposit in an exploratory section provided by a drill hole, test pit or trench is the simplest, since the deposit is exposed and sampled throughout its thickness. Neither extrapolation nor interpolation are necessary. The contact of the deposit with the wall rock and interlayers is plotted directly in the geological log. If the transition between the deposit and the wall rock is gradual, the deposit is delineated on the basis of sampling. The boundary is defined by samples with a previously determined economic material content, designated as the "cutoff grade".

2. The delineation of the deposit in an exploratory profile involves the determination of the boundary in a horizontal or vertical section of the deposit. The most accurate limits are determinable in continuous profiles consisting of drifts and inclined workings on deposits of small or medium thickness. The upper and lower contacts of the deposit can be plotted directly, but the qualitative boundaries must be drawn according to sampling results. Figure 152 shows several methods of constructing the key points for qualitative boundaries:

a) interpolation to the half-distance between two samples $(B - D)$ is simple and sufficiently precise and is therefore the most frequently used procedure;

b) linear interpolation between a pair of samples $(A - B)$ can be solved graphically or by computation;

c) linear extrapolation beyond the last sample (D) can also be solved graphically or by computation.

The c_h values are the cutoff grade of the useful mineral and the $c_A - c_D$ values are the contents of the samples.

The use of methods b) and c) is limited to very regular deposits since qualitative changes in most ore bodies do not occur linearly.

A horizontal section provides data for the construction of a transverse profile $(1-2)$ and a longitudinal profile of the deposit $(3-4$ Fig. 152).

Discontinuous exploratory profiles are obtained by linearly distributed workings. The geological situation in the profile plane is reconstructed using interpolation of data between neighbouring sections or by extrapolation. Either geological or linear interpolation and extrapolation are used for drawing the boundaries. In the former case, the boundaries are drawn through the key points as continuous curves corresponding to the likely shape of the deposit. In the latter case, the boundaries pass through the key points as straight lines to ensure an unbiased delimitation of the ore bodies. Linear interpolation and extrapolation are chiefly used in calculating ore reserves. Discontinuous exploratory profiles are less accurate; they are placed in accuracy class II—III (Table 54).

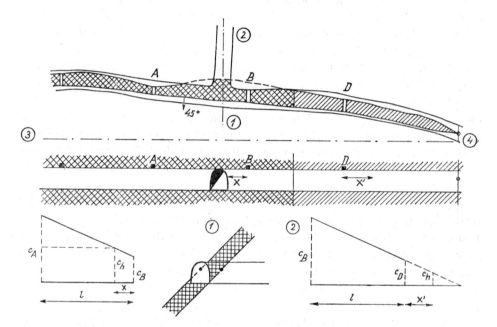

152. Delimitation of a deposit in a continuous horizontal section by a drift, in a cross-section (1, 2) and vertical projection (3, 4).

Figure 153 illustrates the exploration of a lens-shaped body by trenches, pits, cross-cuts and drill holes laid out along the lines. The figure shows the delimitation of the deposit in cross sections ($A - B, C - D$), the construction of a horizontal profile (a "derived" profile, since it is constructed from cross-sections) and the construction of a vertical profile. The derived profiles are less accurate than the primary ones. Here, this can be classed with category IV (Table 54).

3. A complete delineation of the deposit within the exploratory grid is based on data yielded by sections and profiles. This is more complicated since it is not a planar

TABLE 54

Scheme showing the relative accuracy of exploratory profiles

Class of accuracy	Characteristics of exploratory profiles based on their construction
I	profiles constructed on the basis of the continuous tracing of the deposit outline
II	profiles constructed from univariant geological interpolation
III	profiles constructed from univariant linear interpolation
IV	profiles constructed from univariant twofold interpolation ("derived" profiles)
V	profiles with elements of multivariant interpolation
VI	derived profiles which do not enable the geological pattern between exploratory works to be linked unequivocally
VII	profiles in which the geology of only individual exploratory works is plotted; their linking is impossible
VIII	profiles constructed with the use of limited extrapolation
IX	profiles constructed with the use of unlimited extrapolation

but a three-dimensional problem. The boundaries are constructed by transferring key points from exploratory sections and profiles into a horizontal or vertical plane or a plane parallel to the dip or strike of the deposit. Generally, the internal boundary is constructed first. It connects marginal positive exploratory works, enclosing the more thoroughly explored parts of the deposit within the grid. The external boundary (of zero thickness) is then constructed, using various extrapolation methods. The following procedures are used for limited extrapolation:

1. the external boundary is drawn midway between positive and negative exploratory works;

2. the key points for the external boundary are determined in exploratory profiles, according to the character of the wedging out of the deposit (Fig. 153);

3. the construction of key points for the external boundary using the average angle of wedging out. The average angle is calculated as the ratio of the average half-thickness in the marginal exploratory works to the average exploratory works half-spacing (Fig. 154):

$$\tan \frac{\alpha}{2} = \frac{\dfrac{m_p}{2}}{\dfrac{l_p}{2}} = \frac{m_p}{l_p},$$

$$l_p = \sum_1^n \frac{l_i}{n} \qquad m_p = \sum_1^n \frac{m_i}{n}.$$

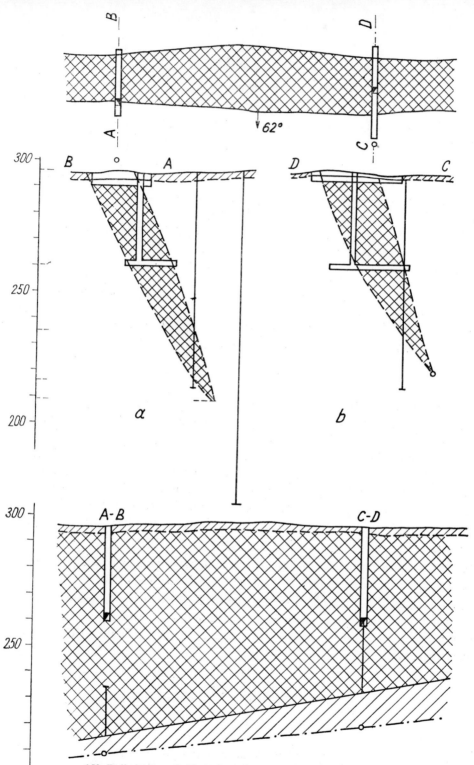

153. Delimitation of a lens-shaped deposit in vertical sections using limited extrapolation (a) and unlimited extrapolation (b) depending on the morphology of the deposit.

From the average angle of wedging out, the distance of the point of wedging out (x) on the lines connecting each positive and negative work can be calculated, taking into consideration the thickness in the marginal exploratory work.

$$\tan\frac{\alpha}{2} = \frac{\dfrac{m_x}{2}}{x} \qquad x = \frac{m_x}{2\tan\dfrac{\alpha}{2}}$$

The points of wedging out can also be constructed by plotting the thickness of the deposit and the angle $90° - \dfrac{\alpha}{2}$, as shown in Fig. 154. This last procedure is seldom used; it is applicable only to lenticular bodies which wedge out regularly. All three methods of constructing the external boundary are suitable for limited extrapolation.

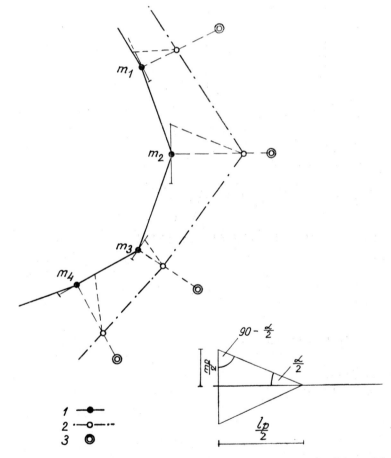

154. Delimitation of a lens-shaped deposit from an average angle of the wedging-out of the deposit. 1 — inner boundary with exploratory drill holes, 2 — outer boundary, 3 — negative drill holes.

The external boundary beyond the outline of the exploratory grid is constructed by unlimited extrapolation unless negative exploratory works occur. Exploratory profiles or parts thereof constructed in this way are the least precise (Table 54). Several methods are used for the delineation of a deposit by *unlimited* extrapolation:

1. Construction of the external boundary based on geological, i.e. tectonic, stratigraphic and facies records, usually yields the best results. Figure 155 shows the delineation of the economic parts of the Rudňany deposit (Slovakia) at the Carboniferous-Permian boundary deduced from the assumption that Permian plastic rocks are unfavourable to the extension of ore-bearing structures.

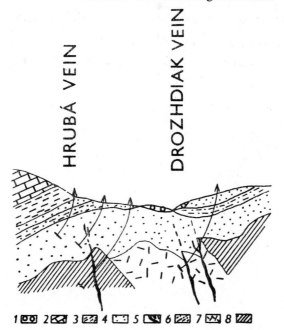

155. Prospecting for concealed ore veins near Rudňany (Slovakia). Economic mineralization is localized in pre-Permian formations. 1 — alluvium, 2 — Mesozoic dolomites, 3 — Mesozoic shales, 4 — Permian conglomerates, 5 — ore veins and indications, 6 — Permian variegated shales, 7 — diabase, 8 — Carboniferous phyllites (after Čillík—Ogurčák, 1964).

2. Construction of the external boundary from the morphology of the deposit also provides good results. The points of wedging out are established in the profiles with due respect to the mode of deposit wedging-out within the exploratory grid. These points are then transferred to the profile lines in a horizontal or vertical plane, on which the projection of the whole deposit is reproduced (Fig. 153).

3. If neither of the two methods mentioned above is applicable, the external boundary is constructed in a conventional way according to the density of the exploratory grid and the dimensions of mining blocks and of the deposit.

a) The external boundary is drawn parallel to the internal boundary at a distance equal to the spacing or half-spacing of exploratory works in the grid. The distance is chosen according to the complexity of the deposit structure.

b) The external boundary below the lowest horizon is drawn at a distance equal to the dip length of a mining block. In extrapolating along the strike of the deposit, the external boundary is drawn at a distance equal to the strike length of a mining block. Depending on the character of the deposit, these distances can be extended to double or reduced to one half the original value.

c) The external boundary of a deposit located on a fracture or fault system is extrapolated in the form of a triangle whose height equals one half of the strike length of the deposit, or as a rectangle whose shorter side equals one quarter of it. The extrapolated volume of an isometric deposit has the shape of a hemisphere or a cone.

The various methods applicable to the construction of the external boundary leave a number of possibilities of delimitation. The reserves between the internal and external boundaries are therefore not established very accurately and are assessed conservatively. The calculation of reserves in the marginal parts is refined with exploration, as the width of extrapolated and interpolated zones diminishes. In addition to geological circumstances and those affecting construction, the distribution of exploratory works in the grid and the time sequence of their performance control the accuracy of the delimitation of a deposit.

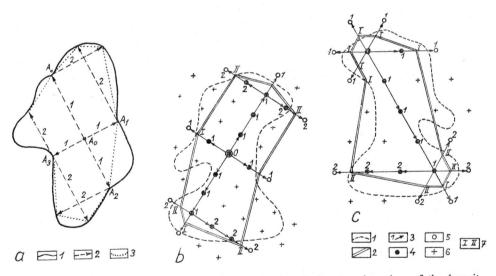

156. a — Scheme showing vectorial delimitation of a deposit. 1 — true boundary of the deposit, 2 — exploratory lines (vectors), 3 — boundary constructed from exploration data, figures denote the order of vectors. b, c — vectorial delimitation of the deposit using a square and a triangular grid; 1 — true boundary of the deposit, 2 — boundary constructed from exploratory data, 3 — exploratory lines (vectors), 4 — positive drill holes, 5 — negative drill holes, 6 — as yet undrilled points of the grid, 7 — order of key points, 0 — the first positive drill hole (after Zenkov and Semenov, 1957).

Zenkov and Semionov (1957) developed a *vectorial delimitation method,* which makes it possible to delimit a deposit in the initial exploration stage using a minimum number of exploratory works (Fig. 156a). Exploration begins at point A_0, which may be an outcrop of the deposit or the first positive exploratory work. Four exploratory lines or first order vectors are drawn from this point towards the periphery of the deposit and a number of second order vectors are drawn from the ends of the first order vectors until the deposit is defined with the required accuracy. The method is convenient particularly in exploration by drill holes or pits, and it has the advantage that most exploratory works of the first stage are used for the delimitation of the deposit. The interior part of the deposit is explored subsequently. For the sake of unifying the works on vector lines and those performed later, an exploratory grid of suitable shape is chosen for laying out the vector lines (Fig. 156b, c). The vector method is best suited for horizontal or moderately inclined stratified and stratabound deposits of coal, manganese, phosphates and for large stockworks and impregnated deposits of non-ferrous metals.

EVOLUTION OF AN EXPLORATORY GRID

The principle of gradual accumulation of data results in multistage exploration. Knowledge of a deposit increases from the propecting stage to the termination of detailed exploration. The density of the exploratory grid is gradually increased both over the whole deposit and parts thereof that have a complicated geological structure. This increase in grid density is especially typical of exploration by drill holes and test pits. In underground exploration, the lay-out of works is largely controlled by the technical conditions.

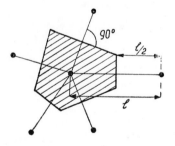

157. Construction of the polygonal area covered by one drill hole or test pit.

The density of an exploratory grid consisting of exploratory sections can be expressed as

$$P_0 = \frac{P}{n},$$

where P — area of the delimited deposit in m^2, P_0 — the area of the deposit covered by one section in m^2 and n — number of exploratory sections in area P.

The density coefficient (Z) can be expressed as the ratio of the area covered by one exploratory section in a lower-order grid (P_1) to the area covered by one section in a higher-order grid (P_2).

The concept and construction of an area covered by relevant exploratory sections are shown in Fig. 157. In this way the degree of density of regular and irregular exploratory grids consisting of drill holes can be given.

The density of geometrically regular grids can be increased in the following ways:

Grid	Density increased by	Coefficient Z
square	one hole in the centre of the square	2
square	one hole in the centre of the square and one hole in the centre of each side	4
rectangular	one hole in the centre of the shorter side	2
rectangular	one hole in the centre of the rectangle and one hole in the centre of each side	4
rhombic	one hole in the centre of the rhombus	2 /
rhombic	one hole in the centre of the rhombus and one on each of its sides	4
triangular (special instance of rhombic grid)	one hole in the centre of the triangle	3

Stammberger (1962) proposed the introduction of a coefficient defining the degree of limitation of a mining block for exploratory grids consisting of underground workings. This coefficient is defined by the ratio of the area of a mining block to the perimeter of the block explored by underground workings. The value of the coefficient K for a group of mining blocks limited to different degrees are given in Table 55.

TABLE 55

Block no.	$K = \dfrac{\text{block area}}{\text{perimeter of the block}}$		Delimitation of sides
1	2400 : 200 = 12	1	2 drifts, 2 raises
2	2400 : 160 = 15	1.25	2 drifts, 1 raise
3	2400 : 120 = 20	1.66	2 drifts
4	2400 : 60 = 40	3.33	1 drift

The coefficient is calculated for a block 40×60 cm in size; it can be expressed in a simplified form when $K = 1$ is used for full limitation. The degree of limitation can thus be compared only in blocks of identical dimensions. Since this procedure cannot be employed for drilling exploration, there is no uniform scale for comparing densities of grids of drill holes and those of underground exploratory works.

The time sequence of exploration is very important for developing an exploratory grid. Exploratory works are performed either successively (each drill hole or pit is begun only after the preceding one is completed and appraised) or simultaneously. In practice, the two procedures are usually combined. Drill holes or pits are sunk synchronously on several lines but exploration along one exploratory line is successive. It is advisable to set out the time sequence of exploratory workings in a working schedule.

It is also important to determine the proportion of exploratory works in each stage. Since exploration of a major ore field or coal basin takes a long time, it is not an easy task and cannot be accomplished by a mere increase of the density of the exploratory grid.

An example of complex geological exploration history on a large ore field in Central Asia has been described by Shekhtman (1963). The schedule of 16 years' drilling investigation is shown in Fig. 158. The deposits are of hydrothermal origin and are located in lithologically favourable horizons of a folded sedimentary complex. The ratios of exploration stages are given in percentages (first column at the bottom of Fig. 158). From the present point of view, the exploration project was deficient in some respects. For example, prospecting was not carried out over the whole ore field and geological exploration was not performed in due stages in many sectors. Prospecting and preliminary exploration were not carried out in sector 11 and in sector 3 preliminary exploration was sparse, while detailed exploration took a disproportionately long time. The volume of works during prospecting and preliminary exploration was too large relative to the detailed exploration.

This example shows that in a given mineralized area, the ratio of exploratory works in the individual stages and their schedule should be designed for a longer period. This general plan is then refined as new records are gradually assembled. If due account is not taken of the overall sequence and reasonable proportions of the exploration stages in examining the whole deposit are not maintained, basic exploration principles can easily be violated.

Exploration, particularly in its later stages, requires a considerable amount of machinery and manpower as well as technical and administrative personnel. The cost of exploration can be reduced by concentrating exploratory works and increasing performance by using adequate equipment. Technical work, however, can be accomplished in a short time only when reliable geological, tectonic, geophysical and metallometric information is available.

Fedorchuk (1964) reported a considerable reduction in the cost of exploration on mercury-antimony deposits in Central Asia by reasonable concentration of drilling

YEARS AND PHASES OF EXPLORATORY WORKS

Deposits and sectors of an ore field		State of exploration
upper level	1	completed
	2	- " -
	3	- " -
	11	- " -
	6	subeconomic
	13	- " -
middle level	NW. field	unlimited
	W. field	- " -
	E. field	completed
	8	unfinished
	SW. field	completed
	S. field	unfinished
lower level	18	beginning of exploration
	12	- " -
	21	- " -

I phase — years 1 2 3 4 5; II phase — 6 7 8 9; III phase — 10 11 12 13; IV phase — 14 15 16

	I phase	II phase	III phase	IV phase
volume of workings in % ①	①33	①11	① 8	①29
	②41	②38	②42	②19
	37 67	61 60	47 40	54 72
number of positive drill holes in % ②	30 92	28 76	45 60	17 57

Legend: prospecting-exploratory stage — preliminary exploration — detailed exploration

158. Schedule of exploratory works on an ore field.

works. The output of one drill rig per month rose from 83 to 140 m, i.e. by 68 per cent, while the cost of boring per one metre dropped by 29 per cent. Working conditions and supply improved and the operation was made cheaper and simpler by, for example, establishing a central transformer station, electrification of drill rigs, pumping station and central preparation of drilling fluid. Complete concentration of drilling exploration on a blind Hg and Sb deposit located in an anticlinal structure is shown in Fig. 159. This is also an example of a sound, geologically substantiated exploration time schedule. With deposits of this type, the fold limbs adjacent to the axial plane and the apex of the fold are most intensely mineralized. The first prospecting holes are situated in these places. If they prove positive, preliminary exploration proceeds in these most promising segments. In the detailed exploration stage, drill holes are sunk to delimit the deposit and examine its entire area. From experience gained on Hg—Sb, usually blind deposits in Central Asia, Fedorchuk reports that continuity in exploration can be achieved by the following drilling programme: 10 % — prospecting, 10 % — deep drill holes, 5 % — deep reference holes, 25 % — preliminary

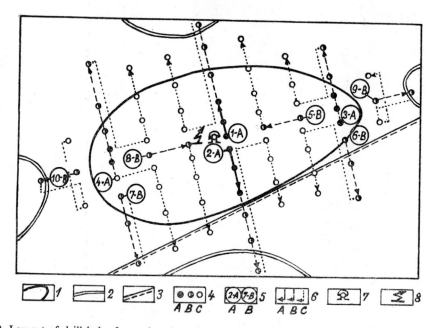

159. Lay-out of drill holes for exploration of a Hg—Sb deposit, based on a structural-prognostic map. 1 — presumed boundary of the mineralized segment in the closure of a brachyanticline, exploratory sector of the 1st order; 2 — ore-bearing structures delimited as sectors of the 2nd order; 3 — reverse fault which acted as the main supply channel of hydrothermal solutions; 4 — drill holes: A — of the 1st order (prospecting), B — of the 2nd order (preliminary exploration), C — of the 3rd order (detailed exploration); 5 — drill rigs operating in the first exploration stage (A — rigs 1A to 4A) and the second stage (B — 5B to 10B); 6 — succession of drill holes on the exploratory lines of the 1st (A), 2nd (B) and 3rd (C) orders; 7 — central preparation of drilling fluid; 8 — transformer station (after Fedorchuk, 1964).

exploration, 50 % — detailed exploration. In our opinion, this division of exploration stages seem deficient in so far as it requires too high a volume of lower exploration stages.

By applying the principles of compound interest reckoning Žežulka (1965) has demonstrated the advantage of shortening the duration of the exploration. He used the formula

$$A = a(1 + r)^p,$$

where A — the total cost of the exploration (here in Kčs), a — the annual investment (in Kčs), r — inverse value of the depreciation time (10 years in this instance) and p — detailed exploration time in years.

If, for example, the annual investment on geological exploration (a) is assumed to be 2 million Kčs, the exploration time (p) 10 years and the time of depreciation (p) 10 years, the cost of exploration during this period will be

$$A_{10} = \sum_1^{10} 2(1 + 0.1)^{10} = 32.856 \text{ million Kčs.}$$

By reducing the exploration time to 5 years at an annual investment of 4 million Kčs the cost becomes

$$A_5 = \sum_1^5 4(1 + 0.1)^5 = 29.700 \text{ million Kčs.}$$

By shortening the time of exploration from ten to five years, 3.1 million Kčs are saved. This example shows how desirable it is to perform exploration in the shortest possible time with due consideration for the correct sequence of exploratory works.

DETERMINATION OF THE OPTIMUM DENSITY OF EXPLORATORY GRIDS

The determination of the optimum density of an exploratory grid, i.e. of the optimum spacing of exploratory works, is extremely important, since it controls the cost and time of exploration and the accuracy and reliability of the results obtained. The grid density should be given particular attention when exploration is performed by drilling or pit excavation. Underground exploration must take into account technical problems connected with the development of deposits and their division into horizons. So far, no universal theory has been developed for determination of the optimum grid density. Research and practice use the following methods: 1. determination by analogy, 2. comparison of exploration data with mining data on the same deposit, 3. gradual decrease or increase in the density of the grid, 4. analysis of the accuracy of exploratory profiles, 5. mathematical statistical methods, 6. modelling and 7. economic methods.

1. Determination by analogy

On the basis of empirical data, convenient spacings of exploratory works have been determined for all normal deposit types. The requirements on the accuracy of exploration for the individual categories of mineral resources have also been unified. The desired grid density for individual materials differs depending on the type of deposit and the category of reserves. The deposit types are determined according to the size, geological structure and morphological and qualitative variation. Table 56 presents empirical average spacings of exploratory works on primary deposits of solid materials (Kreiter, 1961).

Table 57 presents a detailed scheme for the density of exploratory grids on ore deposits, as recommended by the State Commission for Mineral Resources in the U.S.S.R. in 1960–1961 (Kazhdan, 1971). Ore deposits are divided into three groups based on the variation of the deposit:

	Geological structure	Thickness	Distribution of the useful mineral
Group 1	simple	constant	uniform
Group 2	complicated	varying	non-uniform
Group 3	very complicated	highly varying	very non-uniform

If the proposed spacing of exploratory works is to be successful, the deposit must be assigned to the right group. The density of the works is related to the principal bodies of the deposit. In elongated or qualitatively varying deposits the grid should be chosen so that at least 3–4 vertical sections lie in every profile along the direction of greater variation. Over the whole deposit, several dozen positive exploratory sections are required for category C_1. The grid is made denser by exploratory works carried out in successive stages, thus enabling modification of the maps and profiles from one stage to another. The exploratory system and works should be selected with great care. Ore deposits with a complicated structure usually require a certain amount of underground exploration. Therefore, in Table 56 the author augmented the number of deposit types for which underground or combined exploration is recommended. For underground exploration, the recommended spacings are those of drift entries and inclined workings.

The spacing of exploratory works established from long and detailed experience provides a valuable aid, particularly in planning and during the first exploration stages, but it should not be used conventionally regardless of the specific conditions in each deposit.

2. Comparison of exploration data with mining records

Exploration data can be checked with high precision on deposits in operation. A number of parameters are compared, such as the area, average thickness, quality

TABLE 56

Characteristic parameters for exploration of solid-mineral deposits

Group of deposits	Characteristic of the deposit	Principal exploration system	Exploratory works	Spacing of exploratory works for calculation of reserves				Required categories before deposit development
				A	B	C_1	C_2	
a	large deposits, continuous mineralization and morphology, even distribution of useful minerals	drilling	drill holes'	100—200 m	200—400 m	extensive extrapolation		A, B, C, A, B predominant
			underground works	2—4 working levels	extrapolation to larger distance			
b	large, rarely medium deposits, continuous or interrupted mineralization, constant morphology, uneven distribution of useful minerals	drilling + combined	drill holes	60—200 m	sporadic holes			B, C, B predominant
			underground works	2 working levels	3—4 working levels	extrapolation to minor distance		
c	deposits of medium size, discontinuous mineralization, various shapes, uneven and very uneven distribution of useful minerals	combined + underground	drill holes			isolated holes, extrapolation to minor distance		B, C_1, C_1 predominant
			underground works	1 working level	2 working levels	isolated underground works + extrapolation to minor distance		C_1, C_2
d	small or elongated, very discontinuous bodies, very variable distribution of useful minerals	underground	underground works		1 working level, often also sublevels	2 working levels		C_1, C_2
e	very small nest-like deposits	exploitation and exploration carried out synchronously						

predominantly extrapolation

Category A = proved ore, B + C_1 = probable ore, C_2 = possible ore.

TABLE 57

Recommended density of exploratory grids

Metals	Types of deposits
	Group 1
iron	large horizontal or moderately inclined deposits, with constant thickness + uniform mineralization
	large steeply-dipping or stratified deposit of great strike length + constant thickness
manganese	large stratified, horizontal or moderately inclined deposits with fairly uniform ore composition + constant thickness
bauxite	large stratified deposits of fairly constant grade
copper	large stratified deposits + stockworks of simple shape + uniform distribution of useful component
nickel	very large horizontal deposits with simple structure + uniform distribution of useful components
polymetallic deposits	strata-bound, with a simple form and fairly uniform distribution of mineralization
molybdenum	large stockworks + stratified lens-shaped deposits with fairly uniform distribution of mineralization
tungsten	large stockworks + stratified deposits with fairly uniform distribution of mineralization
rare metals	large stockworks + stratified deposits, fairly simple form + fairly uniform mineralization
	Group 2
iron	large stratified deposits, complicated structure + fairly constant ore grade
	strata-bound + lens-shaped bodies, medium size + non-uniform distribution of mineralization
manganese	stratified + lens-shaped deposits tectonically disturbed, rather non-uniform distribution of mineralization + complicated morphology
chromite	vein + lens-shaped deposits, strike length from 300 m to 1000 m and more, constant thickness, broken into blocks of strike length 50 m or more
bauxite	stratified + large lens-shaped deposits of varying quality
copper	large stockworks + lens-shaped deposits of complicated form; mineralization distribution fairly uniform
	vein + lens-shaped deposits of medium size; complicated morphology + non-uniform distribution of useful minerals
nickel	large inclined stratified and lens-shaped deposits and veins with fairly uniform distribution of useful components
	large stratified + sheet deposits with non-uniform distribution of mineralization, veins, lens-, nest-shaped + stratified deposits with complex + very complex structure and very variable distribution of mineralization
polymetallic deposits	large lens-shaped + stratified deposits with complex form and predominant non-uniform mineralization
	lens-shaped + vein deposits of complex form and non-uniform mineralization

Density of exploratory grid category A along: strike	dip	category B strike	dip	category C_1 strike	dip
200	200	400	400	800	800
150	100	300	200	600	300
150—200	150—200	300—400	300—400	600—700	600—700
75—100 75—100	50—75	150—200 100—150	100—150	300—400 200—300	150—200
100	100	200	200	400 — 600	400 — 600
(50—75)	(50—75)	100—150	100—150	150—200	150—200
(50—60)	(60—80)	100—120	100—120	120—200	120—200
(50—60)	(60—80)	(100—120)	(100—120)	120—200	120—200
(50)	(50)	100	100	200	200
—	—	150 (100)	100 (50—100)	300 150	200 100
—	—	150—200	50—100	300—400	100—150
—	—	(40—60)	(20—30)	80—120	40—60
—	—	75—100 75—100	50—75 75—100	150—200 150—200	100—150 150—200
—	—	(50—75)	(50—75)	100—150	100—150
—	—	100—120	50—75	150—200	75—100
—	—	40—60 (20—40)	40—60 (20—40)	80—120 40—80	80—120 40—80
—	—	(50—75)	(50—75)	100—150	100—150
—	—	—	—	(75—100)	(50—75)

Table 57 (continued)

Metals	Types of deposits
tin	large stockworks veins + mineralized zones of great strike length (1—2 km) with non-uniform mineralization
molybdenum	large stockworks + large stratifield, lens-shaped and vein deposits with complicated morphology + non-uniform distribution of mineralization large veins of varying + rather small thickness with non-uniform distribution of mineralization
tungsten	large stockworks and deposits of complicated morphology with non-uniform distribution of mineralization large vein deposits or mineralized zones
gold	large mineralized zones with non-uniform distribution of mineralization mineralized zones of medium size and dykes, lens-shaped deposits with non-uniform mineralization + complex outline large + medium-size veins with uneven + very uneven distribution of mineralization
rare metals	veins or mineralized zones of great strike length and width, complex form or with non-uniform mineralization
mercury and antimony	large stratified deposits and vein zones with non-uniform mineralization
cobalt	large veins and vein zones with non-uniform mineralization

Group 3

chromite	veins, lens- and nest-like deposits, shoots of minor size (from tens of m to 300 m), tectonically broken into blocks
bauxite	lens-shaped deposits of medium size and varying quality
copper	small lens-shaped and vein deposits with complex morphology and very uneven mineralization
polymetallic deposits	veins, lens- and nest-shaped deposits, ore shoots of very complicated form with extremely uneven mineralization
tin	veins and mineralized zones of medium strike length (0.3—1 km) with very uneven mineralization
molybdenum	veins of medium size and minor width with non-uniform mineralization
tungsten	veins of medium size and minor width with non-uniform mineralization
mercury and antimony	stratified and vein deposits of medium size with non-uniform mineralization
rare metals	veins or mineralized zones of medium strike length, relatively small thickness with non-uniform mineralization
cobalt	small ore bodies of irregular lens- or nest-shape, ore shoots with non-uniform mineralization

Numbers in parentheses relate to underground or combined exploration.

Density of exploratory grid category A along: strike	dip	category B strike	dip	category C_1 strike	dip
—	—	(50—60)	(50—60)	100—120	100—120
—	—	(60—80)	(40—50)	120—160	60—80
—	—	50—60	50—60	100—120	100—120
—	—	(60—80)	(40—50)	120—160	60—80
—	—	(50—60)	(50—60)	100—120	100—120
—	—	(60—80)	(40—50)	120—160	60—80
—	—	—	—	(80—120)	(40—60)
—	—	—	—	(60—80)	(30—40)
—	—	—	—	(80—120)	(40—60)
—	—	100	50	200	100
—	—	—	—	(100—200)	(50—60)
—	—	—	—	(80—120)	(40—60)
—	—	—	—	(40—60)	(20—30)
—	—	50—75	25—50	75—100	50—75
—	—	—	—	(50—75)	(50—75)
—	—	—	—	(50—60)	(40—50)
—	—	—	—	(60—80)	(40—50)
—	—	—	—	(60—80)	(40—50)
—	—	—	—	(60—80)	(40—50)
—	—	—	—	(40—50)	(40—50)
—	—	—	—	(100)	(50)
—	—	—	—	(25—40)	(25—40)

of the material, reserves of the raw material and metal, and information on the morphology and tectonics of the deposit. The difference between values assessed during exploration and during exploitation is expressed by the relative error.

The relative computation error for mineral reserves is, for example,

$$Z = \frac{Q_1 - Q_2}{Q_2} \times 100,$$

where Q_1 — reserves calculated from exploration data, Q_2 — reserves calculated from mining records.

Large differences betweeen the exploration and mining records point to insufficient density of the exploratory grid. For similar deposits or for parts of a worked deposit that are still under exploration, a denser grid should be employed. The disadvantage of this method, which yields very accurate results, is its laboriousness and the fact that results are only available long after exploration has been completed.

During exploration the material suffers losses and contamination; it is therefore often more convenient to compare exploration data with the results of exploration carried out during operation than with those of mining. This, however, will be discussed in connection with another method — gradual thinning of the exploratory grid.

3. Gradual thinning of exploratory grid

Using this procedure, the deposit or part thereof should be explored with the aid of a grid that has the necessary density for the relevant exploration stage, or one in excess of it. From the information obtained by all exploratory sections, reserves of raw material, reserves of useful minerals and the average thickness and content of useful components are calculated. The exploratory grid is then "thinned" by omitting every fourth, third or second exploratory work. The compared values and the relative error with respect to the grid of maximum density are calculated from the parameters of the thinned grids. The grid density is satisfactory for exploration purposes when its relative parameters lie within the permissible limits. It should be noted that every degree of thinning provides several variants and that all the alternatives should be compared with the grid of maximum density. Comprehensive numerical calculations are made using automatic computers.

Works carried out on some stockworks of non-ferrous metals in the U.S.S.R. exemplify very detailed study of the optimum density of the exploratory grid (Galkin et al., 1962). For exploration, experimental grids ranging from 400×400 m to 25×25 m were used. Several variants of overlays were put on the plan of the opencast level, which was sampled by drill holes for blasting spaced $4-6$ m apart. The following parameters were computed for every grid in all variants: average content of useful mineral, c_p, area of the industrial ore, P, and the ore reserves, Q. The relationship between the relative errors and the number of points in the exploratory grid on Kounrad (Cu) and Pervomaiskoe (Mo) deposits is shown in Fig. 160. From the analysis of these curves it follows that the relative errors for the deposits' parameters are roughly identical on the two deposits if the number of exploratory sections is the same. For example, 55–65 exploratory sections are necessary for computation of the average amount of useful mineral with a 10 % error. This implies a 100×100 m grid for the Kounrad deposit and a 25×25 m grid for the Pervomaiskoe deposit. The conclusions drawn using mathematical statistics, i.e. that the accuracy of the computed average parameters and total mineral reserves depends on the number of exploratory sections and not on the density of the grid, have thus been confirmed.

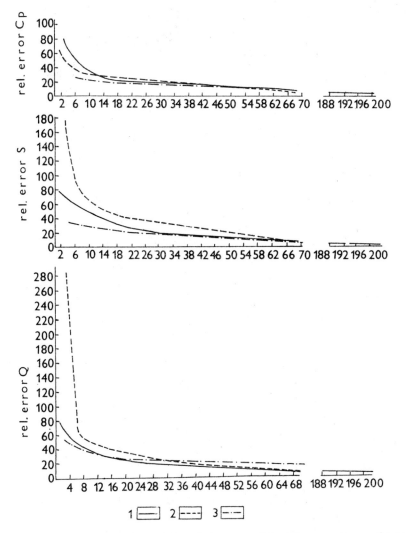

160. Curves showing the dependence of the relative error of deposit parameters on the number of points of the exploratory grid on Kounrad (Cu) and Pervomaiskoe (Mo) deposits. 1 — Kounrad deposit, 2 — Pervomaiskoe deposit, A horizon, 3 — B horizon (after Galkin, 1962).

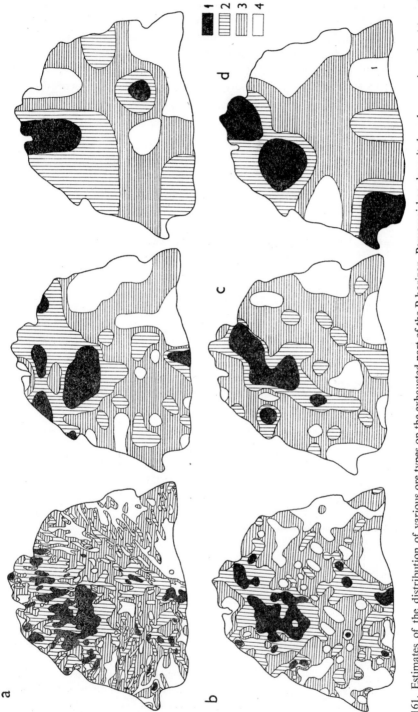

161. Estimates of the distribution of various ore types on the exhausted part of the B horizon, Pervomaiskoe deposit, based on various densities of the exploratory grid. a — delimitation of ore types from sampling during exploration, b — from a 10×10 m grid, c — from a 20×20 m grid in two positions, d — from a 50×50 m grid in two positions. 1 — high-grade ore, 2 — medium-grade ore, 3 — poor ore, 4 — very poor ore (after Galkin, 1962).

Fifteen to 20 drill holes distributed evenly over a deposit would suffice for calculating probable ore reserves (B category) (relative error $\pm 20 - 30\%$) and $30 - 50$ drill holes would allow computation of visible ore reserves (A category) with a relative error $\pm 15 - 20\%$. But average values of parameters and total reserves alone are not sufficient for exploitation requirements. It is necessary to determine the morphology of the deposit, distribution of workable and unworkable ores and of ores having different technological properties, etc. Figure 161 shows that even a 24×24 m grid presents only a schematic picture of the distribution of metal content. Adequate determination of the structure of the Kounrad deposit for extraction purposes would require more than 900 drill holes and more than 400 would be necessary on the Pervomaiskoe deposit. Detailed exploration therefore ends with a grid of 100×100 m or 50×50 m density and the detailed internal structure of the deposit is examined during exploitation.

Biryukov (1965) recommends that the size of the deposit area should be considered in determining the optimum density of the exploratory grid, regardless of the method employed. Figure 162 shows four types of deposit differing in morphology and quality. An important relationship follows from the graphs: the relative error, Z, in calculation of the ore reserves is directly proportional to the complexity of the deposit and indirectly proportional to the areas of the deposit segments. Reserves of the total deposit (P_1) are given with the greatest accuracy, whereas for the smallest part of the deposit, such as a mining block, relative errors are greatest. From the above it follows that, in giving precision to calculation of tonnage, it is necessary to state the area or volume of the deposit to which the relative errors are related. The reserves of individual blocks of a single deposit can be calculated with greater accuracy by increasing the density of the grid. Two curves in Fig. 163 show the dependence of relative error ΔZ on P_1, P_2, P_3 and P_1 areas of various size, P_1 being the area of the whole deposit. Curve γp corresponds to an exploratory grid of the first order with a density of 1.3 sections per hectare, which corresponds to preliminary exploration; detailed exploration was carried out using a second-order grid. The relationship described is expressed by the function

$$\lambda = F\left[\frac{K}{P}\right].$$

The density of the exploratory grid is directly proportional to the increase in the reserves category (category C_2 is the lowest and A the highest) and indirectly proportional to the deposit area (P). These curves are used for the determination of the optimum grid density so that with a deposit explored sufficiently, relative errors of reserves computation (ΔZ_1, ΔZ_2, ΔZ_3) are established for exploratory grids of different densities and for different deposit areas (P_1, P_2, P_3). For each exploration stage the curve and corresponding grid density are selected which ensure the necessary accuracy for computation of reserves for the whole deposit and the parts thereof.

Comparative studies have shown that the relative error in calculating the parameters of a complex deposit can be rather high $- \pm n \times 10\%$ to $\pm n \times 100\%$ — within a mining block. The parameters and reserves of the entire deposit are calculated with relatively high precision of ± 10 to $\pm 20\%$, since positive and negative deviations cancel out.

Determination of the optimum density of exploratory grids according to Biryukov is based on the relationship between the density of the grid, the relative error of reserves calculations and the size of the deposit or the part to which the error is

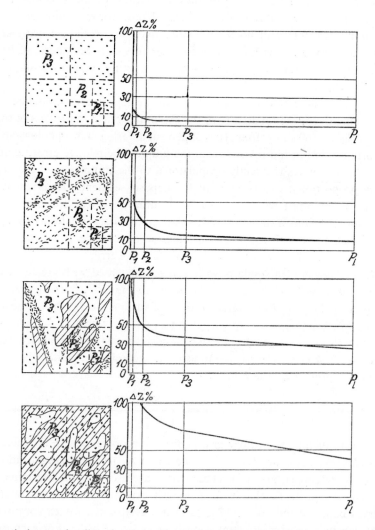

162. Theoretical regression lines for the accuracy of tonnage calculation on deposits differing in morphology and quality of ore. ΔZ — accuracy of calculation expressed by the relative error in per cent, P_1, P_2, P_3 — segments of the deposit of various dimensions, P_1 — area of the entire deposit (after Biryukov, 1965).

related. This is very important since it is desirable to know not only the total relative error but especially the error for individual exploitation units, for both planning and mining.

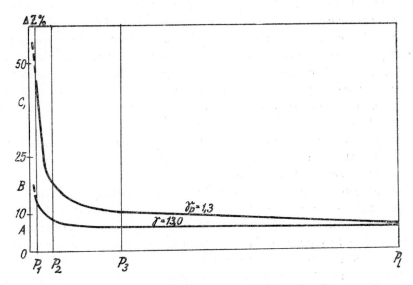

163. Actual regression lines of reserve categories on a stratified deposit with an uneven distribution of metal. γ_p — density of exploratory grid in preliminary stage of exploration, expressed by the number of drill holes or test pits per 10,000 m², γ — density of exploratory grid in detailed exploration stage.

4. Analysis of the accuracy of exploratory profiles

Zenkov (1957) recommends that optimum density of exploratory grids should be defined by analysing the geological structure of the deposit. He recommends that the spacing of exploratory works be such that works along an exploratory line are increased in number until a reliable connection of the geological elements of an exploratory profile is achieved. Sectors of complicated structure in which the geological elements cannot be confidently connected using a dense grid are placed in a lower category. The reliability of the linkage between exploratory profiles is assessed analogously. The degrees of accuracy of exploratory profiles are given in Table 53. Zenkov's method is very ingenious and may be used even when another method is applied to determine the optimum density of the exploratory grid.

5. Mathematical-statistical methods

Although mathematical methods have recently been applied extensively in geology, no universal method for the determination of the optimum density of exploratory grids has so far been developed.

During exploration, however, a large number of quantitative and qualitative parameters of the deposit, such as its thickness, contents of useful and noxious components, volume weight and structural-technical properties, are collected, providing convenient material for mathematical treatment in the form of a mathematical model. This can contribute to the determination of the optimum grid density and to the solution of many other exploration problems.

The selection of a mathematical model will depend on a) the properties of the deposit (the parameters of a deposit may be either 1. a set of independent random variables, 2. a set of random autocorrelated variables or 3. a set of random autocorrelated variables that are functions of the spatial co-ordinates); b) the degree of exploration. In later exploration stages, simple mathematical models can be replaced by more complicated models as the amount of information increases.

Accordingly, the mathematical model will be based on different mathematical methods such as those of mathematical statistics, geostatistics, harmonic analysis and trend-surface analysis (Kazhdan, 1974).

Using mathematical models, the following problems can be solved: the variability of deposits and their random and regular components; the variability along principal structural directions of the deposit; and the number of parameters (thickness, metal content) necessary for ensuring that the mean value will be within the range of the required interval and will be determined with the required probability.

Table 58, given here as an illustration, presents a procedure of statistical analysis that can be regarded as a statistical model for a deposit in the initial exploration stage (Scott, Hazen, 1968). The confidence limit established for the deposit parameters (here the metal content) is a criterion for assessment of a satisfactory number of drill holes.

Of the many recent papers on mathematical methods applicable to the solution of the above-mentioned problems at least those of Koch and Link (1970) and Agterberg (1974) should be mentioned.

In this manual, the classification of mineral deposits based on the coefficient of variation is mentioned in several places (e.g. Table 59). The coefficient of variation controls the density of the exploratory grid and the sampling density. It can, however, be applied only for characterization of random variation of deposits, i.e. of very irregular deposits or of more regular deposits in the initial exploration stages. The coefficient of variation must be computed with respect to the distribution of the studied parameter (thickness, metal content, etc.).

For normal parameter distribution it can be computed using the formula

$$V = \frac{s}{\bar{x}} \times 100,$$

where V — coefficient of variation, $s = \dfrac{\sqrt{\sum\limits_{i-1}^{n}(x_1 - \bar{x})^2}}{n}$ standard deviation,

Assay data

Screen for different assay populations by minerals, grade of ore, and geology

Assay frequency distributions for minerals and grade using unlocated assay values (not considering actual X, Y, Z positions)

Multimodal distributions indicating subpopulations for other minerals or grade of ore

Unimodal distributions indicating single populations of assays

Compute statistics mean, variance, standard deviation, skewness

If relative skewness is approximately 3, test for lognormal assay frequency distribution

Screen for subpopulations (oxide, secondary, primary)

Use log probability paper for cumulative assay frequency distribution plot to screen for subpopulations

If only a small number of assays are available use simulation techniques to generate larger assay populations to form better distributions

If skewness is present in assay frequency distribution try various transformations to normalize the distribution, square root, logarithms, binomial, hypergeometric, etc.

Compare assay frequency distributions with theoretical mathematical distributions

Calculate confidence limits for normal and transformed normal assay frequency distributions

Match computed confidence limits with the acceptable limits specified as the goal for the sampling and evaluation

If confidence limits are not acceptable and more work is needed

Use sample volume-variance relationship $s_1^2\lambda_1 = s_2^2\lambda_2$ and standard error of mean for calculating new sample volumes, or new number of samples of same volume, and number of additional drill holes needed to match specified confidence limits. Calculate new drill-hole-grid spacing, as required, by using standard error of the mean

Phase II of exploration

If confidence limits are acceptable for exploration

Calculate quadratic and cubic regression analysis to determine trends in grade, thickness and mineralization

Compute area of influence of an assay

Compute statistical response surfaces and grade of ore interpolations for planning mining methods and production

Deposit evaluation

TABLE 58

Flow chart of the generalized procedure for the statistical analysis of assay data

TABLE 59

Groups of deposits of solid minerals based on the variation of quality, thickness and reserves

Group	Type of deposit	Variation	Variation coefficient of		
			thickness V_{th}	content V_c	reserves V_r
1	sedimentary: coal, combustible shales, building mat., phosphates, Fe ores. Simple magmatogenic deposits	usually regular	5—50	5—30	30
2	most deposits of non-ferrous metals, some endogenous non-metallic deposits, complex Fe-ore deposits	usually random	30—80	40—100	80
3	most vein deposits of rare and precious metals; dep. of non-ferrous metals with the most complex morphology or tectonically disturbed. Some endogenous non-metallic materials	random	50—100	100—150	130
4	small or very disturbed deposits of rare metals with very intricate distribution of useful minerals	random	80—150	130—300	200

Note: According to Baryshev, the minimum number of parameters needed for the calculation of the variation coefficient is 18 for group 1, 25 for group 2 and 40 for group 3.

x_1 — random variable of deposit parameter, $\bar{x}$ — arithmetic mean of a set of parameters–n; for log-normal distribution:

$$V = (e^{S_{ln}^2} - 1)^{\frac{1}{2}},$$

where e — base of natural logarithm, and S_{ln}^2 — dispersion of the logarithms of the random variable.

6. Determination of the optimum density of exploratory grids by modelling

Kazakovskii (1945) has deduced new expressions for the coefficient of variation and the coefficient of geological assurance of a deposit but his results are not generally valid. The regularities established hold only for a certain type of parameter distribution corresponding to the model studied. Further development of this method is especially difficult since deposit models depict only their areas and thickness; qualitative parameters of their internal structure cannot be expressed by modelling.

7. The economics of exploration costs

This method was suggested by Ograkov and Osetskii (in Pogrebitskii, 1965). The authors begin with the assumption that a deposit consists of parts with a simple geological structure, which they denote as "normal blocks" and parts showing complicated morphology, tectonics and quality – "anomalous blocks". If the latter are not discovered during exploration, the development and extraction of the ore is more costly, since additional exploratory works during mining operations are necessary. For economic reasons, the density of the exploratory grid should be increased to such an extent that the exploration cost does not exceed the cost of additional exploration of anomalous blocks. This method is applied mainly to coal deposits.

The relationships are derived from the coefficient of variation (P_1) of the deposit explored

$$P_1 = \frac{S_1}{S},$$

where S_1 is the total area of the anomalous block and S is the area of the whole deposit. The total number of drill holes in an exploratory grid (n) relative to the complex outline of the deposit is

$$n = 1.5 \frac{S}{l^2},$$

where S is the deposit area in m² and l is the interval between drill holes in metres.

If the area of an anomalous block of roughly isometric shape is expressed by its diameter L, the interval l between drill holes allowing a reliable determination of this block is given by

$$l = \sqrt{\frac{\pi L^2}{4}}.$$

The exploration cost for the given grid density is

$$A = nD \frac{6DS}{\pi L^2},$$

where D is the cost of one drill hole.

The cost of additional unterground-exploratory works on anomalous blocks (B) is calculated using the formula

$$B = B_1 P_1 L^2,$$

where B_1 – cost of exploration of 1 m² of an anomalous block, P_1 – coefficient of variation and L = diameter of an anomalous block not established by exploration.

From the principle of exploration economics it follows that

$$A = B = \frac{6DS}{\pi L^2} = P_1 B_1 L^2$$

or

$$L = \sqrt[4]{\frac{6DS}{\pi P_1 B_1}} \, .$$

The spacing of boreholes is then given by

$$l = \sqrt{\frac{\pi L^2}{4}} = \sqrt[4]{\frac{1.5\pi DS}{4P_1 B_1}}$$

or approximately

$$l = \sqrt[4]{\frac{DS}{P_1 B_1}} \, .$$

The coefficient of variation P_1 is determined during preliminary exploration or on similar deposits. Calculation of the spacing of exploratory works by this method is particularly convenient for detailed exploration.

CATEGORIES OF MINERAL RESERVES AND DEGREES OF GEOLOGICAL ASSURANCE OF THE DEPOSIT

The determination of categories of mineral reserves is included in the computation of reserves and is closely associated with the selection of methods to be used and with the determination of the degree of geological assurance, which can be expressed by the ratio of individual categories.

Since no universally valid methods are available for verifying the accuracy and reliability of estimation of mineral reserves, the categories are defined by requirements on the density of the exploratory grid and on the degree of geological and technological exploration of the deposit.

Classification of mineral reserves on a world scale is not uniform. Designation of ore categories and parameters used for classifying mineral reserves are given in the text below.

Table 60 presents the classification compiled in the Institute of Mining and Metallurgy in London in 1902. It assumes that deposits are explored mainly by underground mining works and the classification is based on the extent of delimitation of blocks by drifts and raises. It does not express the suitability of mineral materials for industrial usage.

Widely known and used is the classification of reserves prepared by the U.S. Bureau of Mines and the U.S. Geological Survey in 1944; it distinguishes measured, indicated and inferred reserves. Progress in geological exploration, drilling techniques and the science of mineral deposits have been considered in this scheme. The most recent classification was set up in 1973, again by the U.S. Bureau of Mines and the U.S. Geological Survey. It has been elaborated with respect to the needs of long-term planning in the exploration, exploitation and consumption of all kinds of mineral

TABLE 60

International Standard Definition of the Institution of Mining and Metallurgy (1902 and 1912)

Designation of reserves	Principal characteristics
visible ore, positive, proved, in sight	1902: ore that has been exposed on 2,3 or 4 sides, which are available to sampling 1912: a) deposit divided into blocks limited on 3 or 4 sides b) reserves that can be rightly presumed even if they are not developed
probable ore	ore limited on 2 or 1 sides, inferred extension of ore in a certain sector
possible ore	ore whose existence can be assumed

raw materials. In addition to the terms "measured, indicated and inferred reserves" which are defined the same as in 1944, a number of new terms are introduced to express the degree of geological assurance and economic feasibility.

A scheme of categories and their definitions follows (according to Geotimes September 1974):

Resources — A concentration of naturally occurring solid, liquid or gaseous materials in or on the Earth's crust in such form that economic extraction of a commodity is currently or potentially feasible.

Identified resources — Specific bodies of mineral-bearing material whose location, quality, and quantity are known from geologic evidence supported by engineering measurements with respect to the demonstrated category.

Undiscovered resources — Unspecified bodies of mineral-bearing material surmised to exist on the basis of broad geologic knowledge and theory.

Reserves — That part of the identified resource from which a usable mineral and energy commodity can be economically extracted at the time of determination. The term "ore" is used for reserves of some minerals.

Measured — Reserves or resources for which tonnage is computed from dimensions revealed in outcrops, trenches, workings and drill holes and for which the grade is computed from the results of detailed sampling. The sites for inspection, sampling, and measurement are spaced so closely and the geologic character is so well defined that size, shape, and mineral content are well established. The computed tonnage and grade are judged to be accurate within limits which are stated, and no such limit is judged to be different from the computed tonnage or grade by more than 20%.

Indicated — Reserves or resources for which tonnage and grade are computed partly from projection for a reasonable distance on geologic evidence. The sites available for inspection, measurement, and sampling are too widely or otherwise inappropriately spaced to permit the mineral bodies to be outlined completely or the grade established throughout.

Demonstrated — A collective term for the sum of measured and indicated reserves or resources.

Inferred — Reserves or resources for which quantitative estimates are based largely on broad knowledge of the geologic character of the deposit and for which there are few, if any, samples or measurements. The estimates are based on an assumed continuity or repetition, of which there is geologic evidence; this evidence may include comparison with deposits of similar type. Bodies that are completely concealed may be included if there is specific geologic evidence of their presence. Estimates of inferred reserves or resources should include a statement of the specific limits within which the inferred material may lie.

Identified-subeconomic — Known resources that may become unrecoverable as a result of changes in technological, economic, and legal conditions.

Paramarginal — The part of "subeconomic resources" that a) borders on being economically producible or b) is not commercially available solely because of legal of political circumstances.

Submarginal — The part of "subeconomic resources" that would require a substantially higher price (more than 1.5 times the price at the time of determination) or a major cost-reducing advance in technology.

Hypothetical resources — Undiscovered resources that may reasonably be expected to exist in a known mining district under known geologic conditions. Exploration that confirms their existence and reveals quantity and quality will permit their reclassification as a "reserve" or "identified-subeconomic" resource.

Speculative resources — Undiscovered resources that may occur either in known types of deposits in a favorable geologic setting where no discoveries have been made, or in as yet unknown types of deposits that remain to be recognized. Exploration that confirms their existence and reveals quantity and quality will permit their reclassification as "reserves" or "identified-subeconomic" resources.

In the U.S.S.R. the first classification of mineral resources was compiled by the Mining Institute of the Soviet Academy of Sciences in 1933 and revised in 1939. Based on the degree of geological assurance and economic feasibility, the reserves were grouped into five or six categories designated by indices A_1, A_2, B (or B_1, B_2), C_1 and C_2.

In a new scheme in 1953, the number of categories was reduced to four — A, B, C_1 and C_2. Each category should comply with contingent requirements on the degree of geological, technological, hydrogeological and mining-technical assurance. The degree of assurance and preparation for the project of a new mine is appraised from the sum of $A + B + C_1$ reserves, which must be present at a certain ratio.

This classification was newly revised in 1960. Other socialist countries use similar classifications; that employed in Czechoslovakia is presented in Table 61.

In addition to the division of mineral reserves into A, B, C_1 and C_2 categories,

TABLE 61

Classification of mineral reserves used in Czechoslovakia

Category	Characteristics
A	Reserves assessed in such detail that their mode of deposition, shape and the structure of the ore body are fully defined; industrial minerals, their natural and technological types, interrelationships and distribution are determined; barren and non-economic sectors are delimited; mining-technical (hydrogeological, engineering geological) factors controlling exploitation and treatment of materials, and their quality and technological properties are determined precisely. Reserves of category A must be outlined by positive drilling or underground workings or other exposures
B	Reserves explored in such detail that the principal peculiarities of their mode of deposition, the shape and structure are known; economic, natural and technical types of mineral materials are determined, as well as regularities in their distribution but not their precise spatial delimitation. Inside the ore bodies the character of barren and unsatisfactory sectors and their relationship to economic sectors are determined; barren and unsatisfactory sectors are precisely delimited. The quality and technological properties of the ore and mining-technical factors (hydrogeological, engineering-geological, etc.) that control the working and treatment are determined. The outline of B category reserves must pass through positive drill holes, mine workings or other exposures. If the thickness and quality of the deposit material is constant, a limited extrapolated zone can also be included in this category
C_1	These reserves are assessed in sufficient detail that the conditions, shape and structure of the ore bodies are broadly outlined; economic types of ores, their natural and technological properties and mining-technical factors determining the exploitation and processing procedures are defined in general. The appraisal of C_1 reserves is established from the results of borings, mine workings, exposures and interpolation and extrapolation based on geological, geophysical and other data
C_2	Poorly explored reserves. The mode of deposition, shape and structure of the deposit are determined on the basis of geological study and geophysical measurements, confirmed by opening of the deposit in several places, or by analogy with the explored sectors. The kind of ore, natural and technological types and the quality are determined from the analyses of samples or from the results obtained in adjacent explored sectors. C_2 reserves are outlined according to the boundaries of favourable geological structures or rock complexes, on the basis of separate drill holes, outcrops, etc., by interpolation and extrapolation based on geological, geophysical and other data
Possible reserves	Unexplored resources presumed to exist on the basis of regularities governing the origin and distribution of mineral deposits and of the results of structural and historical-geological studies of the area in question. Parameters for the assessment of possible reserves (strike length, thickness, average content of useful minerals, etc.) are determined from geological hypotheses or are deduced. Possible reserves are not included in the national budget of mineral reserves; they serve only as a basis for planning geological exploration

316

for which the degree of assurance is the principal criterion, resources in Czechoslovakia are also classed according to other criteria.

On the basis of their commercial importance, *mineral reserves* are classified as *economic* or *subeconomic*. The former are suitable for economic extraction at the present time and techniques for their preparation and treatment are known. The latter are not suitable for exploitation at present due to several factors such as a small amount of ore, a small thickness of deposit, a high content of noxious components, complicated mining-technical conditions or lack of knowledge of their economic preparation and treatment. These reserves are considered to be utilizable in the future.

From the point of their working feasibility, *free and blocked reserves* are distinguished. Ore of safety pillars, for example, is classed as blocked reserves. According to the degree of geological assurance these reserves are also grouped in categories and are divided into economic and subeconomic according to their economic importance.

Commercially profitable reserves in categories A, B and C_1 are designated as *industrial reserves* and the total reserves of a deposit of all categories, both economic and subeconomic, are denoted as *geological reserves*.

According to the stage of development of a deposit, *industrial reserves* are divided into three groups: *opened, half prepared and prepared*.

Figure 164 shows an example of distribution of individual categories ranging from total geological reserves to industrial reserves, which are of principal interest to the miner.

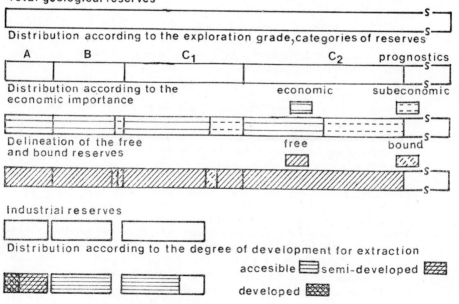

164. Total geological reserves of a deposit.

All classification schemes have some parameters in common and the basic categories can therefore be broadly compared. This comparison is given here for easier orientation, since the designation of categories by indices A, B, and C will be used below.

Comparison of categories of mineral reserves

measured	proved	A
indicated	probable	B, C_1
inferred	possible	C_2

The definitions of mineral reserves lack the determination of admissible errors for individual categories, except the measured reserves (20 %). As the procedures for calculating the reliability interval within which the computed values will vary are not known, a wider complex of requirements on individual categories is preferable, but it is difficult to express mathematically.

Kreiter (1969), for example, proposed the following values of admissible errors:

Category A	$15-20 \%$,
B	$20-30 \%$,
C_1	$30-60 \%$,
C_2	$60-90 \%$.

Of primary importance in determining the degree of assurance of a deposit is the division of mineral reserves into categories, since they are based on an integrated assessment of geological, technological and mining-technical conditions. For geological and economic reasons no deposit can be explored in the same detail throughout. The ratio of categories established in each exploration stage serves as an index of the degree of geological assurance. This relates particularly to the ratio of $A : B : C_1$, as these categories are explored more accurately and allow technical-economic calculations to be made. The importance of category C_2 for the overall appraisal of the deposit must not, of course, be overlooked.

Determination of the degree of geological assurance that should be attained towards the end of detailed exploration of the deposit, when it is to be developed, is most important. This degree must be established in each case, depending on the type of deposit. In principle, higher categories of reserves can be defined both economically and thus also a higher degree of assurance can be obtained on large deposits with slightly varying parameters and a simple geological structure than on small deposits having highly variable parameters.

Therefore, the Soviet classification scheme of 1960 proposed a division of all solid mineral materials into three groups and for each a required ratio of the categories for the project of a new mine has been defined (Kogan, 1971):

Group 1: Deposits of simple structure, constant thickness and uniform distribution of useful minerals. Reserves of A + B categories should cover at least 30 % of the sum of categories A + B + C_1 and category A at least 10 %.

Group 2: Deposits with a complicated structure, varying thickness or uneven distribution of useful minerals. Detailed exploration of category A is useless. Reserves of category B should make up at least 20 % of the sum of categories B + C_1.

Group 3: Deposits with a very complicated structure and highly variable thickness or content of useful minerals. Detailed exploration of category B is unnecessary. A mine project can be based on reserves of category C_1 (see also Table 57).

The optimum ratios of industrial categories are still being studied. As a general rule, a new mining plant must have minimum reserves of the highest category obtainable for the given type of deposit, which will cover several years' exploitation. The remaining reserves are transferred into higher categories on the basis of exploration during mining operations. Assessment of the minimum necessary reserves in the high categories is rather difficult. Approximate values are set out in Table 62 (Pomerantsev, 1961).

The minimum amount of the high categories that will guarantee the return of the investment in the construction of a new plant is calculated as

$$Z_m = a_m A_r,$$

where Z_m — minimum reserves in the high categories, a_m — minimum amortization time and A_r — annual production. If the total industrial reserves of a deposit are Z_c, the required ratio of the high categories to the industrial reserves is

$$\frac{Z_m}{Z_c} = \frac{A + B}{A + B + C_1}.$$

In irregular and small deposits, Z_m can be represented by category B or C_1 and may also include category C_2. Table 62 shows that the amount of reserves explored in detail is chosen to ensure exploitation for 5 to 10 years. The reserves of building materials quarried in opencasts should cover at least 5 years' exploitation.

TABLE 62

Recommended Z_m/Z_c ratios for various groups of deposits

Group of deposits arranged after the complexity of exploration	Total reserves Z_c in millions of tons	Annual production A_r in millions of tons	Minimum time of depreciation A_m in years	Minimum high category reserves Z_m in mil. tons	$Z_m/Z_c \times 100$ in %
a	500—100	6.0—2.0	26—10	60—20	12—20
b	200—50	3.0—1.4	10—8	30—11	15—22
c	100—10	2.0—0.5	10—5	20—3.5	20—25
d	10—1	0.5—0.1	5—3	3.5—0.5	25—30
e	values have not been determined				

The optimum degree of geological assurance can also be derived from the admissible cost of the verification of one ton of reserves. Kaudelnyi (1957) attempted to determine a rational degree of geological assurance on the Kuzbas deposits from the cost of exploration of industrial reserves. He expressed graphically the relationship between the length of the exploratory drill holes (in m) performed for a thousand tons of $A + B$ category reserves and the degree of geological assurance of the deposit, which he gave as

$$\frac{A + B}{A + B + C_1} \times 100.$$

From the analysis of integral curves (Fig. 165) several important conclusions may be drawn:

The same increments in the drill hole length Δl lead to different increases in the degree of geological assurance of the deposit Δa. During reconnaissance exploration, the increase in the hole length leads to a slight increase of the degree and index n is very high. Later on, at about the beginning of detailed exploration, the same increments in the drill hole length produce its rapid increase and the value of n drops gradually to a minimum at point K. Towards the end of detailed exploration (beyond point K) the degree of assurance increases slightly and the value of n again rises. Economically, point K indicates the optimum degree of geological assurance. The shape and position of these curves vary with the type of deposit but the relationship retains the same general trend. A similar procedure can also be employed in underground exploration. Instead of the length of drill holes or underground workings, expenditures on these works can be used for comparison.

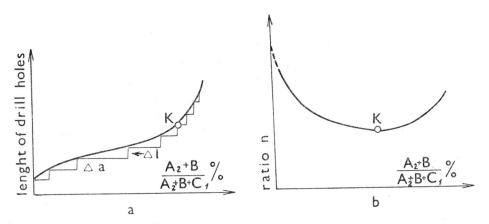

165. a — Curve showing the relationship between the degree of geological assurance and the overall length of drill holes (in m); Δ_a — increase in geological assurance, Δ_l — increase in the length of drill holes; b — curve showing the ratio of drill hole length to 1000 tons of reserves of high categories (n) and to the degree of geological assurance (Kaudelnyi, 1957).

SAMPLING

Sampling of mineral deposits is a very important part of geological exploration. A sample, although it is only a small portion of mineral material taken for laboratory analysis, must be representative, i.e. it must accurately represent all the properties of the material at the sampling site. Another modern method of determining the quality of raw materials is based on examination of their physical characteristics, without sample collecting. These, essentially geophysical methods using luminescence and natural or induced radioactivity, are widely employed. They greatly reduce the expenditure on sampling.

Sampling makes it possible to settle a number of questions as to the qualitative character of a deposit:

1. Determination of the quality of calculated reserves and calculation of the amount of useful components. Delimitation of economic and subeconomic portions of the deposit and delineation of the deposit where its contact with the country rock is indistinct.

2. Compilation of a map showing the quality of raw materials to serve as a basis for planning exploitation and treatment technology.

3. Calculation of losses, impurities and ore recovery.

4. Determination of the quality of broken ore supplied to a dressing plant or a customer.

Chemical and technological sampling are distinguished according to the purpose they are to serve. Chemical sampling comprises various methods of sample collection for analyses; technological samples provide material for technological research. The determination of qualitative parameters of a deposit includes 1. collection of samples, 2. their preparation for chemical assay, 3. chemical assay, 4. treatment of the analytical results and their use for calculation of the reserves. Operations (1) and (2) are related to the geological exploration of a deposit, whereas (3) is of chemico-analytical character and (4) is important in the calculation of mineral reserves.

CHEMICAL SAMPLING

SAMPLING IN UNDERGROUND WORKINGS

There are several methods of sampling in exploratory or mining underground workings. The method is chosen according to the properties of the mineral material and with due regard to technico-economic conditions. The most satisfactory method should ensure that the sample properly represents the deposit and at the smallest cost.

Grab sampling: A sample consists of several pieces of material, 0.5–2 kg in weight, which from visual observation represent the average composition of the deposit. As such a sample is not truly representative, this procedure is not used in systematic sampling.

Chip sampling. A sample consists of pieces taken on a regular grid from the working face, walls or roof of an underground working. Blasted material and broken transported ore can also be sampled in this way (Fig. 166). A piece of material of the same size (approximately the size of a match-box) is taken from each point in the grid. The recommended number of points depends on the variation of the ore: 12 to 15 for uniform to highly uniform deposits, 20 to 25 for non-uniform deposits and 50 to 100 if mineralization is extremely uneven. The shape of the grid is adapted to the morphology and structure of the deposit. The advantage of chip sampling is its high productivity. One man can sample about 50–100 m² of a gallery per shift. Although this procedure is fairly precise, it is recommended that initial results should be checked by, for example, channel sampling (see below).

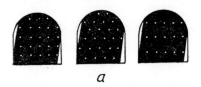

a

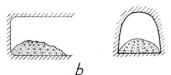

b

166. a — Chip sampling in the face based on square, rhombic and rectangular grids, b — chip sampling of broken ore in the working face.

167. Muck sampling.

Muck sampling is suitable for sampling of broken ore. A sample again consists of pieces taken on a regular grid. At each point of this grid a small pit is sunk through the whole layer of broken ore (Fig. 167). The basic principle is to take equal amounts of fine and coarse material. This procedure can be used only when the broken ore represents the whole thickness of a deposit. The number of sampling points is roughly the same as for the the former methods, as many as 50, if mineralization is very uneven. This sampling method is rapid and cheap.

Channel sampling is the most frequently used for it is very accurate and universally applicable. It consists of cutting a continuous channel in the direction of the greatest variation of the deposit through its whole thickness. For technical reasons, the channel may also be cut along a line making a small angle with the thickness of the deposit. Channel samples are usually collected by hand. The following working procedure should be observed: the exposed rock surface is cleaned and projecting parts are removed; grooves are cut at the borders of a channel; the inner part is excavated keeping a constant channel cross-section. All excavated material is collected and the sampling site is labelled, for example, by a stack with a numbered metal disc. The cross-section of the channel is selected according to the variation in the mineralization quality. In Table 63 the recommended dimensions of the channel cross-sections are given.

TABLE 63

Cross-sections of rectangular channels (in cm)

Character of mineralization	Thickness of ore bodies (in m)		
	2.5	2.5—0.5	0.5
regular	5×2	6×2	10×2
irregular	8×2.5	9×2.5	10×2.5
very irregular	8×3	10×3	12×3

Channel distribution depends on the kind of underground working and on the deposition of the ore body. The bottom and walls of trenches are sampled. In drifts, samples are taken at the heading in the direction of the true, horizontal or vertical thickness. On deposits with random mineralization and small thickness, two to three channel samples are taken at the heading and are grouped into one sample. If there are several ore types at the heading, such as massive and impregnated ores in the hanging and foot walls, sampling is made by sectional channels. Every ore type constitutes a separate sample. Roof channel samples are taken in steeply inclined deposits (Fig. 168). Compared with heading sampling, this technique is advantageous in several respects: it allows a constant spacing of sample points and checking the accuracy of sampling, and the working cycle at the heading is not disturbed by sample collecting. The arched shape of the gallery roof is unsuitable for roof sampling, since the proportions of individual ore types may be distorted and the representativeness of samples reduced. These unfavourable effects can be avoided if the arch is straightened or if the depth of the channel varies along the normal to the periphery of the drift. This method was proposed by Galkin (1950), (Fig. 169). If the ore above the drift is stoped out, samples are taken from the floor. Horizontal wall channels are usually used in cross-cuts (Fig. 170a). Channel samples are occasionally taken in the direction of the true thickness and are shorter than horizontal samples (Fig. 170b). Section channels are generally cut in deposits of medium-to-large thickness, as they allow following of variation in the mineralization in the direction of thickness (Fig. 170c).

Vertical exploratory works, shafts and winzes are sampled by vertical or horizontal wall channels. Raises and offsets are sampled by wall channels in the direction of thickness. In stockworks and massive, relatively homogeneous deposits, channels are oriented arbitrarily.

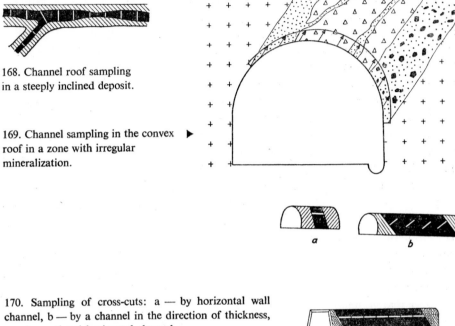

168. Channel roof sampling in a steeply inclined deposit.

169. Channel sampling in the convex ▶ roof in a zone with irregular mineralization.

170. Sampling of cross-cuts: a — by horizontal wall channel, b — by a channel in the direction of thickness, c — by sectional horizontal channel.

The important question of whether or not to sample wall-rock interlayers in a deposit must then be decided. In principle, the sampled thickness should agree with the presumed workable thickness. Therefore, the minimum thickness of unworkable interlayers must be determined by the quantitative and qualitative limits. These are not sampled, as their uneconomical character can be established even macroscopically. Where the contact between a deposit and the wall rock is not sharply defined, the latter is also sampled to determine the amount of impurities.

Channel sampling is laborious and ways of increasing its productivity are sought. On the one hand there is a tendency to reduce the cross-section of samples and, on the other, mechanization of sampling is preferred. Satisfactory results have been achieved on a number of deposits using a channel with a small triangular cross-section and controlling channels of normal rectangular sections. Channel "point" samples along the line of thickness have also been taken, but this procedure cannot be recommended without a very thorough checking by channels of normal cross-sections.

With a channel cross-section of 10×5 cm, one man can sample per shift (Novák 1964):

10 m — in soft rocks (limestone, dolomite, limonite),
6 m — in rocks of medium hardness (siderite, chamosite, hematite),
3 m — in hard rocks (sulphides, quartz, skarn, pegmatite),
1 m — in very hard rocks (quartzite, jaspilite).

Drill-hole sampling. Samples consist of the cuttings from holes drilled by pneumatically operated hammer drills or rotary drills. The reliability of sampling largely depends on catching all the cuttings in special exhausters or in settling tanks, depending on whether air or water is used for flushing. The advantages of drill-hole sampling are: the possibility of taking samples beyond the walls of the underground working, a high labour productivity, fine grade of material (which need not be further treated) and the possibility of accurately maintaining a constant cross-section. On the other hand, the method is inconvenient for the determination of the deposit boundaries and cannot be used in galleries where the deposit shows an inhomogeneous, e.g. banded structure.

Planar sampling is a relatively rare procedure, employed on deposits having a very low content of useful minerals (platinum, gold, cinnabar, cassiterite, wolframite). Samples are taken over a larger area of deposit having a constant thickness of $3 - 5$ cm. In vein deposits of a small width, planar samples can be compared to channels oriented along the dip of the deposit.

In bulk sampling several tons or several tens of tons in weight are extracted. Even here the sample must be oriented correctly to the position and structure of the deposit so that it reliable represents the average composition. This procedure is used mainly for deposits of micas, feldspars, beryl-bearing pegmatite, diamond and other precious stones, platinum and gold. The content of useful mineral is determined from the amount of concentrate obtained by dressing the ore.

SAMPLING OF EXPLORATORY DRILL HOLES

The character of samples from exploratory drill holes depends on the drilling method employed. Exploration of solid raw materials is usually carried out by core drilling. Rotary percussion, auger boring and vibration drilling are less frequently employed.

Sampling of core drill holes. A sufficient core recovery is a prerequisite for satisfactory sampling results. Core recovery varies between 50 and 80 % in many deposits; in others, for example, coal deposits, it drops to zero and it reaches 100 % only in firm, massive ores. Samples are taken from the core, from the cuttings which settle in a bailer, and from sludge which is collected in a sedimentation tank at the collar of the drill hole.

Sampling of drill-hole cores. The core, which has been removed from the core barrel, cleaned and deposited in a special box, is split lengthwise into two pieces;

half is used for laboratory assay and the other half is kept on file for future reference. The core can be split either by hand or mechanically. Using the former procedure, the core is divided into shorter columns, which are halved by a chisel. A mechanical core splitter (Fig. 171) yields better results. Cutting the core with a diamond disc is most accurate. The division of a core should be performed very carefully, as inaccurate splitting leads to the same errors as chang ing the diameter of a channel sample. Albov developed a sampling method using a small-diameter channel. A narrow channel is cut from the core by a disc hardened by metal or diamond. This procedure is still under scrutiny as serious objections have been raised against it. In deposits of medium and great thickness, the core is sampled by subdividing it into smaller sections.

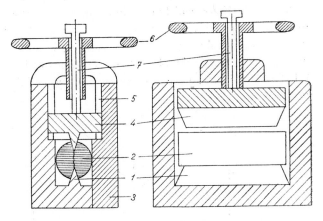

171. A core splitter. 1 — lower bit, 2 — core, 3 — removable part, 4 — upper bit, 5 — slide, 6 — tightening wheel, 7 — driving pin (after Novák, 1964).

Sampling of cutting and sludge. The coarse cuttings settle in a bailer. Sludge is brought to the surface by water, where it is caught in a settling tank. As a rule, the cuttings and sludge must be added to every core run. Since the fragments are brought to the surface with some delay, the drill hole should be flushed after each run until the drilling fluid is clear, or the time necessary for bringing the broken material to the top should be calculated taking into account its size and specific gravity. The cuttings and sludge sample must be cleansed of metallic particles from the drill bit and of shot when the useful minerals are not strongly ferromagnetic. If drilling mud is used, the sample must be freed of mud. A sample of finer-grained cuttings can be taken by quartering, a procedure described below. A sludge sample is used mainly with a minor core recovery, but it is not sufficiently representative. If the core recovery is unsatisfactory, samples can be taken from the hole walls by a lateral core barrel. A set of short core barrels is driven into the wall by explosion of gunpowder.

Poorly cohesive rocks are at present drilled using gas (usually air flushing). The core recovery is generally very good. If necessary, material from a bailer or caught by a filter at the drill hole collar may be used for sampling. Such a sample is representative, as particles are brought up at a high velocity.

Rotary non-core drilling is seldom adopted for exploration of mineralized bodies, but is often used for exploration during mining operations.

Percussion drilling. Rock broken by a chisel is withdrawn by a bailer. The maximum length of individual runs is usually 1 m. The bailer is emptied into a box, from which samples are taken by a sampler after excess water has been carefully poured away (Fig. 172). Each run is analysed separately; only when the deposit has proved to be of fairly uniform quality, are several runs combined into one sample. Percussion drill holes, which have a larger diameter than those drilled by rotary equipment, are particularly convenient for exploration of deposits with an irregular distribution of minerals (stockworks, impregnated deposits, placers).

Auger drilling, widely used in engineering-geological investigation, is employed in economic geology chiefly for exploration of clay, kaolin, bentonite and similar deposits. The length of separate runs is short, about 30 cm. After each run, the helical auger is lifted. Material is cleansed of impurities sticking to the surface and the sample is halved lengthwise; one part serves for analysis and the other is preserved as reference material.

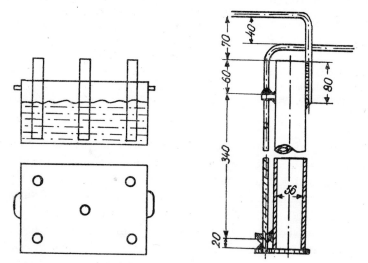

172. Sampling of cuttings obtained by percussion drilling, using a simple sampler (after Smirnov, 1957).

MINERALOGICAL SAMPLING

Mineralogical sampling is essentially determination of the content of useful components on the basis of the mineral composition of the raw material. For this purpose several methods have been developed.

Determination of the quality based on the type of raw material. With some deposits it is possible to define types having a characteristic and relatively stable mineral paragenesis. The average chemical composition of each type is established from the analysis of a sufficient number of samples. The average content in a given part of the deposit (C_p) can be calculated as the arithmetical weighted mean from the average content of the useful mineral in individual mineralogical types and from their thicknesses or surfaces obtained by detailed geological mapping of the deposit (Fig. 173):

$$C_p = \frac{C_1 \times p_1 + ... C_n \times p_n}{p_1 + ... p_n},$$

where C_1, C_n are the average contents of the useful mineral in the individual types of the raw material and p_1, p_n are surfaces or thicknesses of the defined types.

The advantage of this method is the reduction of chemical analyses. Ivanov (1962) used this procedure for the determination of quality on deposits of copper, pyrite, apatite and rare elements. It is also suitable for mica, asbestos, talc and other non-metallic deposits. The use of photography instead of mapping contributes to the refinement of this procedure.

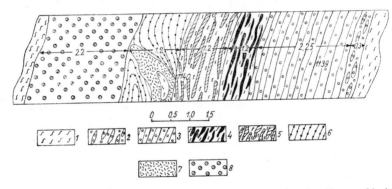

173. Determination of the quality of a deposit on a geological basis. 1 — Quartz-chloritic shales, 2 — faintly banded and unevenly granular Cu ores, 3 — banded Cu ores, enriched with chalcopyrite and sphalerite, 4 — rich impregnated Cu ores, 5 — poor impregnated Cu ores, 6 — banded impregnated pyrite ore, 7 — a quartz lens, 8 — fine-grained and evenly granular Cu ores with a large content of calcite (after Ivanov, 1962).

The planimetric method is discussed in detail in ore microscopy and petrography. The percentage representation of minerals is determined in a thin section or polished section, using a planimetric eyepiece and an integrator. This method is hardly ever used in practice, but effective point integrators permit wider application in special cases.

Determination of the content of useful minerals in artificial heavy mineral concentrates is employed for rapid qualitative evaluation of some types of ore deposits such as auriferous quartz veins and deposits of cinnabar, cassiterite and wolframite. The ore is crushed and heavy minerals are concentrated by washing, elutriation or separation in heavy liquids. The concentrate is then assessed semi-quantitatively or quantitatively under a binocular microscope. This rapid mineralogical sampling allows effective regulation of exploration, since the results are obtained within a very short time.

THE INITIAL SAMPLE WEIGHT AND DENSITY OF SAMPLING

The initial sample must be of the appropriate weight if it is to be a representative one. The choice is influenced by the following factors (Albov, 1961):

1. The initial weight may be smaller on deposits with a regular distribution of useful minerals (massive and banded structures) than on deposits with brecciated and impregnated structures.

2. The coarser the useful minerals, the higher the initial weight of the sample should be and conversely.

3. The larger the amount of mineral grains in a sample, the smaller the initial sample weight can be and the less likely is the error in reducing the bulk of the sample.

4. The higher the specific gravity of a useful mineral, the larger the initial weight of the sample must be.

5. The lower the average content of useful mineral, the larger the initial weight of the sample must be.

The initial weight of a sample should be sufficiently representative, but not too large, since reducing the bulk of a sample for chemical analysis is time consuming and expensive. The necessary weight of samples (Q) is often determined using the Richards-Czeczott formula

$$Q = k \times d^2,$$

where d = the size of the largest grains of useful mineral and k = a constant expressing the qualitative variation of the deposit. Cross-sections of channel samples and their weight calculated on the basis of this formula are tabulated in Tables 63 and 64 (Kreiter, 1961; Albov, 1961) for four groups of deposits.

Group a): Large stratabound sedimentary deposits of Fe, Mn, bauxite, metamorphic Fe-ore deposits with a uniform distribution of the economically valuable component. The variation coefficient of its content does not exceed 70.

Group b): Deposits of magnetite, titanomagnetite, chromite, Cu-bearing sandstone, porphyry ores, polymetallic and copper ores with a less uniform distribution of the economically valuable component. The variation coefficient of the principal useful mineral varies between 70 and 100.

Group c): Major vein deposits of non-ferrous, rare and precious metals and radioactive materials. The variation coefficient ranges from 100 to 150.

TABLE 64

Weights of samples

Group	Character of mineralization	Coefficient k	Sufficient weight of sample (kg) for maximum grain-size (mm)				
			20	10	5	2.5	1.0 mm
a	regular	0.02	8	2	0.5	0.12	0.02
b	irregular	0.1	40	10	2.5	0.6	0.1
c	very irregular	0.2—0.5	80—200	20—50	5—12	1.2—3.0	0.2—0.5
d	extremely irregular	0.5—1.0	200—400	50—100	12—25	3—6	0.5—1

Group d): Small deposits of platinum, diamonds and gold of a very irregular quality. The variation coefficient exceeds 150.

The choice of an adequate initial weight of a sample should ensure that the sample is representative, i.e. that its composition agrees sufficiently with that of the deposit at the sampling site. An adequate sampling density is necessary for determination of the average chemical composition. In principle, more precise average values of the mineral content are obtained by taking a larger number of small samples than with a small number of larger samples.

TABLE 65

Spacing of samples

Distribution of analysed elements	Variation coefficient	Type of deposit	Spacing of samples along the strike of deposit (in m)
regular	5—40	simple deposits of coal, combustible shales, building and cement materials, salts, phosphates, some Fe and Mn ores	50—60
iregular	40—100	hydrothermal, contact-metasomatic, most Cu + polymetallic deposits, some W + + Mo deposits, small part of Au deposits	6—4
very irregular	100—150	some ore deposits, most deposits of tin, W, Mo; many Au deposits	4—2.5
extremely irregular	above 150	many deposits of rare metals, Au, Pt	2.5—2

Note: Values of variation coefficient were calculated from the results of channel sampling.

The density of sample points on a grid consisting of drill holes coincides with the density of the grid, but the number of samples may be higher, when they are subdivided into several portions.

An exploratory grid formed by underground workings following the strike of the body allows for an arbitrary sampling density. A constant spacing of sample points is chosen in underground workings on deposits or their segments having the same qualitative variation. The recommended sampling density based on empirical data is presented in Table 65 (Kreiter, 1961). The optimum spacing of sampling sites in mining exploration of the deposit can be determined by gradually increasing the sampling intervals.

COMBINING BASIC SAMPLES

For determining the quality of raw materials, basic, composite and group samples are employed. Basic samples are collected using one of the methods described above. Composite samples are obtained by combining the material of basic samples before or after quartering. It is recommended that only samples taken by the same technique be combined. Group samples consist of several basic or composite samples and are representative of major parts of a deposit.

Composite and group samples allow 1. the number of analyses and, consequently, their cost to be reduced, 2. integrated chemical analyses, which show the representation of useful, harmful and trace elements in a deposit or part thereof and 3. checking of the calculated average contents of useful components by analysis of an average sample.

Combining of samples is particularly advantageous on deposits where the sampling density is very high (Table 66), as it saves on laboratory analyses. It should be used especially in the advanced stage of exploration, when the variation of deposit conditions is satisfactorily known.

TABLE 66

Number of combined samples

Group	Distribution of metal in ore	Spacing of sampling points (in m)	Number of combined samples
I	very regular	50—15	are not combined
II	regular	15—4	usually not combined
III	irregular	4—2.5	2
IV	very irregular	2.5—1.5	2—3
V	extremely irregular	1.5—1	3—4

Galkin (in Kreiter, 1961) recommended that correlative combination of samples be carried out in addition to combining neighbouring samples from one part of the deposit. Samples are combined on the basis of natural and technological types of raw materials. Within the scope of each type, samples are divided into classes based on the content of the useful component. A group sample then includes basic samples whose mineral content ranges within one class. This method is suitable for the study of the relationship between the principal useful mineral and trace elements (Fig. 174).

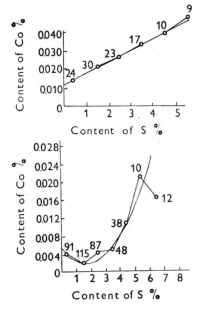

174. Relationship between Co and S contents in pyritized magnetite ores. Example of combining basic samples. Numerals on the graph denote the number of combined basic samples (after Galkin in Kreiter, 1961).

PREPARATION OF SAMPLES FOR CHEMICAL ANALYSIS

The weight of a basic sample varies from several kilograms to tens of kilograms. For chemical analysis about 1 g of material is needed. The final weight of a sample, however, is chosen at $0.5-1$ kg, because a certain number of samples are deposited as duplicates in the chemical laboratory and with the exploration or mining institution. Since samples are generally inhomogeneous, before reducing they must be thoroughly homogenized by milling to a suitable grain-size distribution and by mixing. The grains are thus liberated and distributed evenly throughout the material, but a certain difference between the average chemical composition of the basic sample and a reduced sample remains ("error due to sample reduction").

The bulk of a basic sample can be reduced to weight, Q, which depends on the size of the largest pieces, d, which still enable achievement of the required homogenization. To express this relationship, the Richards−Czeczott formula, $Q = kd^2$, is usually employed; the value of coefficient k is related to the variation in the deposit (Table 64).

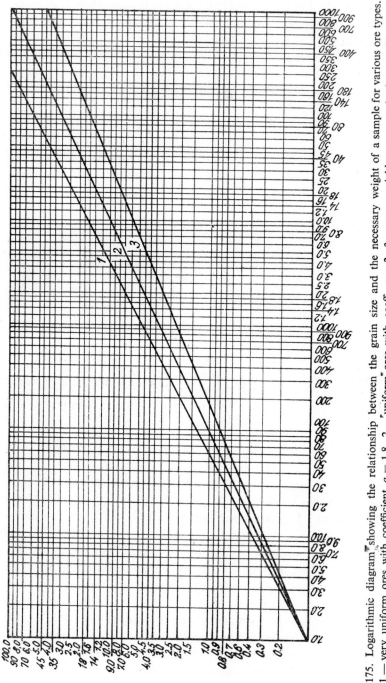

175. Logarithmic diagram showing the relationship between the grain size and the necessary weight of a sample for various ore types.
1 — very uniform ores with coefficient $a = 1.8$, 2 — "uniform" ores with coeff. $a = 2$, 3 — very variable coarse-grained ore with coeff. $a = 2.25$ (after Pozharitskii in Albov, 1961).

Demond and Halferdal define the relationship between the grain-size distribution and the weight to which a sample can be reduced in a more general form: $Q = kd^a$, where the variation of the mineral material is expressed by two independent coefficients — k and a. Pozharitskii gives the values 1.8, 2.0 and 2.25 for coefficient a in deposits with uniform, non-uniform and extremely uneven mineralization, respectively.

Pozharitskii's logarithmic diagram, which expresses equation $Q = kd^2$ in the form of straight lines is very useful. The necessary weight, Q, at maximum dimensions, d, of the pieces can be determined from the diagram with sufficient precision (Fig. 175).

Irrespective of the formula used for the calculation of the weight of the final sample, reduction of samples is performed by the following operations: 1. crushing and milling, 2. sieving and 3. mixing and reduction of sample weight.

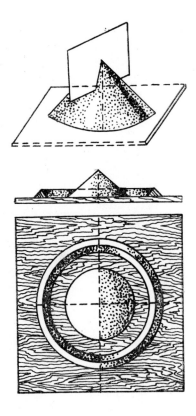

176. Mixing of a milled sample material (after Smirnov, 1957).

Small crushers of various types, such as jaw, hammer, roll and ball mills are employed for crushing and milling. For milling of ceramic materials, ceramic-lined mills with porcelain balls are employed. Twice as many mills as crushers are needed. The milled material is sorted on a set of sieves of adequate mesh size, which determines the dimension, d, of the largest grains in the Richards—Czeczott formula.

334

The milled material is mixed by "coning". Coning consists of piling the ore into a cone, which is then spread into a low, disc-shaped truncated cone (Fig. 176).

The sample is then reduced by quartering: it is divided into four approximately equal portions, and the two opposite portions are combined to give a new sample, the other two being kept as a duplicate. This job is done effectively and rapidly by mechanical samplers.

The process of sample reduction can be carried out to any degree required, each cycle consisting of sorting, crushing or milling, check sorting, mixing and quartering (Fig. 177). At the end of each cycle, the weight of the sample is reduced to Q_n corresponding to grain-size d_n, which can be determined from the mesh:

$$Q_n = k \times d_n^2.$$

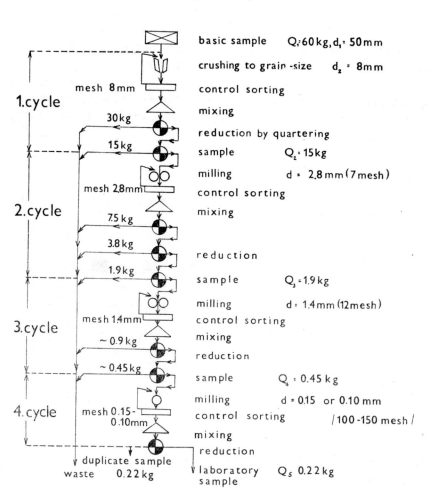

177. Scheme showing reducing of the bulk of sample (after Smirnov, 1957).

Since the calculated weight Q_n cannot be attained precisely, the nearest higher value is considered. Pozharitskii's graph is also a useful tool for reducing the sample volume (Fig. 175).

TECHNOLOGICAL SAMPLING

Technological samples are collected for investigating dressing and processing technology. According to the thoroughness and volume of the technological tests, laboratory, pilot-plant and operational investigations are worked out. The weight and character of the samples depend on the type of technological investigation. The schedule of its individual phases is controlled by the state of the geologico-exploratory works. During the prospecting-exploratory stage, the technological character of normal raw materials can be estimated by analogy with other worked deposits of similar type and on the basis of mineralogical and technological analyses. Laboratory tests are carried out for more complicated types of raw materials.

If raw materials require complicated dressing and processing procedures, technological investigations must be carried out in the full extent during the preliminary exploration stage. Technology then decides whether the deposit can be worked profitably and thus whether detailed exploration is advisable. Unless all problems related to technology and processing are solved fully during the detailed exploration, industrial reserves cannot be delimited.

The weight of technological samples varies greatly, depending on the extent of the investigation. A sample for laboratory analysis should be several to 100 kg in weight, for pilot-plant analysis about 1 ton and up to several tons for investigation at the operational level. The first two are carried out in special research institutes, and the last is most frequently conducted in the dressing plant where a similar raw material is treated or which can be adapted for the proposed technological process. From the rough weights of technological samples it is evident that sampling can be accomplished in different ways. Samples taken from drill cores and channel samples will be adequate for laboratory research. Larger samples can be taken from test pits or underground workings using the planar or bulk method, and samples of several tons weight and of materials that will be extracted underground may be collected only from underground workings.

Technological samples must be taken so that they represent the average character of the material which will be extracted from a major segment of the deposit, i.e. its average mineralogical composition, chemistry, water content, volume weight, mechanical properties, lumpiness, contamination, etc.

Where the sample should have the lumpiness of the ore to be extracted, the same method should be used for both sampling and exploitation. Large technological samples are taken by bulk sampling. Reliable collection of samples requires sound knowledge of the deposit, of the natural and technological types of the raw material and their proportions and distribution in the deposit. Since technological sampling is rather complicated, the co-operation of a technologist and geologist in this job is advisable; in some countries this is incorporated in standards.

CONTROL OF SAMPLING

Chemical analyses which provide a qualitative estimate of the deposit can be imprecise as errors may arise in sampling, in the reduction of samples and in the analysis itself. The magnitudes of these errors in each operation must be determined. Inadmissible errors must be removed, for example, by introducing correction coefficients which would justify the use of the completed analyses.

All types of errors in sampling, sample reducing and analysis can be divided into random and systematic errors. Random errors are either positive or negative so that they cancel out at a sufficient number of values and average values are near reality. The effect of random errors is the greater, the higher they are and the smaller the number of values taken.

Systematic errors have a constant + or − sign so that they do not cancel out and distort the results more than random errors.

The collecting of samples is checked by repeated sampling with the use of more precise though more laborious technique, verified by long practice.

The results of sampling in drill holes are checked by taking samples in underground workings. At the site of the drill hole a channel sample is taken in an underground working and reduced in the same way as the basic sample. A systematic error can also be revealed by study of the core from the deposit, which can show traces of selective breaking-out of softer minerals (cinnabar, galena, sphalerite, calcite) during preservation of hard minerals, such as quartz.

Sampling in underground workings is checked by revising the site of sampling as, for example, the cross-section of channels, the possibility of scatter and loss of friable minerals, and the position of sampling locations with respect to the deposit. Sectors sampled by drilling, chip or muck techniques are checked by channel sampling. Checking of channel samples is done by channels of larger dimensions. The mode of reduction of basic and control samples must be invariably the same.

The reduction of the sample weight is performed under the supervision of a geologist; the remaining portions after quartering are analysed using the same method as in analysing the basic sample.

The quality of chemical analyses is checked by "internal" and "external" methods. In the former case a sample is divided into two parts which, being differently numbered, are sent to the laboratory. Random errors are determined by correlating the results of analyses of the basic and control results. Systematic errors can be established only by "external" control. Samples are sent to another laboratory where analytical methods of a higher standard can be expected. If the differences in the results are too great, control samples should be sent to a third, laboratory. The work of the first laboratory can be controlled simply using reference samples, which are prepared from mineral types of different technological and natural characteristics and are analysed in several laboratories. Average values computed from the analytical results are of high precision. The material of the reference samples is then sent to the basic chemical laboratory at discrete time intervals to check the reliability of its work. Reference samples used in exploration are analogous to standards employed in checking silicate analyses.

In checking sample collection by any method care must be taken that all natural and technological types of mineral material are represented by groups of control samples and that the pairs of basic and control samples can be statistically evaluated. The more irregular and complicated the mineral material, the larger the number of control samples needed. About 20 analyses are thought to be the lower limit for statistical evaluation. The control must be scheduled adequately.

DETERMINATION OF THE QUALITY OF RAW MATERIALS BASED ON PHYSICAL PROPERTIES

Progressive methods are under development to enable the determination of the quality of raw materials without sampling; measurements are conducted directly in bore-holes, mine galleries or surface outcrops. The potentialities of these procedures are steadily increasing as new methods are developed and new devices are constructed.

Geophysical logging allows oil and gas-bearing beds and their properties to be determined. Coal seams in a drill hole can also be identified by these methods and some logging curves are used for the determination of ash, sulphur and volatiles contents.

Reliable magnetic methods are effective in Fe determination in magnetite ores. Measurements are carried out using a kappameter on samples or by the induced magnetization method; in the latter case the induced and natural vertical components of the magnetic field are measured in a mineralized block. The Fe content is directly proportional to the magnetic properties.

Radioactivity methods are widely used on deposits of radioactive minerals. Measurements of natural radioactivity are made in drill holes, galleries or in transported broken ores.

Gamma-gamma ray logging is based on the absorption of soft gamma rays by elements with a high atomic number (Sb, Ba, Pb, W, Hg and others). The quality is determined by recording scattered gamma radiation after irradiation of the ore by Cs^{137}. Sampling of monomineral ores yields the best results.

A number of new instruments have been constructed, such as the X-ray fluorescence analyser employed for the determination of Sn, Sb, Pb, Zn, Ba and other elements. Cd^{109}, Co^{57}, Ce^{130} and Pm^{147} isotopes are used as radiation sources. Another very sensitive instrument is the photoneutron beryllometer using Sb^{124} as a radiation source. It has been developed for the determination of Be in monomineral ores. The Sn analyser is based on the so-called Mossbauer effect; radioactive isotope Sn^{119} in SnO_2 is used as a radiation source.

Using the above analysers, elements can be detected in amounts ranging from hundredths to thousandths of a per cent. The relative error varies within a rather wide range, and can amount to about 10 per cent.

There is little doubt that conventional sampling will be increasingly replaced by "geophysical sampling" methods in the future, but for the present they have to be checked thoroughly against normal sampling and chemical analyses.

GEOLOGICAL DOCUMENTATION OF MINERAL DEPOSITS

The term geological documentation of mineral deposits comprises a complete summing up and recording of all data assembled during geological exploration and exploitation of a deposit. The importance of geological documentation follows from the nature of geological exploration, which often synthesizes an immense amount of data collected gradually over a long period of time. Without adequate geological documentation, exploratory works lose their significance and much costly exploration must be repeated as a result of unsatisfactory or incomplete documentation. Geological documentation preserves the results of geological exploration for the future, just as archive records of past explorations and mining supply us with useful information. Geological documentation is of two kinds:

1. basic geological documentation, including written, graphic records and rock samples gathered during field and technical works;

2. final geological documentation, comprising basic documentary materials in a final form and definitive elaboration of exploration data.

According to the mode of presentation, documentation is written, graphic, or consists of rock samples.

Graphic documentation of trenches, pits, working faces and boring cores is carried out at a sufficiently large scale — 1 : 20 to 1 : 100. Geological mine maps are constructed at scales of 1 : 200, 1 : 500 and 1 : 1000. Geological mine maps and exploratory sections are more deailed than surface maps but since they cover relatively small areas, they are unsuitable for the solution of stratigraphic, magmatic and tectonic problems on a wider scale. Consequently, the detailed data obtained by exploratory and mining works must be combined with the results of surface mapping.

BASIC GEOLOGICAL DOCUMENTATION
OF EXPLORATORY WORKS

Exploratory trenches and pits are documented by plotting the geological features of one wall or, if the geology is very complicated, by plotting them on the developed surface of the exploratory work. The documentation of pits with circular sections is more difficult. The most suitable method is to plot the geology in two profiles perpendicular to each other; the strike and dip of structural elements α and β are easily determined using a simple construction (Fig. 178). The documentation of a heading provides the most detailed picture of the composition, structure and texture of the

340

deposit and of the wall rock (Fig. 179). The headings are plotted on special forms, which also contain columns for recording the results of chemical analyses. Plotting is made at regular intervals, usually at the sampling locations.

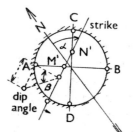

cross sections A B, CD

178. Geological log of a vertical pit by two perpendicular sections and the construction of the strike and dip of beds (after Gerasimenko, 1958).

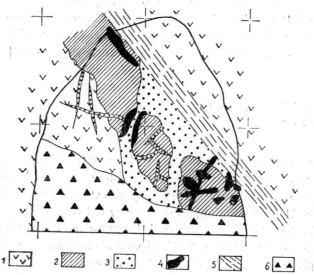

179. Graphic log of the face.
1 — decomposed andesite,
2 — vein quartz, 3 — carbonate,
4 — antimonite, 5 — fault,
6 — broken ore.

UNDERGROUND GEOLOGICAL MAPPING

Underground geological mapping, i.e. plotting of the geological picture of the roof and walls of the work on a mine map, is the fundamental working method of the economic geologist. Other exploratory methods, such as geochemical, mineralogical, petrographic and small-tectonic research, can be interpreted reliably only in relation to the fundamental data which are depicted on the geological-mine map. According to the mode of deposition of the ore body and to the type of mine working, various mapping procedures are used.

1. Mapping on a horizontal plane:

a) The geological features of the roof are projected onto a horizontal plane; the projection is usually complemented by sketches of the headings. This method provides an accurate reduced picture of the geology of the roof without using extra-

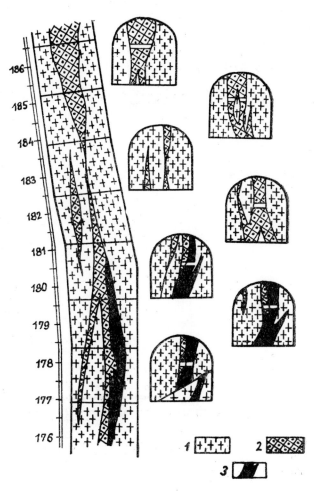

180. Geological logs of a drift roof and of faces with plotted face channels (after Smirnov, 1957). 1 — wall rock, 2 — impregnated wall rock, 3 — high-grade ore.

polation (Fig. 180). This is advantageous, for example, if the locations of channel samples are to be shown precisely on the map. A drawback to this method is the distorted course of the structural elements owing to the convex shape of the roof. This deficiency can be corrected by an alternative mapping procedure on a horizontal plane.

   b) Mapping on a fictitious horizontal plane. This method has been described by many authors (e.g. Forrester, 1946). The geological situation is plotted on a horizontal plane, which is generally selected at shoulder height (about 150 cm). The geological

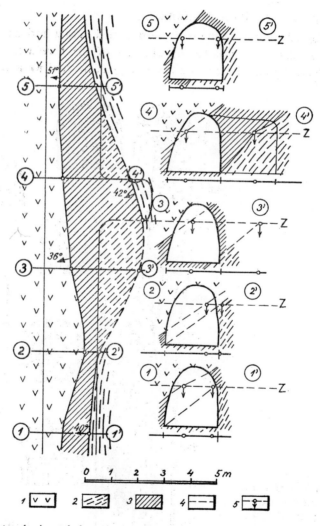

181. Mapping into a horizontal plane. 1 — porphyroids, 2 — chloritic slates, 3 — deposit, 4 — position of projection plane, 5 — key points of structural elements and the direction of their projection into ground plan.

pattern of the walls and roof of a gallery is drawn on the same scale as the map. All tectonic features and geological boundaries are extrapolated or interpolated to determine their intersections with plane Z; they are then projected onto a horizontal plane, in which the walls are plotted. This mapping method is shown in Fig. 181, in which for clarity the geological situation is also plotted in vertical sections and transferred from them onto the horizontal plane. An experienced geologist puts the geological features directly on a mine map without constructing auxiliary sections. These are used only for mapping sectors with complicated geology.

Horizontal or moderately inclined deposits and rock complexes cannot be mapped by this method. Projection onto a horizontal plane is also unsuitable for mapping inclined mine workings.

2. Mapping on a vertical plane is employed in these cases and requires a simple modification of the mine map. A stripe is drawn beside the ground-plan of a gallery, which represents its turned down wall. The set of sections of the gallery is plotted into the map (Fig. 182). If the geological section is complicated, both sides of the gallery are drawn in the same way.

A combined mapping method is used on deposits with a very intricate geological structure. The geological situation is plotted on a developed area of a gallery. Using this method, stripes are drawn beside the ground plan representing the two walls of the gallery, tilted into the roof plane. The combined mapping method requires a sufficiently large scale. It shows the geological setting in the greatest detail but is very time-consuming (Fi g 183).

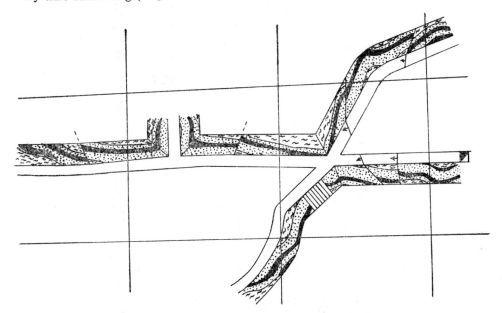

182. Geological underground mapping into a vertical plane (after Pouba, 1959).

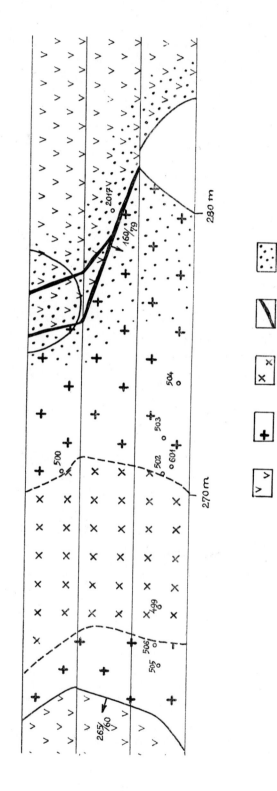

183. A cross-cut mapped by a combined method. 1 — andesite, 2 — pyroxenic diorite porphyrite, 3 — pyroxene-hornblende diorite porphyrite, 4 — ore vein with antimonite, 5 — propylitization; numerals denote sampling sites.

Technique of underground mapping

Rapid and successful underground mapping demands adequate equipment for working in mines. The basis for mapping is a plan of individual mine levels on a suitable scale where a co-ordinate grid, the astronomic and magnetic north and the altitude of the horizon above sea level are denoted. A geologist will further need a map-sheet holder, a notebook, sample bags and labels, a geological compass, a cloth tape, pencils, coloured pencils, an eraser, a protractor, a hammer, a chisel, and a lamp.

The working procedure can be as follows: reconnaissance of the sector to be mapped, washing of the walls, if possible, measuring out of the gallery into segments 2—5 m long (measurement should be checked on survey points, crossings of drifts etc.) and mapping. Mapping of large segments is carried out preferably by two workers—a geologist or technician with a helper.

The geologist plots all the geological elements in a correctly oriented map: contacts of rocks, the course of structural-tectonic features, etc. Proceeding along the gallery, he indicates these features by symbols chosen for the legend and records their strikes and dips. Hydrogeological observations, such as sites of scattered and concentrated water inflows and their discharge, and some technical data (lining or timbering, worked-out spaces) are also recorded. For new unmapped sectors of galleries or old abandoned mines a provisional base map has often to be constructed. A traverse is drawn along the centre line of the gallery, using a tape or cord; azimuths are measured using a compass. In measuring two principles are observed: a) the compass scale should always be set at zero in the direction of measurement; b) azimuths of the polygon sides are recorded on the map clockwise from the north. A hanging miner's compass provides more precise results. The legend to the map should be compiled with great care before mapping begins. The legend is prepared to a detail corresponding to the map scale; it should express all relevant data on the geological setting, deposit bodies and tectonics.

The working map sheet is retraced with india ink every day after completion of the day's mapping and is then redrawn on the original.

The most convenient scales for underground mapping range from 1 : 100 to 1 : 1000, depending on the intricacy of the geological conditions. If general maps of a smaller scale are needed, they are constructed by reducing maps of a larger scale; certain generalization and wider extrapolation of the geological situation are necessary (Fig. 184).

THE USE OF PHOTOGRAPHY IN UNDERGROUND MAPPING

Mapping of underground workings is time-consuming. Mapping of a 150—200 m long gallery at a scale of 1 : 500, including preparatory works, takes one shift. Therefore, photography has recently been introduced for this purpose, which has also the advantage of eliminating the subjective factor. For underground photo-

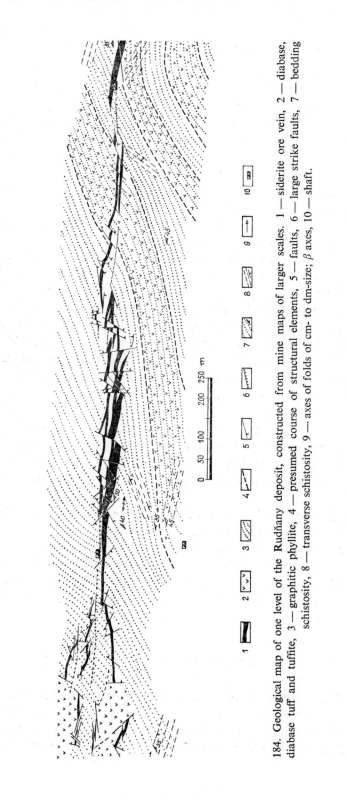

184. Geological map of one level of the Rudňany deposit, constructed from mine maps of larger scales. 1 — siderite ore vein, 2 — diabase, diabase tuff and tuffite, 3 — graphitic phyllite, 4 — presumed course of structural elements, 5 — faults, 6 — large strike faults, 7 — bedding schistosity, 8 — transverse schistosity, 9 — axes of folds of cm- to dm-size; β axes, 10 — shaft.

graphy a reflex camera with short focal length, a negative size of 6 × 9 cm, the use of flash and a film sensitivity of 17/10 DIN are recommended. Photographic material that ensures the best visual discrimination of the deposit from the wall rock should be chosen. Colour photography is convenient for some ore deposits, but usually only developed negatives are used for evaluation. The wall to be photographed is subdivided into a grid of points (e.g. 1 × 1 m apart). The photograph is taken by hand with an open shutter and using flash. The whole procedure takes about 3 minutes.

DOCUMENTATION OF EXPLORATORY DRILL HOLES

Documentation of drill holes depends on the drilling method. Core drilling provides the best results and is therefore used most widely in exploration of solid mineral materials.

Logging of cored holes up to 30 m deep sunk for geological mapping purposes is conducted by a technician at a scale of 1 : 25 or 1 : 50. The section is plotted on a printed form and samples of bedrock are collected.

Deep drill holes are logged in more detail; the records may differ depending on the kind of raw material but the following information should be included in the log: designation of the drill hole, generally after the locality, the serial number of the hole, the district or cadaster, in underground drill holes — the name of the mine, the co-ordinates of the drill hole collar, the dates of the beginning and end of drilling, the names of persons responsible for drilling and logging, the drilling method, the initial and final diameter of the drilling bit and other technical data necessary for assessment of the drill hole, the casing, angle drilling, the kind of drilling fluid, cementation or plugging; the strike and dip of the hole, its length, the co-ordinates of the footwall and other relevant points, the length of runs, the core recovery from individual runs (per cent), records of geology, stratigraphy, petrography, a drawn section, the thickness of the described beds and tectonic data.

The results of hydrogeological and inclinometric measurements, the geophysical survey, chemical and technological analyses, sample sites and gas emanations are also recorded.

The scale of graphic plots depends on the complexity of the geological section and depth of the drill hole. A scale of 1 : 100 is convenient for drill holes of medium depth if the geological conditions are complicated.

Reliable documentation requires correct deposition of the core in a sample box and precise depth records (Fig. 185). An exploratory drill hole is logged in written and graphic form. The positions of geological boundaries are measured along the hole axis with a measuring tape from the beginning or end of a run. The dips of contacts and other tectonic elements are measured either by an inclinometer of the compass, if the core is in vertical position, or by a protractor with a movable strip. In this way, the α' angle ("core dip") of the structural element is determined (Fig. 186);

the true dip $\alpha = \alpha' +$ zenith angle γ, measured by an inclinometer. The orientation (azimuth) of a dip line is determined either from the total geological disposition or — in important cases — using oriented core withdrawal.

from 325.94   to 325.94   from 325.21

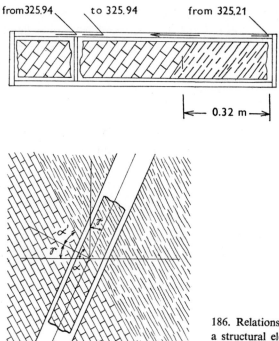

185. Determination of the position of a structural element on the axis of the drill hole. Boring core showing the contact of two rock types, deposited in a core box. Depths of core runs denoted by numerals.

|← 0.32 m →|

186. Relationship between the true and core dip of a structural element. α — true dip, α' — core dip, γ — zenith angle of the hole axis.

If the core recovery is unsatisfactory, some sludge from a bailer and settling tanks at the collar of a drill hole should be used for documentation. In logging a drill hole, the core recovery is systematically assessed from individual runs. If the core is broken and incomplete, the pieces are arranged so that the core volume corresponds to the volume of the hole as closely as possible. The length of the core (l_c) is measured with a tape and the length of a run (l_r) is calculated as the difference between its beginning and end. The core recovery is given by the relationship

$$v = \frac{l_c}{l_r} \times 100.$$

A more precise method for calculating the core recovery was proposed by Petrík and Zeman (1960). The theoretical length of an undisturbed core l_c in a core barrel of radius r for material of unit weight o is calculated on the basis of the weight of material (M) obtained from one run:

$$l_c = \frac{M}{\pi \times r^2 \times o}.$$

If the length of a run is l_r, all necessary parameters for calculation of the core recovery are available.

Documentation of percussion drill holes based on the cuttings lifted by a bailer is far less reliable. Percussion drilling with a reverse circulation provides a better opportunity for documentation. The transporting power of an ascending water current is large and rock fragments are collected on sieves with a very small delay i.e. about several minutes.

Auger and vibration drilling enables relatively precise logging to be carried out. The rotary system employs mainly geophysical methods for determining the lithological section. Only short sectors are explored by core drilling at regular intervals of $20-50$ m.

At present, photographic cameras and television are often used for logging of exploratory drill holes. Since new technical means for geological documentation contributes to precision and economy, they are worthy of attention.

DOCUMENTARY ROCK MATERIAL

In addition to graphic and written documentation of exploration and mining works, rock material should also be collected and stored, i.e. mineralogical, petrographic, palaeontological samples, duplicates of chemical samples and drill cores. These samples are deposited until the calculated reserves are certified. A special commision then decides which material is of permanent value and should be preserved. The samples, arranged and numbered, are of great importance in the study of a deposit; they represent a collection of standards with which new samples may be correlated systematically. Exceptional attention should be paid to the collection of unique mineralogical, petrographic and palaeontological specimens of which the deposit is a type locality. Samples sent for chemical analysis are documented by duplicates, in writing (records of analyses, on punched cards) and graphically (in sketches of headings, maps and sections).

The storage of core samples varies with the kind of drill hole. Samples from reference and deep (so-called structural) drill holes, are usually all preserved. Samples from reconnaissance and exploration holes on a deposit are discarded in two stages. In the first stage after logging and sampling the drill hole, the whole core from the deposit, the hanging wall and foot wall, all rock types and rocks which could be used as reference materials are stored. In the second stage after the exploration works are completed, only samples of rocks constituting the stratigraphical section and at least one section through the deposit and typical parts of all drill cores from the deposit are stored. Storage of samples is expensive and requires a large storing space.

TABLE 67

Determination of the apparent dip of structural elements in geological sections

True dip of the bed	Angle between the strike of the bed and the direction of the section							
	80°	75°	70°	65°	60°	55°	50°	45°
10°	9°51'	9°40'	9°24'	9°5'	8°41'	8°13'	7°41'	7°6'
15°	14°47'	14°31'	14°8'	13°39'	13°34'	12°28'	11°36'	10°4'
20°	19°43'	19°23'	18°53'	18°15'	17°30'	16°36'	15°35'	14°25'
25°	24°48'	24°15'	23°39'	22°55'	22°0'	20°54'	19°39'	18°15'
30°	29°37'	29°9'	28°29'	27°37'	26°34'	25°18'	23°51'	22°12'
35°	34°36'	34°4'	35°21'	32°24'	31°13'	29°50'	28°12'	26°20'
40°	39°30'	39°2'	38°15'	37°15'	36°0'	34°30'	32°44'	30°41'
45°	44°34'	44°1'	43°13'	42°11'	40°54'	39°19'	37°27'	35°16'
50°	49°34'	49°1'	48°14'	47°12'	45°54'	44°17'	42°23'	40°7'
55°	54°35'	54°4'	53°19'	52°18'	51°3'	49°29'	47°35'	45°17'
60°	59°37'	59°8'	58°26'	57°30'	56°19'	54°49'	53°0'	50°46'
65°	65°40'	64°14'	63°36'	62°46'	61°41'	60°21'	58°40'	56°36'
70°	69°43'	69°21'	68°49'	68°7'	67°12'	66°8'	64°35'	62°46'
75°	74°47'	74°30'	74°5'	73°32'	72°48'	71°53'	70°43'	69°14'
80°	79°51'	79°39'	79°22'	78°59'	78°29'	77°51'	77°2'	76°0'
85°	84°56'	84°50'	84°41'	84°29'	84°14'	83°54'	83°29'	82°57'
89°	88°59'	88°58'	88°56'	88°54'	88°51'	88°47'	88°42'	88°35'

Apparent dip of the bed →

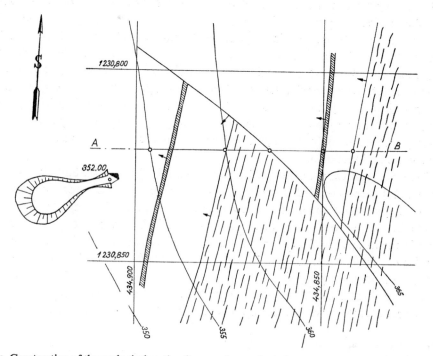

187a. Construction of the geological section from a mine geological map. a — surface geological map.

40°	35°	30°	25°	20°	15°	10°	5°	1°
6°28′	5°46′	5°2′	4°15′	3°27′	2°37′	1°45′	0°53′	0°10′
9°46′	8°44′	7°38′	6°28′	5°14′	3°33′	2°40′	1°20′	0°16′
13°10′	11°48′	10°19′	8°45′	7°6′	5°23′	3°37′	1°49′	0°22′
16°41′	14°58′	13°7′	11°9′	9°3′	6°53′	4°37′	2°20′	0°28′
20°21′	18°19′	16°6′	13°43′	11°10′	8°30′	5°44′	2°53′	0°35′
24°14′	21°53′	19°18′	16°29′	13°28′	10°16′	6°56′	3°30′	0°42′
28°20′	25°42′	22°45′	19°31′	16°0′	12°15′	8°17′	4°11′	0°50′
32°44′	29°50′	26°33′	22°55′	18°53′	14°30′	9°51′	4°59′	1°0′
37°27′	34°21′	30°47′	26°44′	22°11′	17°9′	11°41′	5°56′	1°11′
42°33′	39°20′	35°32′	31°7′	26°2′	20°17′	13°55′	7°6′	1°26′
48°4′	44°47′	40°54′	36°14′	30°29′	24°8′	16°44′	8°35′	1°44′
54°2′	50°53′	46°59′	42°11′	36°15′	29°2′	20°25′	10°35′	2°9′
60°29′	57°36′	53°57′	49°16′	43°16′	35°25′	25°30′	13°28′	2°45′
67°22′	64°58′	61°49′	57°37′	51°55′	44°1′	32°57′	18°1′	3°44′
74°40′	73°15′	70°34′	67°21′	62°43′	55°44′	44°33′	26°18′	5°31′
82°15′	81°20′	80°5′	78°19′	75°39′	71°20′	63°15′	44°54′	11°17′
88°27′	88°15′	88°0′	87°38′	87°5′	86°9′	84°15′	78°41′	44°15′

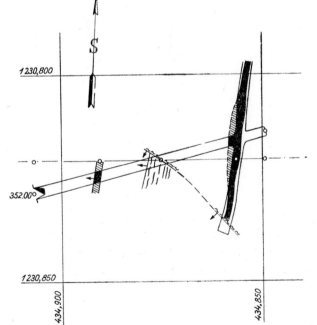

187b. Geological map of an exploratory drift.

CONSTRUCTION OF GEOLOGICAL SECTIONS

Geological sections of a mineral deposit are constructed on the basis of data obtained by underground mapping and drill hole logging. The exploratory workings are generally distributed so that the largest number lie in the plane of a geological section. The construction of geological cross-sections of deposits explored by underground workings differs in detail from their construction in deposits examined by drilling.

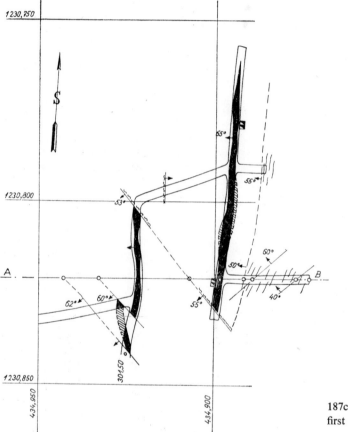

187c. Geological map of the first level (301.50 m a.s.l.).

1. The construction of a cross-section through mine workings will be shown using the example of a vein deposit located near the tectonic contact between porphyroids and chloritic schists (Fig. 187a—e). The deposit was explored on levels 1 and 2 and by a short trial drift and the outcrop is plotted on the geological surface map. In this case the scale of all map sheets is the same, which in reality is very rare. Generally, it is necessary to bring the different scales of maps to the scale of the geological section.

In constructing a section from a map, the section lines in geologically complicated parts should be drawn through as many exploratory works as possible. If several sections of a deposit are to be constructed, their spacing should be roughly the same. The course of structural elements is interpolated and extrapolated on all maps up to

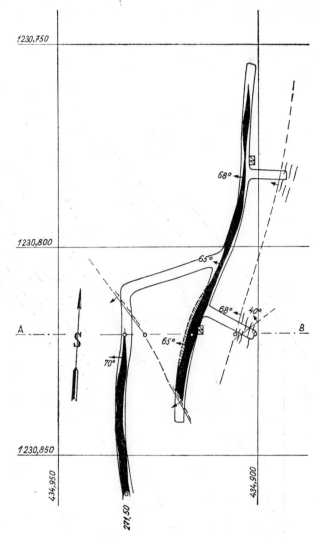

187d. Geological map of the second level (271.50 m a.s.l.).

the section line. All key points (mine workings) and geological features which the section line cuts across are transferred from the maps into the section at the given elevation above sea level. The mine workings located near the section line are projected onto it and depicted by a dashed line. Key points of the same geological unit are connected at all levels and on the surface. Structural features that are encountered

354

only at one point are plotted in the section according to their dip. If their strike is not perpendicular to the section line, an apparent dip is plotted. This can be found in tables (Table 67) or nomograms. The construction of geological sections is not only a matter of geometry; it also requires geological knowledge of the deposit in order to reconstruct correctly those parts of the section which do not enable unequivocal interpretatiton.

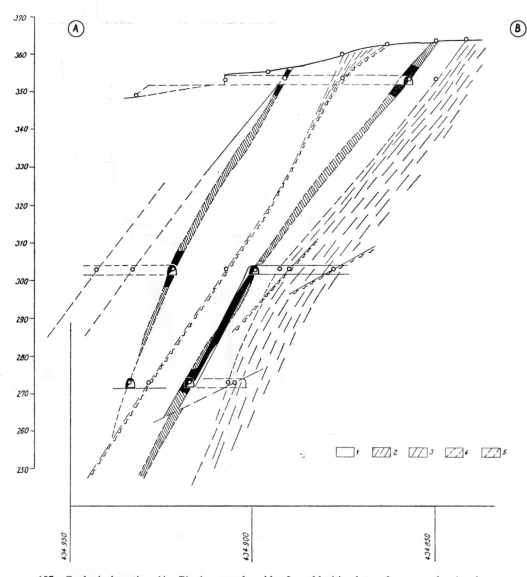

187e. Geological section (A—B). 1 — porphyroids, 2 — chloritic slates, 3 — ore vein, 4 — interpolated and extrapolated segments of the ore vein, 5 — faults with clay gouge and a mylonitized zone, faults without clay gouge.

2. The construction of a geological section of a brown-coal deposit from a row of a drill holes is shown in Fig. 188 (Pekárek 1960). The process is quite simple: a) plotting of the section line on the map; b) drill holes whose distance from the section line is not greater than half the spacing between drill holes are also projected on the section line; c) orientation of the profile, uniform for the whole system of sections; d) construction of the topographic profile; e) plotting of the position of drill holes and the geological situation from basic documentation; f) drawing of the key horizons and the course of other beds.

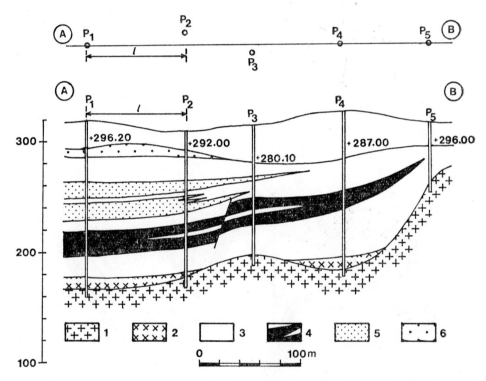

188. Construction of a geological section from boring data. 1 — crystalline complex, 2 — kaolin, 3 — clay, 4 — coal, 5 — sand, 6 — gravel (after Pekárek, 1960).

3. The construction of a geological section from a row of exploratory drill holes in a folded sedimentary complex is more difficult. The method developed by Skok (1959) advantageously uses measurement of bedding along the whole axis of a drill hole. This method can be employed where the thicknesses of beds at the site of the section are constant and changes in the dip of beds are not sharp, but gradual. The principle of construction is shown in Fig. 189.

The MN line is the axis of the drill hole. The dips of a bed, α and β, were measured at points A and B. The true thickness of the bed should be determined in the segment

356

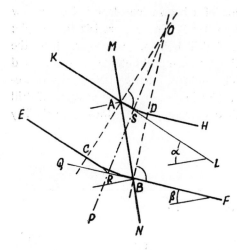

189. Construction of a boring section using the bisector method based on the measurement of two angles of dip (after Skok, 1959).

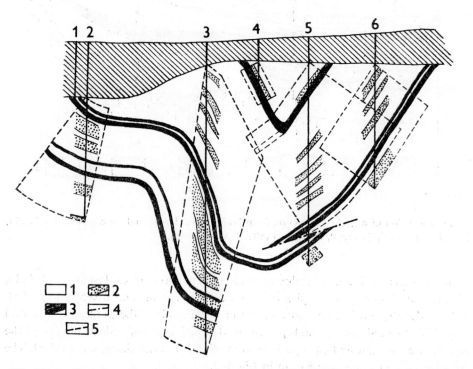

190. Construction of a section along an exploratory line using the bisector method. 1 — overburden, 2 — sandstone, 3 — coal seam, 5 — limitation of areas adjoining the axis of the drill hole (after Skok, 1959).

between points A and B and the course of it should be drawn. A normal to the line of dip KL is constructed at point A and to the line of dip QF at point B. The two normals meet at point O. The points where the dip of the bed changes are points R and S, at which the bisector $(O - P)$ of the angle formed by two normals intersects the lines of dip. The change in the dip is gradual and can be indicated by arcs circumscribed from point O. This construction is suitable only for areas adjoining the axis of the drill hole, which are limited by the upper boundary of the folded complex determined in the drill hole, by the depth of the drill hole and the line of dip as measured in the hole (Fig. 190). The key horizons and coal seams between the individual drill holes are then connected by a continuous line. As seen from Fig. 190, the profile is constructed on the principle of geological interpretation, using the measurement of all structural elements to the greatest degree. Application of linear interpolation to connect the key horizons established in drill holes by straight lines, regardless of the dips of beds, would lead to distorted results.

SPATIAL ILLUSTRATION AND MODELS OF MINERAL DEPOSITS

Mine maps, vertical sections, borehole logs and the geological surface map make it possible to illustrate a mineral deposit spatially in the form of block diagrams. Perspective, axonomic, affine and vector projections are used for the construction of block diagrams. At present mechanical equipment makes possible the drawing of the spatial projection of a geological body from its plan and front view.

Models of mineral deposits are also highly instructive. A great variety of techniques is used in their construction. The most widely used models illustrate the geological situation by means of horizontal, vertical or combined sections drawn on perspex or glass plates. Models made of plastic substances are less frequently employed. Mineral deposits explored by vertical drill holes or pits can be shown by a peg model. Pegs illustrate vertical holes and intersections with the deposits are interconnected by coloured threads.

HYDROGEOLOGICAL AND ENGINEERING-GEOLOGICAL INVESTIGATIONS OF MINERAL DEPOSITS

The study of the hydrogeological and engineering-geological properties of solid economic minerals is an important part of the integrated investigation of deposits in prospecting, exploration and exploitation. The objective of such studies is to determine data necessary for the drainage of deposits, for forecasting changes in the ground-water regimen in their environs (with respect to the utilization and protection of water sources) and for the design of mining works. Hydrogeological and engineering-geological investigations should therefore be carried out during early exploration.

Since the hydrogeological and engineering-geological conditions control the extent and methods of hydrogeological and engineering geological investigations as well as the possibility, method and economy of exploitation, the basic criteria of the hydrogeology of solid mineral deposits are discussed in detail below:

Every mineral deposit should be assessed from two points of view: 1. hydrogeological conditions, 2. mining-technical drainage conditions.

The factors characterizing the hydrogeological conditions are as follows:

a) boundary hydrological conditions (ground-water table, proximity to a river, inflow from other aquifers);

b) the hydrogeological character of rocks considering their permeability (impermeable rocks, rocks with porous or fissure permeability, karstified rocks and their distribution);

c) the tectonic structure of wall rocks in relation to the hydrology (the character of the perviousness of faults, their distribution and possible water content);

d) the presence of drinking, industrial or mineral water within reach of the deposit drainage (effects on existing or future ground-water sources).

With regard to mining-technical drainage conditions, deposits are characterized on the basis of the following factors:

a) rate and regimen of ground-water inflow,

b) physico-chemical properties of subsurface water (corrosive, thermal, chemically polluted, etc.),

c) engineering-geological properties of rocks in the deposit,

d) method of deposit drainage.

HYDROGEOLOGICAL INVESTIGATION OF VARIOUS TYPES OF DEPOSITS

1. *Deposits in loose, unconsolidated, non-cohesive rocks*

In exploration and working of deposits situated predominantly in loose, unconsolidated rocks, drainage and control of running sand is of primary concern. Running sands occur especially in coal deposits. The coal seams are often under- or overlain by clay beds, which prevent the water from percolating through. Where they are of small thickness and especially if the water is confined, it breaks through into the mining works, often with a large amount of running sand. Hydrogeological investigation is directed at determining the principal parameters of water-bearing beds and at locating the sites of potential water inrush. Points to be considered include drainage of the rock complex, reduction of hydraulic pressure to a safe height and connection between the individual aquifers. In surface deposits, the uplift and velocity of ground-water flow causing the liquefaction of sands and piping are determined. The maximum permissible angles of slopes are proposed as well as the necessary lowering of the ground-water table to maintain their stability. The basic permeability parameters of aquifers should be established for the computation of inflow and the drainage design.

2. *Deposits in fissured solid rocks*

In mineral deposits of this type, heavy flows of fissure waters are of primary importance. If the deposit is situated at a greater depth, the increased temperature and often strongly corrosive character of the fissure water must be considered, as it is then deleterious to both the building materials and the rock itself, which promptly loses its strength. After tapping a major fracture, the inflows are at first considerable, but decline with time and occasionally may gradually cease altogether. Tiny hair-thin joints of weathering and other origin are usually of no practical importance. The amount of water pumped from these deposits is generally not large. Hydrogeological investigation is especially concerned with the water-bearing capactiy of major faults and the chemistry, temperature and corrosive character of the water. It should also be determined whether the major fissures extend to the surface and whether they are accessible to surface water. The possibility that working by caving methods could produce fractures extending to the surface at the sites of water courses or reservoirs should also be considered. A secondary increase in the water inflow into exploited parts of the deposit would then be expected.

The bulk of water is drained from deposits of this type into collecting shafts and is pumped to the surface. Special drainage of the rock itself is carried out only occasionally.

3. *Mineral deposits in karst areas*

In karst areas, the main water inflow into deposits is derived from fissures and karst cavities. When they are tapped, large inflows, often in catastrophic amounts, gush into the deposit but subside in the course of time. The total water inflow depends on the amount and distribution of precipitation, the karstification of the carbonate rocks and the depth of the deposit below the base of erosion.

Hydrogeological investigation is directed at 1. the determination of faults and their water-bearing capacity; 2. determination of the occurrence, depth and presence of water in karst cavities; 3. measures to be taken for reducing water inflows and 4. discovering places where there is a hazard of water inrush and proposals for controlling it. It is important to decide where the water should be discharged to prevent its flowing back into the deposit. Working over the largest possible area at approximately the same depth is usually recommended.

4. Salt deposits

Owing to their high solubility, salt deposits have been preserved only where they were buried by impermeable rocks. The most dangerous are aquifers overlying the deposits, which must be crossed in mining.

Hydrogeological investigation is related to 1. the determination of the parameters of aquifers in the neighbourhood of the deposit, 2. the choice of mining technique, 3. the establishment of changes in the ground-water regimen and 4. the determination of the thickness of salt safety pillars. Salt deposits should never be drained! The invasion of water can be prevented by cementation of the deposit, back-filling of worked-out spaces, or construction of hydraulic closures and other sealing methods. In boring exploration, normal drilling fluid can occasionally be used outside the deposit when the leaching of salts is not imminent. In the deposit itself, a salt-water-based drilling fluid must be employed. After the termination of exploratory work, the holes are liquidated by careful plugging with cement under pressure. During mining it is recommended to leave safety pillars around the boreholes.

5. Water, oil and gas deposits

Prospecting for and exploration and exploitation of water, oil and gas deposits, as well as calculation of their reserves, are carried out by special procedures depending on their physical properties. The methods used have been amply described in the literature and the subject has developed into a separate scientific branch – the geology and hydraulics of oil and gas deposits. Hydrogeology and ground-water hydraulics deal with prospecting, exploration and exploitation of ground-water sources.

METHODS OF HYDROGEOLOGICAL INVESTIGATION OF MINERAL DEPOSITS

The objective of the hydrogeological investigation of deposits is to determine the ground-water conditions and the drainage possibilities, to estimate the changes in the hydrogeological regimen caused by mining, and to assemble the necessary data for planning the drainage of deposits. Successful and rational drainage requires:

1. Sound knowledge of the geological structure of the aquifer and its neighbourhood, including the supply and drainage conditions, its shape, the distribution and hydraulic properties of faults, changes in the thickness of the aquifer, location of more permeable or impermeable beds and others.

2. Determination of the hydraulic parameters of the aquifer and of the over- and underlying beds.

3. Establishment of ground-water flow conditions from the hydraulic point of view and observations of the water regimen before, during and after exploratory works.

4. The determination of the ground-water table before drainage and of the depth to which it should be lowered by draining.

From the above discussion it follows that the success and economic realization of the drainage depends on full knowledge of ground-water flow in the deposit area, and on an adequate project. In complicated hydrogeological conditions it is advisable to base the drainage project on a simulation method using analogue or digital computers or other physical methods.

In order to determine the data listed under 1−4 hydrogeological investigation uses the following methods:

1. Hydrogeological mapping, which involves an inventory of surface and ground waters, large springs and wells and the determination of their effect on the deposit; differentiation of rocks according to their permeability; characterization of faults and fractures and the establishment of ground-water flow and recharge.

2. Tests on hydrogeological wells to determine basic parameters such as the coefficients of hydraulic conductivity, permeability, transmissibility, storage, porosity, yield, and undersaturation. The water content of individual beds, their hydraulic interconnection, chemical composition of waters and piezometric surfaces of separate horizon are also established. These parameters are found by pumping tests on separate or combined, especially arranged wells. In fractured solid rocks, water-pressure tests are occasionally carried out to ascertain whether the inflow can be reduced by cement or clay grouting.

The permeability parameters of an aquifer are usually determined in hydrogeological wells The tests can be divided into three groups:

a) Pumping tests and pumping-in tests with a constant pumped or pumped-in water amount or with a constant change in water table (head) in the borehole.

b) Tests of pressure equalization in a hole after pumping or pumping-in of water; these are termed tests of water head rise or decline.

c) Special short-term tests on aquifers, for example, with the use of a "tester", when water is allowed to flow into the borehole or directly into the tester for a few minutes. These tests can be applied during boring or in producing boreholes.

The boring tests are evaluated using the methods of steady and unsteady flow. Detailed descriptions of evaluation methods are given in textbooks on ground-water hydraulics.

3. Hydrogeological observations and measurements within the basic exploratory grid provide information on the extent of piezometric water surfaces for the construction of a chart of hydroisohypses or hydro-isopiestic lines, and on the direction of ground-water flow. They also enable approximate determination and delimitation to be made of the basic permeability parameters.

4. Observations on the water regimen serve mainly as basic data for the estimation of the ground-water table after drainage works are begun, for the determination of the time-advance of drainage operations and for calculations of some hydraulic parameters.

5. Special exploratory methods involve geophysical measurements for the determination of the depth and shape of aquifers, faults, logging, etc., artificial raising of drainage well discharge, taking of special water samples, determination of the gas/oil ratio, etc. The interconnection of aquifers, perviousness of faults and the principal direction and velocity of ground-water flows are established using the dye method.

6. Laboratory tests are applied to the determination of the grain-size distribution in rocks, their permeability, ground-water discharge and the physico-chemical properties of rocks and waters. Laboratory analyses and hydrogeological computations are performed by various methods, e.g. the analogue method.

Hydrogeological measurements and observations in mine workings should verify the preliminary computations. At this stage, the hydrogeologist is mainly concerned to determine where and when the ground water was encountered, its physico-chemical properties, the discharge and its changes in time, and the number and thickness of water-bearing beds. For mineral deposits which are explored by comprehensive mine workings, a special hydrogeological sheet is compiled as a supplement to the mining-geological map. When the conditions are simple, the hydrogeological data are plotted directly on the geological map.

ENGINEERING-GEOLOGICAL PROBLEMS IN PROSPECTING, EXPLORATION AND WORKING OF MINERAL DEPOSITS

Excavation and earth works are an inherent part of exploration and exploitation of mineral deposits; they include digging of test pits, trenches and shafts, driving of galleries and other underground works as well as the construction of cuttings, slopes, rock walls and pit heaps. Adequate planning and design, and compilation of a schedule for all these works demand a sound knowledge of the engineering-geological properties of the rocks, in particular:

1. physical properties (specific weight, volume weight, grain-size distribution, porosity, conductivity),
2. behaviour of the rocks on contact with water (absorption capacity, permeability, resistance, filtration anisotropy),
3. mechanical properties (strength, deformability),

4. technological properties (workability, drillability, resistance, to excavation, rock pressure),

5. rheological properties (changes in stress/deformation relations of rocks in time).

In the prospecting, exploration and exploitation of mineral deposits, the engineering geologist studies the engineering-geological properties of rocks and rock masses, classifies them, investigates dynamic processes with regard to mine workings and predicts their further development.

Another task of engineering-geological investigation relates to the stability of mine workings. This demands a detailed study of the fabric of the rock mass, i.e. its texture and structure. Particular attention should be paid to surfaces of discontinuity such as planes of bedding and schistosity, tectonic fractures and joints resulting from weathering or pressure release. The principal characteristics of these planes of weakness (orientation, length, density, openness, filling) are assessed both quantitatively and qualitatively.

For the purposes of underground and opencast mining, the following classification of rocks is recommended:

1. hard rocks, 2. weak rocks, 3. cohesive rocks, 4. non-cohesive gravelly (psephitic) rocks, 5. non-cohesive sandy (psammitic) rocks, 6. unsuitable rocks.

Some engineering-geological properties of these rock classes are summarized in Table 68.

Should the mining works meet all requirements they must also be designed with due regard to the technological properties of the rocks, particularly their resistance to excavation, rock pressure and deformability. Classifications of rocks based on their workability are given in standards and norms which are specific for individual countries.

By excavating a shaft, gallery or other underground opening, part of the rock supporting the load of the overlying rock is removed, which may cause deformation of the worked-out opening. The size and form of the deformation depend mainly on the engineering-geological properties of the rocks, the size and shape of the mine working and the pressure of the overlying rocks.

On the basis of their technological properties, rocks are divided into the following four categories:

1. Hard compact rocks which provide suitable conditions for underground mining. Only at exceptionally high pressures does the profile of the worked-out space not correspond to the rock arch; if the rock is disturbed, spalling of walls ("bursts") may occur. Supports are only necessary occasionally and the roof may be opened to a greater breadth.

2. The behaviour of hard stratified and schistose rocks is controlled by the mode of deposition, engineering-geological properties of individual beds, and the density of joints and their filling. Light-to-medium type supports are usually necessary and heavy suports must be placed in some sectors.

3. Densely jointed clastic and sandy rocks are incapable of taking up the arch

TABLE 68

Some engineering-geological properties of rocks

Rock	Cohesion	Compressive strength $(kp/cm)^{-2}$	Compressibility of rock sample	Permeability	Rock pressure
hard rocks	crystalline, cementation-crystalline	500—4000	practically nil, $E_0 \sim 100,000$ kp/cm^2	low fissure-type	low
weak rocks	crystalline-cementating, water-colloidal	10—500	relatively small, $E_0 \sim 20,000$ to $100,000$ kp/cm^2	low fissure and porous $k = 1 \times 10^{-4}$ to 1×10^{-8} m/s	medium, locally high
non-cohesive gravelly	none	none (shear strength given by φ)	small $E_0 \sim 500$ to 5000 kp/cm^2	very high $k = 1 \times 10^{-1}$ to 1×10^{-4} m/s	low, locally high
non-cohesive sandy	none	none (shear strength given by φ)	usually medium $E_0 \sim 100$ to 1000 kp/cm^2	medium to high $k = 1 \times 10^{-3}$ to 1×10^{-6}	various, mostly high
cohesive	water-colloidal	< 10 (shear strength given by $\varphi + c$)	usually high, long-term, $E_0 \sim 10$ to 200 kp/cm^2 (except. greater)	very low $k = 1 \times 10^{-6}$ m/s	high to very high
unsuitable (organogenic) collapsible loess with large water content, waste dumps	none	very low	very high, irregular	various, depending on composition	usually very high

pressure and carrying the whole weight of the overburden. Special driving methods are used and the excavations require continuous lining and thorough backpacking of the natural arch space.

In sandy water-bearing rocks the stress conditions are aggravated by the pressure of the water flow. Special procedures such as shield tunnelling, grouting, ground freezing and drainage are necessary.

4. Rocks with clayey cement, or clayey joint fillings and clays possess low cohesion and are liable to swelling. The pressures exerted on the supports are therefore very great, approaching the weight of the overlying rock mass. Driving in these rocks does not generally present great difficulties but the openings must be provided with linings of the heaviest type.

EXPLORATION DURING MINING OPERATIONS

Exploration during mining operations begins with the opening of the mine and lasts until its exhaustion. In the interval between the end of detailed exploration and the beginning of exploitation, when surface and underground mine constructions are carried out, the geological service is concerned with detailed geological recording of the principal works in mine opening and development, since they are only temporarily accessible before being provided with lining. After the economic reserves are exhausted, additional exploration is conducted where there is a possibility of the extension of the deposit along the strike or in the hanging and foot walls. The aim of this exploration is to find out whether part of the deposit remains at depth after liquidation of the mine. From what has been said it is apparent that this exploration stage may be of long duration and that it is specialized according to the kind of raw material. In addition, it should help in solving the basic problems in exploitation and dressing of the material. Exploration during operation involves the following principal tasks:

1. the refining of exploratory data in all mining sectors down to the smallest mining units (blocks);

2. systematic study of the mineralogy, chemistry, structures and textures of the raw materials, with special emphasis on their dressing;

3. increasing the precision of information on the mechanical properties and hydrogeology of rocks within the individual mining sectors;

4. refining of data on the reserves. Experience has shown that exploration during operation can find new ore bodies which, due to faulting, were displaced to a greater depth or which were not large enough to be encountered by an exploratory grid of the density employed. If, however, exploration during operation shows that the reserves much differ from those suggested by detailed exploration, some deficiencies in exploration, such as insufficient density of the exploratory grid must be presumed;

5. continuous calculation of reserves from records gathered during operation and from the amount of materials recovered;

6. determination of losses and impurities in individual sectors and blocks.

A brief survey of particular features of exploration performed during operation is given below.

The technique and system of this exploration stage essentially depend on the mode of division and preparation of the deposit for exploitation. These works can furnish better geological information than working faces whose geological logging is difficult or impossible when working methods providing large open spaces are used.

In underground mining, exploration during operation uses drifts, cross-cuts and other mine workings combined with underground drilling. In opencast mining, drill holes are sunk from the top of quarry benches or in front of the working face.

Exploratory works are laid out as in detailed exploration, in horizontal profiles on single levels and sublevels. Horizontal profiles consist of drifts, cross-cuts and horizontal drill holes. Raises, inclines, offsets and fans of small-dimension underground drill holes constitute vertical profiles. The exploratory profiles thus obtained are linked up with or laid between detailed-exploration profiles.

During mine operation, exploration should be very detailed when effective underground-mining methods are being employed (e.g. sub-level stoping, top slicing, longwall method), since the design of a mining block must be based on detailed knowledge of the morphology, tectonics and composition of relatively small parts of the deposit. When geological conditions differ from those assumed, losses are caused by ore degradation or by incomplete exploitation of the deposit.

Exploration can be less detailed with other mining methods, such as cut-and-fill stoping, as these can be adapted to changes in the geological conditions.

The thoroughness of exploration carried out during operation is also predetermined by the dressing technique of raw material. It can be less detailed, for example, for magnetic separation of magnetite, which tolerates a higher impurity content.

Marginal and those parts of an ore body which are not examined by detailed exploration, as well as newly found bodies and parts displaced by faulting, are also subject to exploration during operation. If the results are positive, the reserves are transferred to higher categories.

The exploration of minor ore bodies must be carried out in due time. If an ore body situated in the hanging wall of the principal deposit is threatened by subsidence because of undermining, its extraction is difficult if not impossible. In opencast mining, exploration is aimed at determining the detailed morphology and quality of the deposit well in advance so that operation of the mining equipment can be regulated.

If raw materials of rather complicated composition are extracted in an open pit, drill holes for blasting on quarry levels and exploratory holes in front of working faces should be evaluated. These drill holes need not penetrate the whole deposit, as in all foregoing exploration stages, but are sunk to a depth equal to the height or twice the height of the bench height.

The project of this exploration stage is compiled in connection with but in advance of preparation of parts of the deposit for exploitation. It is particularly important to maintain the right proportions of opened, half-prepared and fully prepared reserves.

It should be stressed that the density of the exploratory grid is the greatest in the exploratory stage under discussion; this fact is taken into account in determining adequate grid dimensions in earlier exploration stages. A thorough exploration can throw light on the genesis of a deposit and this knowledge, in turn, enhances the effectiveness of geological exploration.

Technological investigation should be completed during exploratory stages;

major changes and new technological studies are performed only when a sudden unforeseen change in the material character occurs.

An important task of geological service is to supervise complete extraction of economic parts of a deposit and to erect a reliable system for continuous reserves determination. The amount of reserves changes as the results of operational exploration are refined and the raw materials are being extracted.

The geological service provides all information necessary for the calculation of losses and ore degradation. Under "losses" are included the economic reserves which remain permanently unextracted in pillars or at the contact of a deposit with the wall rock. Degrading is caused by mixing and depreciation of economic reserves with subeconomic reserves or by the wall rocks.

Very small and irregular mineral deposits pose special problems. Since not even detailed exploration can provide a precise picture of the amount of reserves and the economic importance of such deposits, exploitation is carried out together with exploration with minimum capital expenditure on mining installations. If the reserves eventually prove to be larger, the capacity of the plant can be increased by reconstruction.

PROSPECTING AND EXPLORATION OF PLACER DEPOSITS

Placer deposits originate by mechanical concentration of economically important minerals in clastic sediments of various origin, which resulted from weathering of rock complexes and primary ore deposits.

On a world scale, placer deposits are still industrially important types of deposits for many minerals. The number of minerals recovered from placers is steadily increasing with the rising consumption of trace elements. They comprise minerals containing heavy metals — Os, Ir, Pt, Au, cassiterite, wolframite, scheelite, magnetite, cinnabar; minerals with rare and trace elements: containing Ti as the essential element — ilmenite, rutile, perovskite; minerals containing Ta or Nb — columbite, tantalite, pyrochlore, fergusonite, euxenite, samarskite; Th-containing minerals — xenotime, monazite, bastnaesite, and Zr-bearing minerals — baddeleyite, zircon. Many of these minerals contain several useful elements. Monazite, pyrochlore and zircon, for example, also possess uranium in industrially significant amounts. A group of minerals extracted from placers are used as abrasives, precious or semiprecious stones such as diamond, sapphire, ruby, garnet, topaz, tourmaline, beryl and amber. Piezo-electric quartz is also obtained from placers.

Data from the U.S.S.R. well illustrate the present importance of placers. It is is reported that $65-67\%$ Au, 100% Pt, 26% tin, 40% Ti, 85% Zr, 25% diamonds, 12.9% Ta and 100% amber are exploited from placers.

Of many regions rich in placer deposits only the best known can be mentioned here. Large gold deposits occur in the drainage basins of the Lena and Amur rivers, in Kolyma, in California, in the basin of the Sacramento River (whose primary sources are in the Sierra Nevada), in Alaska, mainly in the basin of the Yukon River, in Australia, in New Guinea and in the drainage basins of many Indian rivers.

The largest platinum placer deposits occur in the Central Ural, in Africa and in Colombia. There are rich cassiterite placers in Tasmania, Malaya, Indonesia, Thailand, China, and in the Victoria region of Australia. Rich diamond placers were discovered in the drainage basin of the Vilui river in Siberia and in Africa (Sierra Leone, Guinea, Congo, Angola, Transvaal). Diamond placers are also exploited in South America, Australia and Borneo.

The increase in the importance of placer deposits at present is also due to technical progress enabling their exploitation from the sea floor, particularly on the continental shelf, which occupies an area equal to about $15-20\%$ of the landmass. It contains appreciable resources of gold, diamonds, manganese, titanium, and other minerals.

GENETIC TYPES OF PLACER DEPOSITS

As with other deposits, the amount of prospecting and exploration works for placers is controlled by the size, shape and attitude of the deposit, by the content and distribution of economic minerals, which in turn are predetermined to a large extent by the genetic placer type. Consequently, one of the first tasks for the prospector is to define the genetic type. Irrespective of the kind of minerals, the following genetic types are differentiated:

Eluvial placers are connected spatially with the eluvium over the primary source, which may be a deposit, an impregnation or accessory minerals in the rock. Mineral association of a placer is similar to that of the primary source. Economically important minerals preserve their original habit; intergrowths with other minerals in the deposit or with rock-forming minerals is frequent so that useful minerals are not loosened completely. An eluvial placer does not usually possess a markedly enriched zone. It may be enriched under favourable conditions when, for example, the thickness of the eluvium is reduced by leaching or washing out of fines, or when the ore-bearing bed is softer than the neighbouring rocks and forms depressions in which useful minerals have accumulated over a long period. If the primary source protrudes above the terrain, the minerals become dispersed.

Colluvial placers are related to clastic rocks of the weathering mantle mixed with clay, which move downslope under gravity. They occur at a certain distance from the primary source, from which they are continuously supplied. Colluvial placers do not provide suitable conditions for major concentrations of useful minerals unless there are depressions in the slope surface; the deposit extends along the inclination of the slope and the fan of scatter is minimum. Loosening of useful minerals is again insufficient; it increases on moderate slopes exposed to long weathering under favourable climatic conditions.

Alluvial placers are the most widespread and of the greatest economic value. Their origin is due to river activity. Of several subtypes, placers in river beds and in valley alluvium are generally most important.

Placer deposits in river channel. Ore minerals are concentrated in or immediately below the bottom of the channel. They originate where the transportation power of the stream decreases for some reason.

Placer deposits of sand and gravel banks. They develop in river channels by aggradation of fluvial sediments giving rise to shoals, sand and gravel banks or islands. During floods they are inundated, but are usually exposed at a normal water level. Placers of alluvial plains occur in river banks flooded recurrently by high water. Facies of river alluvium and types of placers are shown in Figs. 191, 192 and 194.

Valley placers are economically the most important of all alluvial placers. They are not related to recent river channels but to earlier alluvium filling the valley above the bedrock. They are in fact old placers of the river bed, formed during its migration in the valley. The normal section of a valley placer from the base upwards is as follows:

1. bedrock, 2. productive layer of sandy gravel, 3. non-productive sand-gravel layer,
4. another productive bed resting on a false bottom may occur in multi-level placers,
5. sandy-loamy bed occurring mainly on the alluvial plain inundated during floods,
6. humus (Fig. 193). The significance of the bedrock character for catching heavy

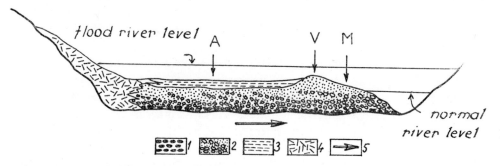

191. Distribution of alluvial facies. 1 — proluvial sediments, 2 — sediments of river channel, 3 — sediments of flood plain, 4 — eluvium, 5 — shift of the river bed; A — flood plain, V — aggradation ridge, M — shoal (after Smirnov, 1965).

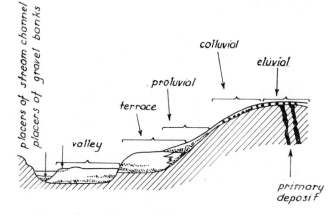

192. Distribution of placers of various genetic types in the section across the valley (after Smirnov, 1965).

minerals is universally known. Schistose rocks dipping downstream are good collectors as are carbonate rocks, in the depressions of which heavy minerals accumulate, forming rich nests and pockets. A smooth valley floor made up, for example, of massive magmatic rocks is unfavourable. A bed of valley alluvium that prevents vertical displacement of heavy minerals which concentrate on it is called a false bottom. It may be a clay seam or sandy gravel cemented together by Fe hydroxides or carbonate. The position of valley placers in the cross-section of a valley is shown in Fig. 192.

Terrace placers occur in alluvial sediments deposited in earlier stages of evolution of river valleys. They were preserved from denudation on the valley slopes after the downcutting of the river channel. Their formation is connected with the lowering of the base of erosion and with vertical river erosion. Part of the river alluvium preserved on the valley slopes may be categorized as terrace placers. Since the origin of river terraces is often caused by regional epeirogenetic movements, their character is uniform even over extensive areas. As a result of interrupted movement several successive terraces can develop, the oldest terrace standing in the uppermost position. Terrace placers are shown in Figs. 192, 194.

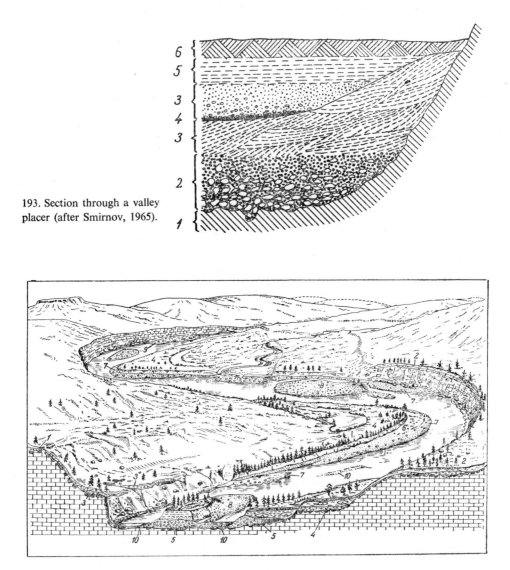

193. Section through a valley placer (after Smirnov, 1965).

194. Geomorphological scheme showing diamond placers of the river Markha (Siberia). Terrace placers of the 5th terrace (1), fourth terrace (2), 3rd terrace (3), 2nd terrace (4), 1st terrace (5); valley placers of: flood plain (6), aggradation ridges and sand bars (7), washed terraces (8), islands, shoals (9) and river bed (10) (after Bobrievich et al., 1957).

Placer deposits of ravines or dales are localized in depressions on valley sides. For most of the year, the depressions are dry. Streams flow through them only intermittently.

Coastal placers form on the coasts of seas and lakes. Deltaic placers, which are transitional between alluvial and coastal deposits, are classed with this group.

Glacial placers occur in glacial sediments. The transportation of clastic material by a glacier does not provide suitable conditions for the concentration of useful placer minerals. The moving glaciers often "scoured" placer deposits of other genetic types and caused their dispersion. Moraines may be sources of economically valuable minerals which concentrate in an aqueous environment to form alluvial or lacustrine placers.

Eolian placers occur mainly in desert regions. They are not of great industrial value. Rich mineral concentrations are produced by deflation, i.e. by winnowing of the light fractions from ancient fluviatile or lacustrine sediments.

CLASSIFICATION OF PLACERS ACCORDING TO THEIR AGE

The largest amount of placers is found in Late Quaternary and recent sediments, but concentrations of useful minerals also exist in clastic rocks of older geological formations. On the basis of age, placer deposits are divided into young, Quaternary to recent (associated with the present drainage pattern and sea and lake coasts) and old, fossil placers. These are subdivided into a group of Tertiary and Mesozoic placers and a group of placers occurring in Palaeozoic, Proterozoic and Archean rocks.

The Tertiary and Mesozoic placers fill the depressions of the ancient drainage pattern, which usually differs from the present one. The old placers are unconsolidated or cemented, tectonically disturbed and may be affected by karstification when deposited in carbonate rocks.

Placers in Palaeozoic and older formations are invariably strongly lithified and related to the facies of conglomerates and coarse-grained sandstones. They are often tectonically disturbed and metamorphosed.

The genetic division given above can be applied to all these groups of placer deposits.

Placers either occur on the surface or are buried by younger rocks. In subsiding areas they may be covered by sedimentary rocks, glacial sediments, colluvium or by volcanic products, which preserved the placers from denudation or exploitation in the past. On the other hand, they make prospecting and exploration difficult as little or nothing on the surface suggests their presence, and geomorphological criteria cannot be employed.

GEOMORPHOLOGICAL AND GENETIC CLASSIFICATION OF PLACERS (WITH A VIEW TO PROSPECTING AND EXPLORATION)

Placers of individual genetic conditions occur under various geomorphological and geotectonic conditions. This circumstance must be considered in the choice of exploratory methods.

The following natural and technical-economic factors should be taken into account in placer exploration: 1. the mode of deposition and morphology of the deposit, 2. the lithology and grain-size distribution of clastic sediments, 3. hydrogeological conditions, 4. the depth of the placer, 5. characteristics of economic placer minerals, especially their specific weight, dimensions and morphology of grains, 6. distribution of useful minerals within the placer, which controls the qualitative variation of the deposit, 7. dimensions of the placer with regard to the conditions of exploration, 8. the average content of useful minerals, 9. the required degree of exploration accuracy and 10. the presumed mining method.

Because of these factors, placers of the same genetic type must often be explored using different methods. Bozhinskii (1965) proposed a classification of placers for prospecting purposes, based on their position and geomorphological conditions. This classification is presented in Table 69.

1. PLACERS IN THE VALLEYS OF RECENT DRAINAGE PATTERN

Placers of this group are related to recent or Quaternary geomorphological features, particularly the young drainage pattern. From the point of view of exploration, placers of this type are the simplest and in many long-exploited areas they have already been depleted. The evolution of these placers depends on the evolution stage of streams, expressed particularly by the character of the cross and long profiles of the valleys.

Young valleys with steep slopes, in which vertical erosion predominates (e.g. as a result of tectonic uplift) do not provide favourable conditions for the development of major placers. The thickness of the alluvium is small. Placers occur in river beds or in narrow floodplains. Heavy minerals as well as alluvium migrate downstream, particularly during floods, but some minerals with a high specific gravity, such as gold and platinum, especially their coarser-grained fraction, cannot be transported very far by water and are often concentrated in placers of mountainous areas. Rich concentrations of heavy minerals may occur in depressions, fissures, potholes or sinkholes in a rocky stream bed. Nor does rapid modelling of slopes afford suitable conditions for the origin of placer deposits. The weathering mantle on steep slopes is of small thickness and useful minerals are not released completely from the enclosing material so that their higher specific gravity cannot play a significant role. They are transported in pebbles by water and scattered over a long reach of the stream. In narrow valleys with steep slopes not even terraces are preserved and the steep gradient

TABLE 69

The classification of placer deposits on the basis of their location
(with a view to prospecting and exploration)

Position of a placer in the relief + geol. section	Elements of the present and buried relief	Type of placers genetic	morphological
I. Placers in valeys of recent drainage pattern			
1. V-shaped valleys in the stage of vertical erosion	river bed and adjacent part of the valley	alluvial	river-bed placers
	surface of the valley	alluvial	floodplain placers
2. U-shaped valleys in stage of equlibrium. Lateral erosion predominates	river bed and adjacent part of the valley	alluvial	river-bed placers
	gravel banks and islands in the river	alluvial	gravel-bed placers
	valley terraces and accumulation terraces	alluvial	valley placers
	erosion-accumulation terraces	alluvial	terrace placers
3. broad valley of old evolution stage; no erosion	erosion-accumulation terraces	alluvio-colluvial	slope placers
	valley floor	alluvial and eluvio-colluvial	valley placers
	valley slopes	alluvial	slope placers
4. young valleys developing in valleys of an older cycle	river bed and adjacent parts of the valley	alluvio-colluvial	river-bed placers
		alluvial	river-bed placers
	erosion terraces	alluvial	terrace placers
5. valleys in subsiding areas, filled with sediments deposited during several cycles of erosion	floodplain and accumulation terraces	alluvial	multistage valley placers
6. valleys buried by thick sedimentary cover of non-alluvial origin	recent floodplain	alluvial	floodplain placers
	buried rocky stream bed	alluvial	buried valley placers
7. valleys with frequent karst features	buried terraces	alluvial	buried terrace placers
	sinkholes in the valley floor	colluvio-alluvial	karstic placers
8. valleys with intermittent streams	large karstic depressions	colluvio-alluvial	karstic placers
	narrow ravines	colluvio-alluvial	ravine placers
II. Placers of the previous drainage pattern, occurring on elevated parts of the terrain			
9. placers on the bedrock surface on elevated broad, slightly inclined watersheds	fossil valleys perceptible in the relief	alluvial	fossil valley placers
	fossil buried valleys imperceptible in the relief	alluvial	buried fossil valley placers
	karst sinkholes	colluvial and colluvio-alluvial	karstic placers
10. strongly dissected watersheds and plateaus	peneplaned parts of land surface	alluvial	peneplaned watershed placers
	surface of a dissected peneplain	alluvio-colluvial and colluvial	slope placers
	summits and slopes of elevations	alluvio-colluvial and colluvial	slope placers
11. fossil placers in Mesozoic and Cenozoic stratigraphic horizons, forming upper struct. platform stage	broad watershed and uplands	marine, lagoonal, lacustrine and alluvial	fossil buried or exhumed placers
III. Placers of piedmont and intermontane depresions			
12. valleys of buried stream pattern	buried valleys and terraces	alluvial	buried valley and terrace placers
13. recent valleys of piedmont depressions	floodplain, gravel banks and islands	alluvial	river-bed and floodplain placers
14. transitional area between mountainous terrain and a plain	talus fans	colluvial	talus fans
IV. Placers of sea coasts			
15. recent beaches	beaches, spits, bars	coastal	beach placers
16. elevated beaches, marine terraces	marine terraces	coastal	placers of marine terraces
	coastal dunes	eolian	dune sands
17. submerged beaches	sediments of sea floor	marine	placers of sea floor
18. submerged deltas	deltaic sediments below sea level	sediments of submerged delta fans	delta placers
19. old alluvial river channels at the sea coast	marine sediments covered by swamps and alluvial-lacustrine sediments	coastal and alluvial	placers of plains

of the river does not enable sorting and concentration of heavy minerals. Placers of this type are not systematically explored, chiefly because of their impersistence; only heavy mineral analyses of the alluvium and "pockets" in the rocky valley floor are carried out.

The largest placers of various morphological types originate in mature valleys with a graded profile. In the river bed, placers occur in gravel beds and islands. They are, however, of minor importance since they are displaced during floods and only fine fractions such as "floating gold" accumulate. Enriched alluvial beds form small and thin lenses. The relatively small size of these placers is partly controlled geomorphologically. Enriched beds can lie above or below the mean water level. Placers of the river bed are located on the floor or at a small depth below it. The thickness of the productive bed and barren alluvium is small. A small clay fraction makes the washing of placer material simple. Placers of this type invariably occur below the water level and exploration is conducted by grabs mounted on pontoons, along lines perpendicular to the axis of the river channel.

Valley placers located in the lower part of the alluvium, near the bedrock or on a valley terrace are the most valuable. The thickness of the productive beds and frequently even of barren alluvium is greater. The total thickness of the alluvium varies between 5–10 and 20–30 m. It increases on tectonically subsiding blocks. The step in the valley floor at the crossing with a fault forms a barrier, in front of which heavy minerals are often accumulated. Valley placers form strips elongated parallel to the valley. Their width amounts to several tens or hundreds of metres and their length to several hundreds of metres to tens of kilometres. For exploration of these placers, which are located in water-bearing alluvial sediments, drilling along lines normal to the axis of the valley is used.

Terrace placers can also be preserved on valley slopes. They are usually less important than valley placers, especially when they are strongly disturbed by denudation. Placers on lower terraces are generally larger, because younger terraces were not so long exposed to denudation. The composition and technological properties of undisturbed placers do not differ from those of alluvial placers. Terraces that suffered denudation or creep, are mixed with colluvial rocks; the content of useful minerals decreases and clay fraction increases, which makes the treatment of material more difficult. Terraces affected by slope movements are often divided into separate parts, for which individual exploratory grids should be used. Generally, test pits prove suitable for exploration of terrace placers lying above the ground-water table.

Figure 194 shows an example of a mature valley with a typical group of alluvial placers.

Valleys of an old evolution stage in peneplaned terrains are generally less favourable for the origin of placers. Valley placers can form there if eluvio-colluvial placers occur on the river banks or if tectonic zones connected with primary mineralization extend in the rocky bed of the river. The alluvium is thus enriched and eluvio-alluvial placers originate. Exploratory works must be sunk to the bedrock in order to determine the trend of the mineralized zone. Colluvio-alluvial placers can also originate by destruction of terraces developed on the valley slopes.

Young valleys deepened in valleys of the older evolution stage due to the lowering of the base level of erosion are narrow, canyon-shaped. Old alluvial sediments are preserved on the valley slopes in the form of long, broad terraces. Since they are only slightly disturbed, terrace placers can be regarded as a continuous deposit and a uniform exploratory method (most frequently test pits) can be chosen. River alluvium is enriched by washing of terrace placers. As conditions are not suitable for the formation of large deposits in these narrow valleys, river-bed placers determinable by panning are indicators of terrace placers.

Valleys in subsiding areas are filled with a thick alluvium, which can contain multistage valley placers lying on the bedrock and on the "false" bottom in the upper part of the alluvium. If the river shifts its channel, several strips of placers can develop in the valley alluvium. Regular synsedimentary subsidence of the area leads to the dispersion of useful minerals to a great depth in alluvial sediments. It is important to explore the whole thickness of the alluvium in the cross-section of the valley.

Both valley and terrace placers can be preserved in valleys that have been buried by glacial, lacustrine, marine and eolian sediments or by volcanic products. Exploration of these placers is difficult.

Karst placers occur in the valleys of rivers flowing through karstic areas. Concentrations of useful minerals are confined to sinkholes and joints, often extending up to several tens of metres into the bedrock.

Placers of another type occur in ravines and gullies that dissect valley slopes and are usually dry during several months of the year. Heavy minerals can be derived either from local colluvium and eluvium or from river terraces if a ravine cuts across them. In the latter case, small but rich placers often originate.

2. PLACER DEPOSITS RELATED TO THE FORMER DRAINAGE PATTERN

Placer deposits of all genetic types can be found in these areas. Topographic features to which the placers were related are often faded, causing difficulties in choosing a suitable exploratory grid. Exploration should begin with the delimitation of old topographic features (fossil valleys, sinkholes, river terraces) and proceed with detailed exploratory works. Old coastal placers associated with a definite stratigraphic horizon and coastal sedimentary facies differ sharply from old eluvio-colluvial and alluvial placers. The prospecting method must be chosen with a view to the fact that this group includes mainly fossil placers. Principal data are obtainable from a geological map, while geomorphological studies are less significant. Panning of unconsolidated rocks, which is largely used for recent placers, is replaced by metallometric analysis of fragmentary aureoles for fossil outcropping placers.

3. PLACER DEPOSITS OF PIEDMONT AND INTERMONTANE DEPRESIONS

A common feature of placer deposits in this tectono-geomorphological position is their location in a tectonically unstable region, far from their sources in mountainous areas. Valley and terrace placers occur below the sedimentary filling of depressions, which protects them from erosion and displacement. Exploration of these buried placers is difficult and expensive. First, drilling and geophysical methods are employed to determine the course of buried topographic features (buried valleys or slopes with terraces); they are followed in several sections and only then are detailed prospecting and exploration of thus localized placers carried out. Owing to the great distance from the primary source, minerals with a high specific gravity occur in recent streams in a very fine-grained form, such as fine-flaky "floating" gold, which is difficult to dress. At the contact of mountains with a lowland, placers can originate in talus fans and sheets but the concentration of useful minerals is usually low. These deposits can serve as sources for alluvial placers in the foothills.

4. PLACER DEPOSITS OF SEA COASTS

This group of placer deposits, which differ greatly from those described above, include placers of recent and old sea and lake coasts, and deltaic placers. Coastal placers are confined to a narrow strip of recent beach sediments, which are deposited on terraces both above and below the sea level.

Narrow strips of recent beach deposits may extend over several tens or even hundreds of kilometres. Their width depends on the slope of the coast, height difference between high and low tide and the height of surf, but usually does not exceed several tens of metres. Productive beds have the form of thin lenses or beds dipping seawards lying either on solidified clayey sediments forming a false bottom or directly on the bedrock. Heavy minerals are supplied either by coastal rock complexes disintegrated by abrasion, or by rivers. Minerals transported by rivers are distributed by currents along the shore. The mineral composition of placer deposits of the former type differs little from that of the rocks building up the coast; minerals that do not survive long transport are also present. Beach sands containing magnetite, pyroxene and olivine (Haiti) or feldspar, hornblende and garnet (the coast of the Baltic Sea) exemplify this type of placers. Placer deposits supplied by river-transported material are characterized by their simple mineralogical composition. Highly resistant minerals, such as diamond, zircon, rutile and monazite, predominate in quartz sand.

Topographically, beach placers differ from terrace placers which occur on elevated terraces, and from placers lying below the sea level. Terrace placers are occasionally enriched by wind winnowing (deflation).

Deltaic placers were thought to be economically unimportant, but many industrial placers have shown that this is not so (e.g. gold in New Zealand or platinum in

British Columbia). Useful minerals occur in deltas mainly as a fine fraction. Figuratively, deltaic placers can be compared to tailings from alluvial placers. The variation of sedimentary regime in deltas causes a lenticular, highly variable form of deltaic placers.

Extensive placer deposits can originate on coastal plains where intensive sedimentation takes place. The concentration of economically valuable minerals, however, is not as high as on stable coasts, where the amounts of material supplied and removed are roughly identical and where rock material is thoroughly sorted according to its specific gravity and grain-size distribution.

Placers located in non-consolidated sediments are usually not explored systematically, only productive beds found in single exploratory works are sampled. Since systematic exploration is useful only in sediments which are not subject to displacement, it is also not carried out in eolian placers of coastal dunes.

Exploration of terrace placers is generally conducted using test pits laid out in lines perpendicular to the elevated shoreline. Different methods are employed for exploration of submarine placers distributed on the continental slope. The geology of the sea floor is derived from geophysical data and from drill holes sunk from pontoons or special vessels and by extrapolating the geological conditions of the coastal area. At a smaller depth the topography of the sea floor can be determined from aerial photographs or by echo-sounding devices. Samples are taken from drill holes or using grabbing equipment.

The above outline of the great heterogeneity of the geologico-geomorphological disposition of placer deposits show clearly how important it is to choose adequate methods for prospecting and exploration.

METHODS FOR PROSPECTING AND EXPLORATION OF PLACERS

The search for and preliminary estimate of placer prospects must be based on a number of geological criteria and indications. Stratigraphic-lighological criteria must provide detailed data on the distribution and stratigraphy of clastic rocks which act as accumulators of economic placer minerals. Great attention should be paid to the delimitation and stratigraphy of Quaternary sediments. Structural criteria enable the tectonic position, geomorphological history and neotectonics of the area to be determined. According to magmatic criteria, the sites and character of primary endogenous sources controlling the mineral associations of placers can be established.

Sources of gold placers, for example, are auriferous quartz veins, fault and metamorphosed zones with dispersed mineralization. Sulphidic or polymetallic deposits usually containing submicroscopic grains of gold cannot develop major placers. In metallogenic provinces with tin-tungsten mineralization, quartz-cassiterite deposits and zones of tin-bearing greisen are primary sources. Tin placers call for integrated mineralogical exploration, since they can contain additional minerals such as wolframite and tantalo-niobate. Of magmatic rocks, acid hypabyssal granitoids, tin-bearing pegmatites and granite dykes should be thoroughly studied.

Platinum placers occur in areas containing ultrabasic rocks, particularly dunites and pyroxenites. Platinum-palladium deposits associated with polymetallic mineralization in ultrabasites do not develop large placers.

In platform areas, diamonds accompanied by pyrope, phlogopite and kyanite are associated with kimberlite pipes.

Areas with granitoid rocks and pegmatites of nepheline-syenite are promising for placers containing niobium-tantalum minerals i. e. minerals of the columbite-tantalite, euxenite samarskite and pyrochlore groups.

Primary sources of Ti placers are diverse, as concerns the distribution of Ti minerals. Attention should be concentrated on gabbroic rocks, nepheline-syenite with ilmenite and rutile impregnations, and primary deposits of ilmenite-magnetite and apatite-ilmenite-magnetite.

The areas of potential placer occurrence can be delimited preliminarily from these geological criteria and metallogenic analysis. Geomorphological study of the area will assist the prospector in deciding whether placers actually developed in the given area and where they are likely to be found. The formation and development of placers in space and time can be very complex. Economic minerals may be derived from ancient periods of weathering and after multiple redeposition concentrated in Quaternary sediments. The multistage evolution of placers following the scheme: primary source — collector I (clastic rocks of various origins) — disintegration and transport — collector II — disintegration and transport — Quaternary or recent accumulator — is widely distributed and it must be taken into consideration. The number of transitional collectors varies. According to the above scheme, the age of placers can also be estimated, i.e. when economic minerals were loosened from the primary sources and when transitional collectors originated. The age of clastic rocks containing useful minerals should also be established. It is possible to determine which topographic features a placer is related to by means of detailed geomorphological study.

No less important are palaeoclimatic criteria. On the basis of palaeoclimatic analysis, warm and humid periods of intensive and long-lasting chemical weathering of rocks and mineral deposits can be determined.

Promising areas are explored by panning in streams to a detail corresponding to the scale of the geological map. Methods of panning exploration have been described in previous chapters.

PRELIMINARY AND DETAILED EXPLORATION OF PLACERS

Exploration is carried out in areas where the presence of placer deposits has been determined by prospecting works; it generally includes detailed geological mapping, geomorphological study, mapping of Quaternary complexes, geophysical survey, hydrogeological research, sampling, laboratory analyses and evaluation.

The area is mapped at scales of 1 : 25,000 to 1 : 5,000, depending on the size and composition of placers to be explored. In the detailed exploration stage, geological maps at scales 1 : 2,000 to 1 : 1,000 are often prepared for small deposits. Maps of the Quaternary cover and of placers based on geomorphology are constructed at the same scale. Aerial photographs are often useful in geomorphological studies.

The principal subjects of geomorphological study are the following: 1. determination of morphogenic features of the relief, 2. the history and age of the relief, 3. the relationship between the distribution of placers and topographic features, 4. collecting of complementary data for the geological map, such as the extent of bedrock below the sedimentary cover or records of neotectonic movements and 5. division of the Quaternary based on genetic types and stratigraphic horizons and determination of the thickness and position of productive beds.

A map of placer deposits should contain 1. classification of deposits according to age, genetic and morphological types, 2. relationship of placers to primary sources and types of mineralization and 3. degree of exploitation and exploration of various placer types.

Topographic mapping should be very detailed in order to plot as many geomorphological features important for localization of placers as possible. For maps at a scale of 1 : 5,000 to 1 : 10,000 contours of 2−5 m interval are required and of 1−2 m spacing for maps at scales 1 : 1,000 to 1 : 2,000. A topographic plan should cover a somewhat larger area than the designed exploratory grid. The laying-out and surveying of exploratory works are carried out continuously.

Geophysical prospecting and exploration methods are direct and indirect. Determination of the depth to the bedrock and of its configuration below superficial clastic rocks is an example of indirect geophysical method. Electric logging and, if the cover is of greater thickness, seismic measurement are employed for this purpose. Magnetometric methods are helpful when economic placer minerals such as magnetite or ilmenite show a high susceptibility (direct use of geophysics) or when they are associated with other useful minerals, for example, gold and platinum. Good results are obtained particularly when the placers lie on rocks with low magnetic parameters (limestones, acid magmatites). Geophysical measurements are conducted along lines perpendicular to major geomorphological features.

Hydrogeological and hydrological research works, which are described in a separate chapter, provide information on surface and ground waters, necessary for placer mining.

EXPLORATORY SYSTEMS

The following systems are employed for preliminary and detailed exploration:
1. shallow vertical drill holes (Fig. 141-1),
2. exploratory trenches (Fig. 142-6),

3. exploratory pits (Fig. 142-7),

4. shallow drill holes and exploratory pits (Fig. 143-9).

The exploratory grid is laid out depending on the morphology of placers and their variation. Characteristic configurations of individual mineral placers can be used to advantage for the distribution of exploratory works. Alluvial gold placers, for example, form very narrow strips so that exploratory holes must be closely spaced — 10 to 20 m apart — within lines at 400—800 m intervals. Tin placers are distinguished by their greater width and smaller length; therefore exploration works are located approximately 20—40 m apart and the lines are not more than 400—600 m apart. Ilmenite placers are usually developed in the whole valley profile and over a considerable distance. Intervals between exploratory works are about 40—80 m and the lines are spaced about 600—800 m apart.

Similar to other deposits, placers are divided into several groups according to their characteristics.

Very persistent placers are characterized by a relatively uniform distribution of useful minerals in a productive bed of a comparatively constant thickness. The bottom of a placer is relatively even and slightly inclined. The productive bed differs markedly in lithology from barren alluvium. Valley and terrace placers of large rivers are classed with this group.

Persistent placers are continuous in both longitudinal and transverse directions but the distribution of economically important minerals is less regular. The bottom of a placer is generally uneven and steeply inclined. The productive bed does not differ from the surrounding alluvium, so that it can be delimited only on the basis of sampling. Mineral grains are worn to different degrees, large crystals and intergrown aggregates are frequent. Alluvial placers of minor rivers and streams are of this type.

Irregular placers are distinguished by random distribution of useful minerals with exceedingly rich concentrations locally. The thickness of productive alluvial beds, colluvium or eluvium and their aggregate thickness are highly variable. The productive bed can only be delimited by sampling during exploration or mining. The placer bottom is usually uneven, steeply dipping with scours and pockets. Mineral grains are poorly segregated, while intergrowths and nuggets are frequent. Eluvio-colluvial placers and alluvial placers of ravines, dells and short brooks belong in this class.

Figure 195 shows schematically the distribution of useful minerals in the placer types described above.

Eluvio-colluvial placers of irregular form are explored using a grid of any type; those located in elongated slope depressions are explored along lines drawn at right angles to their longer axis. Since these placers mostly occur above the ground-water table, exploratory trenches (to a depth of 3 m) and pits (to a greater depth) are used. Because of the small size and irregular form of placers, exploratory works are spaced about 20 m apart.

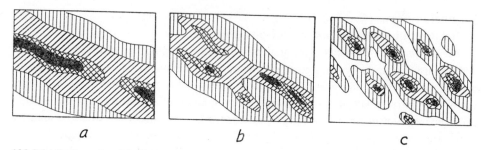

195. Distribution of useful minerals in an alluvial placer: a — very regular, b — regular, c — irregular.

The following principles were proposed by Bozhinskii (1962) for spacing exploratory works, particularly in valley placers:

1. The spacing of exploratory lines is controlled by the length of the placer, category of reserves and variation of the placer.

2. The spacing of exploratory works along lines is controlled by the breadth of the economically valuable part of the placer and by its variation but not by the category of the reserves. From this it follows that the final density of exploratory works is already achieved in the preliminary exploration stage.

Determination of the spacing of exploratory lines is based on the assumption that several complete lines (4—6) are necessary for assessment of the geological structure and mode of deposition of a placer and of the distribution of useful minerals. The size of a block for the calculation of reserves, which is limited by two lines, is also taken into consideration. This block should be of approximately the same size as

TABLE 70

Spacing of exploratory lines recommended for various reserve categories of placer deposits

Degree of variation (group of deposits)	Length of placer to be explored (distance between boundary lines in km)	Spacing of exploratory lines (in m)			Minimum number of expl. lines for category C_1
		A	B	C_1	
very persistent	more than 10	200—300	400—600	800—1200	8
	4—10	150—200	300—400	600—800	6
	less than 4	150	300	600	4
persistent	more than 3	100—150	200—300	400—600	6
	less than 3	100	200	400	4
impersistent	more than 2	—	200	400	6
	less than 2	—	100—200	200—400	4
very impersistent	more than 1	—	—	200	6
	less than 1	—	—	100—200	4

a mining block, corresponding roughly to the annual operating capacity of the extraction equipment (e.g. bucket ladder excavator). Its capacity is chosen with due regard to the size of placers. The spacing of exploratory lines is thus governed by several natural and technical factors. The spacing of exploratory lines based on these principles is presented in Table 70.

The broader a placer deposit, the greater the spacing of exploratory works can be. Since the relations of placers to the geomorphological features are determined mainly in cross-sections, the necessary density of exploratory works is established already during preliminary exploration. By increasing the density of exploratory lines, valley placers are explored in greater detail and the reserves can be placed in higher categories. The recommended distances of exploratory works on the lines are listed in Table 71.

TABLE 71

Spacing of exploratory lines and the minimum number of exploratory works recommended for the lines of placer deposits

Width of economic part of placer (in m)	Very persistent placers		Persistent placers	
	spacing of exploratory works (in m)	number of expl. works	spacing of expl. works (in m)	number of expl. works
up to 60	—	—	10—20	3—6
60—120	20—40	3—6	20	3—6
120—250	40	3—6	20—40	6—10
250—500	40—80	6—10	40	6—12
more than 500	80	more than 6	40—80	more than 8

Width of economic part of placer (in m)	Impersistent placers		Very impersistent placers	
	spacing of expl. works (in m)	number of expl. works	spacing of expl. works (in m)	number of expl. works
up to 60	10	3—6	10	2—6
60—120	10—20	6—10	10	6—12
120—250	20	6—12	10—20	8—12
250—500	20—40	more than 10	20	more than 12
more than 500	—	—	—	—

METHODS OF PLACER EXPLORATION

Test pits are very suitable for placer exploration. They are widely used since they enable reliable sampling to be carried out. Where the deposit lies above the ground-water level and at a small depth, pits are preferable to drill holes. If they are dug rapidly and effective pumps are employed, they can be used even in rocks with a minor water content. Special devices have been constructed for mechanical digging of pits in water-bearing rocks. Hand-digging of pits in water-bearing rocks in northern countries can be facilitated by freezing the pit walls. For shallow exploratory holes percussion drilling is employed.

Hand-drilling rigs are widely used particularly in inaccessible areas where the transport of heavy drilling rigs is impossible, and in countries where human labour is cheap. A suitable hand-drilling device which was used for many years for placer exploration in the U.S.S.R. was the Empire rig; similar systems were also employed in other countries. The rig was developed especially for drilling in gravel, gravelly sand and sandy clay. Casing pipes, $4'' - 6''$ in diameter, are provided with a sawtooth bit. A casing string is usually driven into clastic rock in advance of drilling at least $5 - 10$ cm below the bottom of the hole. A casing string driven in advance ensures reliable sampling and prevents mixing of samples during bailing.

Special drilling rigs for placer exploration are now constructed on the same principle, as, for example, the UBR-1 percussion drilling rig in the U.S.S.R. Casing pipes are driven either by ramming or rotation; the maximum depth attainable is 15 m, the diameter of drill hole is 121 mm and the weight of the rig is 745 kg. The rig is operated by two workmen. Light, mobile rigs are constructed by the Conrad – Stork company in the Netherlands and the Kirk Hillman Co. in the U.S.A.

Experimental drilling was carried out by ZIF-550 core drilling rigs. A casing string is provided with a metal bit; runs $0.5 - 1$ m in length are drilled without using drilling fluid until the core is fixed in the casing. It is then easily removed, being either knocked out by hand or forced out by water pressure. Up to a depth of $5 - 10$ m the hole is drilled using a casing with a diameter of $219 - 230$ mm. This is then reduced and the wider pipes serve as casing of the hole. This method provides a very good record of geological conditions over the cored sector.

SAMPLING OF PLACER DEPOSITS

The purpose of placer sampling is to obtain complete qualitative and technological information on the deposit: 1. the presence of economically valuable minerals, their size, shape and chemistry are determined; 2. the extent of the productive bed is delimited; 3. technological properties of clastic rocks, mainly the grain-size distribution and clay content are established. The following kinds of sampling may be distinguished:

1. Mineralogical sampling determines the content of useful minerals in samples collected during exploration. The content of useful minerals in 1 cu. metre of rock, which is one of the basic parameters in the determination of reserves, is calculated from their weight in a sample of known volume. The useful mineral concentrates are also examined by sieving and chemical analysis. The latter is used, for example,

for determination of the fineness of gold, the Sn content in cassiterite, the Ti content in ilmenite and contents of other trace elements.

2. Technical sampling determines the coefficient of bulking of the productive bed by measuring the volume of the sample in a graduated vessel. Using this coefficient, the content of a useful mineral can be recalculated to its content in the original state. Grain-size distribution in clastic rocks and the percentage of pebbles over 20 cm in diameter are also established. Larger fragments are usually eliminated from samples before their treatment but they must be taken into account in calculating the content of useful minerals in a unit volume. The technical properties of minerals are studied in diamond, ruby, topaz, beryl and similar placers with regard to their suitability as abrasives or as jewelry.

3. Technological investigation is carried out in economically valuable placers to develop a technological scheme for sand treatment. The possibility of complete usage of placers is examined, i.e. recovery of all valuable minerals and the feasibility of using gravel and sand in the building industry.

Mineralogical sampling embraces the following operations:
a) collection of samples and determination of their volumes,
b) washing of samples — coarse concentration,
c) fine concentration,
d) mineralogical evaluation of heavy mineral concentrates and weighing of useful minerals.

The sampling method is chosen depending on the kind of exploratory works and on the required initial sample volume.

Channel or planar sampling is used in test pits and trenches. Bulk sampling is employed when a large initial volume is required. The material of a productive bed can be deposited on heaps around a pit according to depth intervals and then either treated as bulk samples or only part of this material is selected for further treatment by hand picking or with a sampler.

If the percussion drilling procedure is not strictly observed, the original composition of the rock sample can be greatly disturbed. Correct data on depth and composition cannot be obtained if, for example, the clay component and heavy minerals are washed out, part of the hole wall is pushed back by chiselling, a strong inflow brings rock particles into the drill hole or the drill bit becomes blunt. It is necessary that the casing be driven to the bottom of the hole in advance. In bouldery rocks, where this is not practicable, a short run is drilled by chiselling and the casing is sunk into the broken rock. During drilling a run, all data necessary for determination of the ideal run volume and the position of the end of the casing string are recorded. If there is a large difference between the ideal volume, which is calculated from the length of the run and the inner diameter of casing, and the measured volume, the reason for this difference must be discovered and corrected.

If placer deposits lie under the water level of rivers, seas and lakes, drilling rigs have to be mounted on pontoons or special boats. River-bed or sea-floor placers

TABLE 72

Relationship of the initial sample volume to the dimensions of grains and average content of useful minerals in placer deposits

Useful minerals	Distribution of useful minerals (in %) in grain-size fractions (in mm)										Number of grains of useful minerals im 1 m³ sand at the limiting content	Initial sample volume according to the formula $v = Kd/c$		
	< 0.07	0.07 to 0.15	0.15 to 0.25	0.25 to 0.5	0.5–1	1–2	2–4	4–8	8–16	> 16		number of grains in sample (K)	mean grain weight (d) m?	sample volume (V) m³
diamonds	—	—	—	—	0.2	2.8	30	60	7	—	0.05	2	70	35
	—	—	—	—	8	30	40	20	2	—	2.14	5	9	2
gold	—	—	0.5⁺	1.2⁺	2.4	7.3	13	40.6	35	—	145/10	5	14/200	0.5
	—	—	14.9	25.6	42.5	10.5	4.5	2	—	—	900	100	0.6	0.12
	—	1.4	37.4	21.8	36.3	—	2.9	0.2	—	—	770	100	0.3	0.15
cassiterite	—	2⁺	4⁺	12	8	10	18	15	—	31	$246 \times 10^3/22 \times 10^{3-}$	1000	1.6/17	0.42
	—	13	17	23	30	10	5	2	—	—	700×10^3	10 thousand	0.3	0.015
columbite	—	—	10⁺	14	24	30	17	5	—	—	$54 \times 10^3/10 \times 110^{3-}$	1000	0.4/1.8	0.045
	—	39	25	29	7	—	—	—	—	—	1400×10^3	10 thousand	0.014	0.007
zircon	5	35	27	31	2	—	—	—	—	—	360×10^6	1 million	0.008	0.004
	67	33	—	—	—	—	—	—	—	—	1560×10^6	1 million	0.0015	0.0008
ilmenite	—	3	12	27	45	13	—	—	—	—	520×10^6	1 million	0.06	0.002
	6	85	9	—	—	—	—	—	—	—	$10{,}000 \times 10^5$	10 million	0.003	0.001

Note: In the nominator – total number of grains, in the denominator – number of grains without the smallest fractions denoted by +.

are also sampled using small excavators along trenches extending at right angles to the direction of the placer deposit.

The necessary initial sample volume is established mainly on the basis of the useful mineral concentration in the placer, the grain size and specific gravity, Bozhinskii (1965) recommends deriving of the necessary initial sample volume from the formula

$$V = K \frac{d}{c},$$

where V — initial volume,

K — coefficient of probability expressed by the number of mineral grains which the sample should contain,

c — average or limiting content of useful mineral in placer,

d — average weight of a grain.

Volumes of samples as recommended according to the above formula are set out in Table 72.

The distribution of useful minerals in placers of the same genetic type is controlled chiefly by their specific gravity. This determines the length of a suitable sampling interval. Gold and platinum which because of their high specific gravity concentrate in a productive bed near the floor, should be sampled in very short core runs or very short sections of about 0.2 m. Cassiterite and columbite whose specific gravity varies around 6 — 8, concentrate farther from the placer bottom, so that core runs and sections of about 0.5 m are usually sufficient. Core runs or sections of up to 1 — 2 m are used for sampling of ilmenite and monazite, which have a specific gravity of about 4 — 5 and are dispersed almost uniformly within a large volume of clastic sediments.

The volume of a large sample is measured in a calibrated vessel (Fig. 196). The heavy minerals are separated and concentrated using a pan, a trough or a simple sluice (Fig. 197). Samples are washed directly in the field. A typical equipments for percussion drilling is shown in Fig. 198. The coarse heavy mineral concentrate is dispatched for detailed study and mineralogical analysis. Mineralogical analytical methods are described in special text-books (e.g. Rost, 1956; Kukharenko, 1961). A well-equipped laboratory for heavy mineral analysis needs a magnet, an electro-magnetic or electrostatic separator, an elutriation tube or other similar device, a concentrating table for gravitational segregation, heavy liquids, an ultraviolet lamp, a binocular and polarizing microscope, immersion oil, reagents for microchemical analysis, a blow pipe, a set of laboratory sieves and an analytical balance.

An alternative treatment for tin-placer samples was proposed by Janečka (1966). The sample is divided into a coarse fraction (above 2 mm), subsizes and a fine clay fraction, which is removed by settling and decantation. The volume of the coarse fraction and the subsizes is measured in calibrated vessels. The mesh size is chosen depending on the assumed size of the useful minerals. The volume of the fine fraction is determined from the difference between the total volume of the sample and the sum of the coarse-fraction and subsize volumes. In the prospecting stage, or in the initial stage of exploration, the whole association of the heavy mineral concentrate is evaluated: economic minerals are evaluated quantitatively and the remaining minerals semiquantitatively. In advanced exploration stages only economically valuable minerals are appraised quantitatively. A complete analysis of a heavy mineral concentrate is made on key drill holes, for example, on one hole in every boring profile.

For mineral analysis, the concentrate is divided into grain-size fractions. Each of them is weighed and part of the sample is separated by quartering for further analyses. Magnetic minerals separated from each fraction are weighed. After separa-

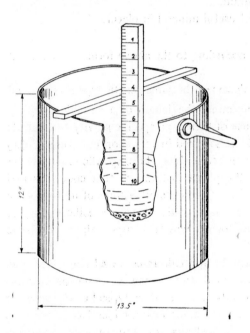

196. Vessel for measuring the volume of samples (after Harrison, 1954).

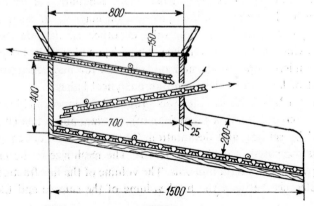

197. Nedelyaev's rocker. Dimensions in millimetres (after Azhgirei et al., 1954).

tion in heavy liquids, the heavy fraction is segregated into electromagnetic fractions and non-electromagnetic fractions and both are weighed. Average samples (200—500 grains) are selected from the two fractions and the content of the useful mineral is determined quantitatively and that of remaining minerals semiquantitatively. Before microscopic quantitative determination, HCl is poured on cassiterite grains deposited on a zinc plate for easier identification. A tin film exsolved on the grain surface makes them more striking. The content of the useful mineral can also be read from the curves of its specific gravity/content ratio, or by X-ray spectrographic analysis.

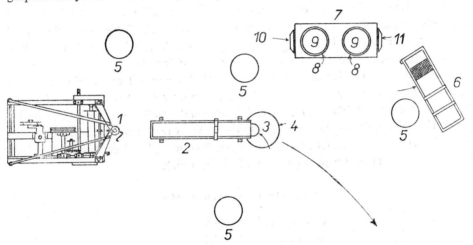

198. Arrangement of a drill rig and installation for washing samples. 1 — percussion drill rig, colla, of a drill hole, 2 — trough for emptying bailers into, 3 — vessel for measuring the sample volumer 4 — rounded vessel, 5 — barrels, 6 — rocker capable of transverse motion, 7 — washing table, 8 — vessels for waste, 9 — pan for waste catching, 10 — screen, 11 — washing pan (after Harrison, 1954).

For checking the results of mineral analysis, part of the sample (about 2—3 %) is examined by chemical analysis of the concentrate. Experience has shown that gravimetric concentration methods make it possible to obtain cassiterite of about 0.2 mm grain-size. Finer-grained fractions pass into the tailings. Therefore, grain-size analysis of useful minerals in the concentrate is also part of mineralogical analysis.

Gold and platinum are readily separable from the concentrate because of their high specific gravity. Primary panning is more thorough; it proceeds until a black concentrate is obtained, since the possibility of losses is minimum. Minerals are separated from the black concentrate by magnetic and electromagnetic separation. The remainder is spread on a piece of black paper with the edges bent up, and segregated by air current. Gold and platinum do not move on the inclined paper sheet, while other minerals are carried away. The concentrate is then sorted under a binocular microscope. A more suitable method for concentration of very fine-grained gold is amalgamation. The concentrate is put into a porcelain plate coated with mercury. The amalgam is then dissolved in nitric acid or sublimated by heating. Gold obtained by this separation procedure is then weighed with an accuracy of 0.1 mg. During mineralogical examination, the properties of gold and platinum that may affect their dressing are also studied. The percentage of free metal and of metal intergrown with rock minerals is determined; the content of metal bound in quartz, pyrite and chromite is defined by chemical analysis of these mineral concentrates.

Because of the losses which are due particularly to panning, the waste accumulated in the tank is washed once more. A control is carried out either for every sector of the drill hole or test pit or for the whole hole at once. In the former case the useful mineral content is added to that for the respective interval; in the latter case it is added to the average content for the whole drill hole or pit.

Despite this control, it has been ascertained that owing to losses during sampling and handling of samples and to natural conditions only about 60 per cent of actual reserves are identified. Therefore, correction factors depending on specific conditions in individual deposits are introduced. On the basis of experience from the exploitation of placers in the U.S.S.R. the following correction factors have been determined for the calculation of Au placer reserves (Albov, 1961):

Type of placer	Correction factor
very persistent	1.1—1.2
persistent	1.3—1.4
irregular	1.4—1.6

The value of a correction factor is also controlled by the exploration procedure. With a certain placer type, it is lowest when exploration by test pits was used and highest when drilling was employed.

It will be useful to record two instances of placer exploration (V. I. Smirnov, 1955) to conclude the chapter on placers. Figures 199, 200 show exploration of an aluvial gold placer in western Siberia. The placer extends over more than 25 km, from the middle course of the river to its mouth. The exploratory polygonal grid has a side 12,450 m long. The gradient of the valley is 5—6 m/km and the breadth of the valley varies between 1 km and 60 m. River terraces are indistinct; the thickness of alluvial sediments is 1.5—9. 4 m on average. The construction of excavators enables operation to a depth of 1 to 6.82 m. The productive bed does not differ in lithology from other alluvial sediments. Gold is concentrated mainly in the lower part of the gravel but is occasionally dispersed throughout or constitutes seams lying over one another. The gold content varies considerably.

Exploration was carried out by drilling with an Empire rig (inner diameter 93 mm). In digging test pits (1.4 × 1.8 m) natural freezing of the walls was utilized. Pits that could not be dug to the desired depth for technical reasons were completed by drilling. Exploratory lines were spaced $250-500$ m apart, and in exceptional cases 1,000 m. The distance between drill holes and pits was 20 m and only rarely 40 m. The lowest ore-bearing interval in exploratory works was regarded as the lower boundary of the placer. Reserves calculated by the method of vertical geological sections were classed as probable ore (C_1 category) with the spacing of the lines of 600 m.

Figures 201 and 202 illustrate the exploration of an irregular gold placer situated in a river terrace $0.5-2$ m above the flood plain. The placer was 2.1 km long and $450-1,150$ m broad. It has an alluvio-colluvial character, since the terrace is disturbed

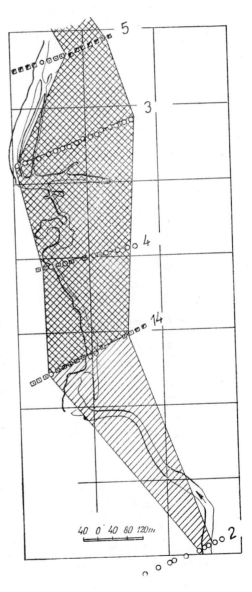

199. Schematic representation of the classification of reserves in a regular gold placer. Circles — exploratory drill holes, squares — test pits, squares with dots — pits combined with drill holes; cross hatching — reserves of B category, diagonal hatching — reserves of category C_1 (after Smirnov, 1950).

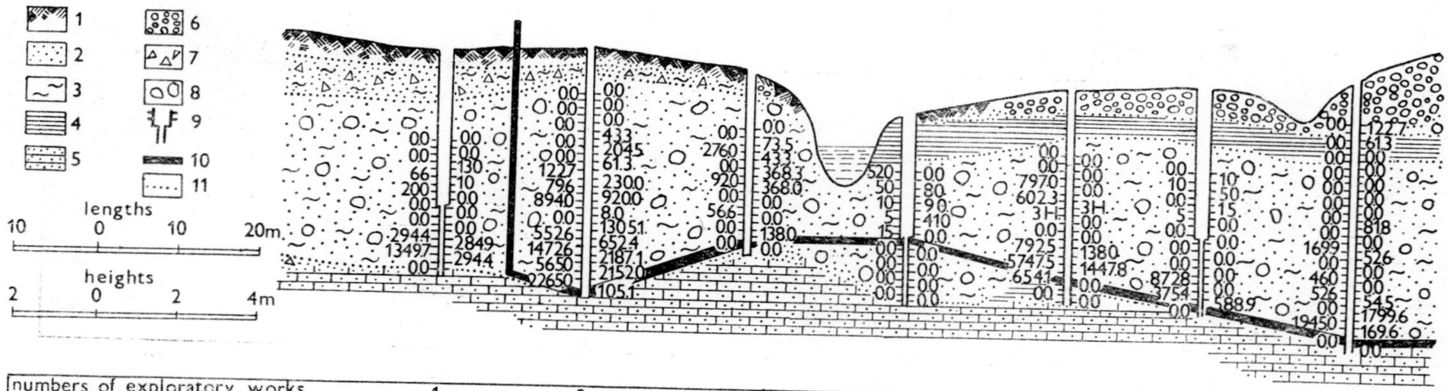

numbers of exploratory works	1	2	1	3	6	4	7
spacing of exploratory works m	18		20	20	20	17	19
altitude above sea level	4462	4473	4418	4315	4400	4414	4440
depth of exploratory works, m	580	640	480	500	560	600	720
depth of excavation, m	560	620	440	320	500	580	880
thickness of barren layer, m	480	440	260	140	400	500	620
thickness of productive layer, m	080	180	060	160	100	080	040
average content of productive layer, mg/m³	556	1252	276	146	4362	444	1872
average content of total volume, mg/m³	94	446	64	74	935	64	142

200. Part of the exploratory line no. 4 from Fig. 199 (a section). 1 — humus layer, 2 — sand, 3 — clay, 4 — sandy clay, 5 — sandstone, 6 — gravel, 7 — debris, 8 — pebbles, 9 — test pit and drill hole, sampling site denoted as well as the content of useful mineral, 10 — boundary of mining, 11 — geological boundary (after Smirnov, 1950).

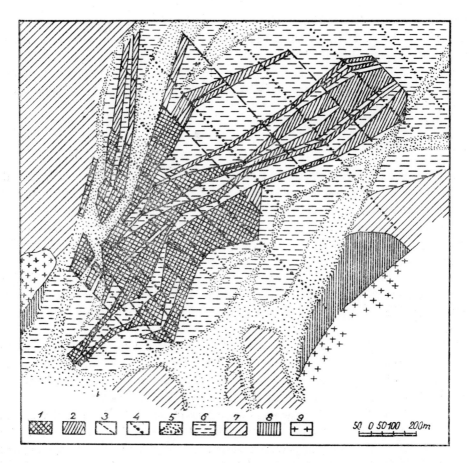

201. Schematic representation of exploration and delimitation of an irregular gold placer. 1 — reserves of B category, 2 — reserves of C_1 category, 3 — boundaries of blocks, 4 — test pits, 5 — flood-plain sediments, 6 — terrace sediments, 7 — loam, 8 — dells, 9 — basalt (after Smirnov, 1950).

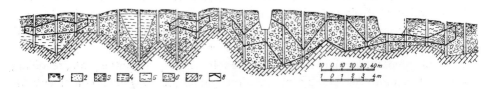

202. Cross-section through the placer shown in Fig. 201. 1 — humus layer, 2 — sand, 3, 4 — sandy clay, 5 — clay, 6 — pebbles, 7 — bedrock, 8 — limit of calculated reserves (after V. I. Smirnov, 1950).

by slope processes. The thickness of the terrace varies between 3 and 6 m; its average thickness is 3.5 m. The productive bed consists of coarse-grained sand and gravel with sandy clay, and its thickness ranges from several decimetres to 4−5 m. Gold occurs in sandy gravel. The deposit was explored by test pits, 10 m apart, along exploratory lines spaced from 100−200 m to 300−400 m apart (probable ore, categories B and C_1).

PLACER MINING

Since mining methods for placer deposits differ in many respects from those employed for working solid mineral resources, they are briefly mentioned below. With regard to the great variety of placer types, a number of methods has been developed.

Buried placers that cannot be worked by open mining are extracted by underground procedures which are chosen according to the same criteria as with stratified deposits in sedimentary rocks.

Placers situated above the ground-water table and covered with a thin overburden are worked by open mining. Slightly solidified rocks of placer deposits do not require blasting and excavators of various types are adequate to the stripping of overburden and mining.

Hydraulic mining is very suitable for placer working. Jets of water are passed through nozzles to break down the rock and wash the material through sluices to the preparation plant. This method is cheap and simple but a large volume of water is required, which can be provided by ponding the stream by earth dams.

Many placers lie below the ground-water level or below the level of rivers, seas and lakes. These are worked by dredgers, which also provide for separation of useful minerals and haulage of tailings onto a waste tip. Dredgers are of suction, grab, and most often of ladder types. A dredger, being an expensive machine, can be employed profitably only on rich placers, where there is a possibility of long exploitation allowing for amortization. Dredging requires a moderate land surface and bedrock configuration, a sufficient width of the placer, a satisfactory content of useful minerals and a small content of clay, alluvium free of boulders and tree trunks, and easy transportation. Dredgers constructed for placer mining in remote regions (e.g. of Australia, Canada, Siberia) consist of segments of the minimum possible weight, transportable by air. The principal technical parameters of dredgers are the bucket volume, which controls the mining capacity and the height and depth of dredging. Heavy types of dredgers developed for exploitation of placer deposits at great depths below the sea level are shown schematically in Fig. 203.

A simple preparation plant has usually the following equipment: a grate, revolving screen, installation for gravity preparation (tables or jigs) and equipment for amalgamation on gold placers. Often only a coarse concentrate is obtained on a dredger and final preparation is carried out in plant operating for several mining sectors.

Placer deposits provide favourable conditions for both mining and preparation of metals. They are located in slightly solidified rocks, which need not be broken by blasting; useful minerals are most commonly loosened, so that costly process of dressing, crushing and milling is unnecessary. The high specific gravity of most minerals allows for simple gravity dressing.

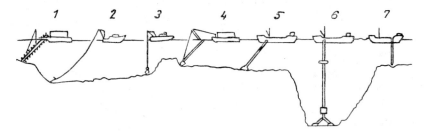

203. Various types of dredgers used for placer mining in near-shore water. 1 — bucket ladder dredge for working rocks of medium hardness, 2, 3 — clamshell dredger suitable for unconsolidated sedi, ments and loam, 4 — hydraulic dredger with cutting head suitable for rocks of medium hardness- 5 — hydraulic dredger with a bin for excavated material, suitable for unconsolidated sediments, 6 — hydraulic dredge for excavation at a great depth with equipment for stripping overburden from the sea floor, 7 — dredger constructed on the principle of an air pump (after Krukshenk, 1967).

The relatively low capital expenditure on exploration, mining and dressing makes it possible to work placers profitably, even if relative to primary deposits they have an appreciably smaller content of useful minerals. The determination of economic standards depends on several factors, particularly on the content of useful components and on the conditions available for mining and dressing.

ECONOMIC EVALUATION OF MINERAL DEPOSITS

The complete cycle of exploratory stages also includes calculation of the reserves and economic evaluation of the deposit. Since special papers are usually devoted to these two problems, they are discussed here only in general in order to present an overall picture of exploration tasks.

Calculation of mineral reserves comprises not only calculation of the amount of useful minerals and determination of the quality and economic importance of the deposit, but also drawing up of a final report summarizing all information obtained in the individual exploration stages.

The report contains the following chief items of information:
1. The amount of industrial raw materials and distribution of the individual kinds in the deposit,
2. the quality of the mineral materials,
3. geological conditions that are decisive for selecting the mode of opening the deposit and the mining method,
4. reliability of the calculation of reserves and the degree of geological assurance for a confident estimation of the industrial value of the deposit.

At present seven methods are commonly used for the calculation of mineral reserves. These are the 1. geological block, 2. exploitation block, 3. triangular block, 4. polygonal block, 5. geological section, 6. the isoline and 7. structure-contour methods. Calculation of solid materials is simple, consisting of calculation the deposit volume V (in m^3), of reserves Q in tons and of the reserves of useful components S (mostly in tons)

$$V = P \times m_p, \qquad Q = V \times v_p,$$

$$S_1 = \frac{Q \times C_{1p}}{100} \qquad S_2 = \frac{Q \times C_{2p}}{100},$$

where P is the area of deposit in m^2, m_p is the average thickness, v_p is the average unit weight in tons/m^3 and C_1, C_2, C_n are the average contents of useful components in the raw material (in per cent).

The above formulae indicate that the basic parameters for the calculation of solid mineral reserves are the area and thickness of the deposit and the unit weight and content of useful components. These parameters must be determined with the highest possible precision, since the accuracy of the calculation depends entirely on their reliability. Some methods of calculation differ from the above procedure in the determination of the deposit volume; using the method of geological sections, it is calculated as the volume of a block between a pair of sections; using the method of isolines it is computed from the plan of thickness isolines, and from the structure-contour map of the deposit when the contour method is employed.

The set of exploration data and calculation of reserves form the basis for estimating the economic value of the deposit.

Economic evaluation of a deposit is of a complex character and is thus referred to as geologico-economic or technical-economic evaluation. A deposit must be evaluated using a complex of parameters since no reliable synthesizing index for determining the economic value of a deposit is available.

The maximum fulfilment of economic needs using the minimum labour and financial outlay per production unit is a general criterion.

Indices characterizing the value of a deposit are given in Table 73 (from Kobak-hidze, 1973, slightly modified). These indices should be recalculated for one ton of reserves, one year of exploitation and the total economic reserves of the deposit. The individual indices are briefly described below:

The natural value of a deposit is given by the explored reserves of mineral raw materials, their quality, the amount of useful principal and subsidiary components in both the ore mass and the final product. The reserves are the main index of the industrial value of the deposit. In evaluating a deposit in lower exploratory stages, potential reserves should be considered in addition to industrial reserves. Stress should be laid upon complete utilization of the raw material. Although expression of the value of a deposit by its price is not a decisive factor, a simple formula can be employed:

$$C = A \times B \times p \times q \times r,$$

where C — price of the deposit, A — workable reserves of metal, B — cost of metal, p, q, r — yield coefficients in extraction, treatment and smelting.

Annual working capacity of the mine: Correct determination of the optimum working capacity is decisive for the determination of many other economic indices. Technical feasibilities, the volume of the reserves and the life of the mine should be taken into consideration in the determination.

For inclined deposits, Agoshkov recommends determining the annual output from the advance to depth during one year.

$$A_{max} = S \times V \times d \times \frac{k_0}{1 - r}.$$

A_{max} — maximum annual output, S — area of a horizontal section through the deposit in the pertinent section (in m^2), V — advance to depth in one year (in m), d — unit weight, k_0 — coefficient of recovery, equal to $0.8 - 0.95$ in underground mining, r — coefficient of loss and depreciation, $1 - 20\%$, depending on the working method.

If the deposit is horizontal or moderately inclined, the annual advance of the face can be determined analogously according to mining-technical conditions.

TABLE 73

INDICIES OF INDUSTRIAL VALUE OF A DEPOSIT

NATURAL VALUE OF A DEPOSIT

Reserves of mineral raw material (in tons)
Average content of useful component in the ore (in %)
Reserves of useful component (in tons)
Amount of useful component utilized in the end product (in tons)

MINING EFFECTIVENESS

Annual working capacity of the mine (in tons)
Production of the principal useful component (in tons)
        subsidiary useful components (in tons) expressed
        as their price

Costs of geological exploration and mining:
        geological exploration
        mining
        dressing
        smelting

Profit from exploitation of the deposit
        during exploration
        exploitation
        dressing
        metallurgical treatment

Differential rent from exploitation of the deposit
Profitability of mining in relation to
        operational cost (in %)
        production cost (in %)

EFFECTIVENESS OF INVESTMENTS

Sum of investments on the construction of industrial plant, including
        mine
        dressing plant
        smelting plant

Specific (unit) costs of investments to:
        annual production price on:
        annual production unit
        cost of 1 ton of reserves
        cost of 1 ton of annual ore production

Returnability of investments
        of mine
        dressing plant
        smelting plant

The annual working capacity can be determined from the optimum life of the mine taking into account depreciation of the investment and full utilization of important installations.

$$t = \frac{Q}{A} \qquad A = \frac{Q}{t},$$

where t — life of the mine (in years), Q — workable reserves in the deposit (in tons), A — annual production (in tons of ore).

The annual production is sometimes determined by the interdependence of the mine and preparation plant. In this instance

$$A = \frac{Q_k \times C_k}{r \times k}.$$

A — annual production (in tons), Q_k — annual capacity of the dressing plant of concentrate (in tons), C_k — metal content in the concentrate (in %), r — metal content in the ore (in %), k — yield coefficient for the metal from the ore to the concentrate ($n \times 10^{-1}$).

Annual mine production is determined by ore production but most frequently by the amount of concentrate or metal,

$$n = \frac{A \times r \times k}{C},$$

n — amount of concentrate or metal (in tons), A — volume of dressed ore (in tons), r — content of useful component in ore (in %), k — total coefficient yield ($n \times 10^{-1}$), C — content of metal in final product (in %).

If the prices are known, production can also be expressed in financial terms.

Production cost is the principal indicator of economic evaluation. In deciding on future mining, those deposits which enable all operations up to the final product to be carried out at the lowest cost are to be preferred. The production cost can be determined using Rachkovskii's formula:

$$C = \frac{R_3 \times (C_p + C_{ob})}{f \times i_0 \times i_m \times (1 - q)} + \frac{R_3 \times (c_t + c_m)}{R_f \times i_m},$$

C — production cost of 1 ton of product; R_3 — content of useful mineral in the final product; C_p — cost of extraction and transport of 1 ton of ore; C_{ob} — cost of treatment of 1 ton of ore; f — content of useful mineral in the ore, based on sampling of the deposit; i_0, i_m — coefficients of yield of useful mineral in the dresing and smelting plants; q — coefficient of ore depreciation during mining; c_t — transport cost from the dressing to smelting plant; c_m — cost of smelting 1 ton of concentrate; R_f — content of useful mineral in the concentrate.

Profit and profitableness: the profit can be determined as the difference between the cost of exploration, working and treatment and the price of the product. All factors — exploration, mining and treatment — share proportionately in determining the profit.

Besides the above-mentioned profit index, the differential mining rent, which expresses the natural value of the deposit in terms of its price, is also an important factor.

Volodomonov recommended the formula

$$V = (Q - q) \times P \times p \times r \times f,$$

for determination the mining rent (V), where Q — the limiting cost of production; q — the cost of production of useful component of a deposit; P — mineral reserves (in tons); p, r, f — coefficients of yield in working, treatment and smelting. Use of this kind of differential rent is hampered by the fact that the cost depends not only on natural conditions but also on technical and organizational work on individual deposits.

The profitableness, or degree of effectiveness of the utilization of mineral raw materials, is expressed by the formula

$$N_r = \frac{R}{P} \times 100.$$

N_r — rentability standard (in %), R — total profit of the mine; P — operational resources.

Costs and their effectiveness are expressed as

$$E = \frac{C_r}{k} \qquad E_1 = \frac{k}{C_r},$$

where E — proportion of annual production cost per investment unit; C_r — annual production cost; k — investments on the construction of the mine and plants for the production of the end product; E_1 — proportion of the investment per annual production unit.

In some case the relations are useful;

$$T = \frac{K}{P} \qquad E = \frac{1}{T} = \frac{P}{K}$$

T — time of return; K — capital investments; P — annual gains; E — coefficient of effectiveness.

These indices enable the explored deposit to be compared with similar deposits and determine its importance and position in the country's raw-material reserves. The explored deposit should also be compared with deposits that are being worked. Practice has shown that a greater effectiveness of investment can be achieved on reconstructed than on new deposits. Similarly, mining in the neighbourhood of worked deposits requires lower investments than do so far unexploited areas, since existing mines can be at least partly used.

In addition to the above indices of the economic value of a deposit, several other factors controlling the economy of mining should be taken into consideration:

1. natural conditions in the area: geographic position, orography, hydrography and climate;

2. geological and mining-technical conditions: position of the deposit, physico-mechanical properties of the deposit and rocks, etc.;

3. technological conditions: quality of raw material, technological types, possibilities of dressing, etc.;

4. general economic characteristics of the deposit area, sources of energy, man-power, possibility of co-operation with industrial plants;

5. industrial requirements on the mineral raw material (standards): minimum average contents of useful components, minimum thickness of the deposit, maximum thickness of rock interlayers in the deposit;

6. the requirements of mineral raw material in the national economy with regard to the planned development of other industrial branches.

The sum of all these parameters determines the economic value of the deposit, but this overall evaluation requires a great deal of data that are available only in the advanced stages of exploration. If, however, the risk of useless investment on exploration is to be reduced, it is necessary to develop unbiased, even if more or less approximate evaluation methods also in the lower exploration stages. In principle, the estimate of the economic value of a deposit must also be carried out in stages.

In Czechoslovakia the following methods of economic evaluation of deposits are used:

In the prospecting-exploration stage: 1. general qualitative parameters and 2. evaluation of deposits by point-alloting method.

In preliminary and detailed exploration: 3. special qualitative parameters, 4. evaluation of deposits using geological variants, economic estimate, and 5. direct calculation methods.

The principal parameters determining the economic profitableness of the deposit are, in our concept, only a selection of indices, which are part of the total economic evaluation of the deposit. They are used particularly for the appraisal of the economic profitableness in calculating the reserves.

1. General parameters are not based on the whole complex of data, since some of them are known with insufficient accuracy in the initial exploration stages. Instead, average values valid for a given raw material, for deposits of a given type and for a certain common dressing technology are used.

General parameters of exploitability must obey the equality or inequality of total expenditure and cost (B. Soukup 1966)

$$n \leqq c.$$

Costs of 1 ton of concentrate depending on the principal factors are expressed by the equation

$$n = \frac{100 \times (n_1 + n_2)\,\beta}{\alpha_1 \times \varepsilon}$$

or with two-degree dressing

$$n = \frac{100\beta_1 \times (n_1 + n_2)}{\alpha_1 \times \varepsilon_1} + n_3 \frac{100\beta_2}{\beta_1 \times \varepsilon_2},$$

where n — total cost of 1 ton of concentrate; c — cost (maximum admissible total expenditure) of 1 ton of concentrate; n_1 — cost of extraction of 1 ton of ore and transport to the dressing plant; n_2, n_3 — cost of first and second degrees of preparation of the ore; ε — yield of metal to concentrate in per cent; ε_1, ε_2 — yields of metal to concentrate of the first and second stages of dressing; β — metal content; β_1, β_2 — metal contents in the concentrate of the first and second stages of dressing; α_1 — metal of the extracted ore.

Total cost will depend on the following indices: Z_0 — geological reserves of the deposit, m — thickness of the deposit, s — ratio of the deposit to the overburden (in opencast mining), α_0 = content of the useful component in geological reserves.

The total expenditure is a function of these indices:

$$n = f(Z_0, m, s, \alpha_0).$$

The individual parameters cannot be determined independently. Since calculation of individual variants of exploitability are time-consuming, nomograms are useful, particularly in the early exploration stages. An example of such a nomogram is given in Fig. 204 (Soukup, 1966). A nomogram is most frequently used for determining the minimum average metal content that makes the deposit workable at a given thickness and a given amount of reserves. Another problem which can be solved using a nomogram is determination of the necessary minimum reserves for exploitability of a deposit with a given content of useful mineral and a given thickness. In using a nomogram, the direction of the procedure is not important, but three variables out of four must be known. The price of the raw material or maximum admissible costs are plotted as the limit of exploitability in the nomogram. Deposits that lie on the line ($n = c$) or below it are workable.

This method is generally accepted as convenient since it allows rapid if not definitive determination of parameters decisive for the workability of the deposit. The only deficiency is that limiting conditions for the use of these nomograms have not been published and sets of nomograms usable for various groups of industrial deposits have not been compiled.

2. Evaluation of deposits by alloting points. In this preliminary evaluation of deposits, the parameters are assessed by points according to an adequate scale. The sum of points then expresses the economic value of the deposit. Several proposals of such evaluation have been put forth but all give only approximate data, since the scoring of parameters is subjective.

A very precise objective scoring method was developed by Žežulka (1965) for iron ore deposits in Czechoslovakia, and could also be applied to mineral deposits of other types.

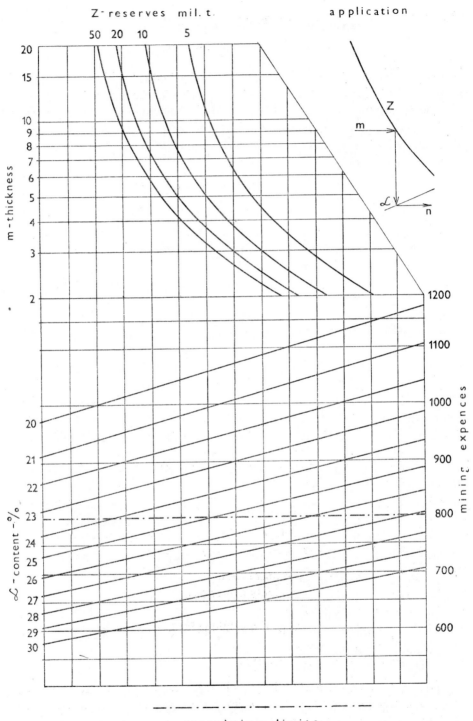

204. Nomogram for the calculation of mining expenses (after Soukup, 1966).

The economic characteristic of a deposit is expressed as

$$\delta = [(m_v \times A) + B + C + D \times (Mr_h, m_v) \pm E] \, \alpha \times \beta,$$

where m_v — amount of ore (in tons) necessary for the production of 1 ton Fe; A — cost of extraction of 1 ton of ore; B — cost of ore dressing for the production of 1 ton Fe; C — cost of smelting; D — cost of transport; Mr_h — amount of one charge of prepared product necessary for producing 1 ton Fe; E — effect of useful and harmful admixtures; α — coefficient characterizing the volume of the deposit; β — coefficient characterizing the cost of the construction of a plant for the given raw material.

The expression in brackets gives the total cost of the production of 1 ton of pig iron from the deposit.

The parameters of evaluation in the above formula are expressed by points, which the author arranged in tables for the industrial types of iron-bearing deposits in Czechoslovakia, after a thorough analysis of a large amount of data provided by exploration and exploitation. He divided these deposits into five groups, characterized by economic value, which can serve as a basis for exploration of analogous deposit types. A great advantage of this method is that, from the number of points, deposits can be ordered on the basis of their δ value.

Both these methods present the sum of parameters which the mining geologist does not have at his disposal "coded" in the form of nomograms or point systems. It is necessary to limit the use of methods according to the type of deposit and time interval in which they are valid, as conventional application cannot provide reliable results.

3. Special parameters are used in computing reserves assessed by detailed exploration and are included in planning the exploitation of a deposit. Computing of parameters for the evaluation of a deposit is based on concrete conditions.

4. Evaluation of deposits using geological variants and economic estimate was suggested by Žežulka (1965). The optimum utilization of the deposit is computed gradually from the assessment of geological, technical and economic parameters. Of several possible variants the geologist tries to select that variant which enables profitable exploitation, taking into consideration non-economic factors. Unfortunately, much necessary data are often unavailable. These were therefore compiled in the form of indices to be employed prospectively.

The computation consists of five operations:
a) determination of limiting exploitability conditions,
b) analysis of non-economic factors,
c) determination of optimum geological conditions for the utilization of the deposit,
d) calculation of true profitability,
e) characterization of subeconomic reserves.

5. Direct calculation methods provide the most accurate results but need precise and comprehensive basic data, which are usually determinable only after completing detailed exploration and during projection works for the mine.

Special qualitative parameters are based on a complex of natural and technical-economic indices and serve mainly for delimiting economic and subeconomic ores in the calculation of reserves, planning of mining and checking of recovery. The established parameters are valid for a given mining technology and dressing, and for given costs of mineral raw materials; they must be corrected if these conditions are changed. They usually comprise the following indices:

1. minimum average content of useful components in each geological block, from which the reserves are computed.

2. limiting content of useful components (cutoff grade) in a marginal sample, which defines the boundary between economic and subeconomic ores. This is particularly important where the ore-rock contact is indistinct;

3. the cutoff grade in a marginal sample for the determination of the boundary of subeconomic ores;

4. the maximum admissible content of harmful components in a block and in a marginal sample for delimiting the economic ore;

5. differentiation of technological types of the mineral raw material;

6. differentiation of useful subsidiary components whose reserves are also calculated, determination of their minimum content in a block or in the technological type;

7. the minimum coefficient of ore productivity for irregular mineralization where the computation of reserves is not based on delimitation of single ore nests but on statistical data (the coefficient can be computed as a ratio of mineralized sectors of galleries to their total length driven in an ore-bearing structure);

8. minimum (limiting) thickness included in the economic reserves;

9. maximum thicknesses of interlayers of gangue or subeconomic ores in the bulk of economic reserves;

10. maximum depth of opencast mining and the limiting overburden coefficient.

The requirements can be modified according to the character of the deposit. Most of the factors are determined with regard to the mining technology employed. The minimum average content of economic ores is determinable from the equality of production costs and the price of useful components (Kobakhidze, 1973):

$$r_{\min} \times e \times k_d \times k_m \times \frac{C_m}{v_m} = a + b + d,$$

$$r_{\min} = \frac{a + b + d}{e \times k_d \times k_m \times C_m} \times v_m,$$

where $r_{\min.}$ — minimum content of useful component (in %), a = cost of the extraction of 1 ton of ore and its transport to the preparation plant; b — cost of dressing of 1 ton; d = cost of smelting 1 ton of ore, including transport to the smelting plant; e — coefficient of recovery $\frac{100 - P}{P}$, where P — depreciation during extraction in %; k_d — coefficient of yield in dressing; k_m — coefficient of yield in metallurgical treatment; C_m — price of metal in the final product, and v_m — metal content in the final product %.

The formula is simplified if the price of the concentrate is given as

$$r_{min} = \frac{a + b}{e \times k_d \times C_k} \times v_k,$$

where C_k — price of the useful component in the concentrate, v_k — metal content in the concentrate (for the other symbols see above).

In the ore which contains several useful components, with contens c_1, c_2, c_n all must be converted to one element on the basis of their price ratios and yield coefficients. An example of recalculation given by Kobakhidze (1973) is given in Table 74

TABLE 74

Transfer coefficients for the determination of the average contents of multi-component ores

Product	Cost per product unit	Cost ratio	Yield of metal to end product	Ratio of yield coefficients	Transfer coefficients
1	50	1	80	1	$K_1 = 1 \times 1 = 1$
2	100	2 : 1	72	0.9 : 1	$K_2 = 2 \times 0.9 = 1.8$
3	150	3 : 1	88	1.1 : 1	$K_3 = 3 \times 1.1 = 3.3$
4	300	6 : 1	80	1 : 1	$K_4 = 6 \times 1 = 6$

The content of "one element" in 1 ton of ore:

$$c = c_1 \times k_1 + c_2 \times k_2 \dots + c_n \times k_n.$$

The above-mentioned relations show that the minimum average content of economie ores is not constant even within one deposit. It is controlled by natural conditions, mining technology applicable in various parts of the deposit, and other factors.

This short survey indicates some of the problems bearing on the economic evaluation of mineral resources included in geological exploration. The theory of economic evaluation should facilitate the development and extension of the raw material base which is so important for the contemporary economy.

REFERENCES TO PROSPECTING

ABDULLAEV, KH. M. (1957): Dykes and mineralization.
(Аблуллаев, Х. М.: Дайки и оруденение. — Гос. научнотех. изд. Москва, 232 стр.)

ACKERMANN, E. (1937): Die Aufeinanderfolge von Prospektionsmethoden erläutert am Beispiel von Südzentralafrika. — Geol. Rundschau, 28, no 3—4, 259—281, Leipzig, Berlin, Stuttgart.

AGRICOLA, G. (1556): De re metallica libri XII. — Basel.

BADHAM, J. P. N. (1974): Plate tectonics and metallogenesis with reference to the silver, nickel-cobalt arsenide ore association. — Symposium Metallogeny and plate tectonics, St. John's, Newfoundland.

BAIN, G. F. (1936): Mechanics of metasomatism. — Econ. Geol. 31, p. 505—526, Lancaster, Pa., USA.

BATEMAN, A. M. (1950): Economic mineral deposits. — John Wiley and Sons, New York; Chapman and Hall, London, 916 pp.

— (1951): The formation of mineral deposits. — John Wiley and Sons, New York; Chapman and Hall, London, 371 pp.

BELEVTSEV, J. N. (1961): Structural conditions of formation of ore deposits.
(Белевцев, Я. Н.: Структурные условия образования рудных месторождений. — Изд. АН УССР, Киев, 232 стр.)

BERNEWITZ, M. W. von (1943): Handbook for prospectors and operators of small mines. — McGraw-Hill Co., New York and London, 547 pp.

BILIBIN, Y. A. (1955a): Metallogenic provinces and metallogenic epochs.
(Билибин, Ю. А.: Металлогенические провинции и металлогенические эпохи. — Госгеолтехиздат, Москва, 86 стр.)

— — (1955b): Principles of geology of placers.
(Билибин, Ю. А.: Основы геологии россыпей. — Изд. АН СССР, Москва, 471 стр.)

BLANCHARD, R., BOSWELL, P. F. (1934): Additional limonite types of galena and sphalerite derivation. — Econ. Geol. 29, 671—690, New Haven, Conn.

BOYLE, F. W. (1970): Regularities in wall-rock alteration phenomena associated with epigenetic deposits. — In: Problems of hydrothermal ore formation, pp. 233—260. Schweitzerbart, Stuttgart.

BULGAKOV, Y. I. (1970): The method of transitional fields and its application in prospecting and exploration of mineral raw materials.
(Булгаков, Ю. И.: Метод переходных процессов и его применение при поисках и разведке полезных ископаемых. — Сборник Состояние и задачи разведочной геофизики, 270—276, Недра, Москва)

CLARK, A. H., CAELLES, J. C., FARRAR E., HAYNES S. J., LORTIE R. B., McBRIDE S. L., QUIRT S. J., ROBERTSON R. C. R., ZENTILLI M. (1974): Longitudinal variations in the metallogenetic evolution of the Central Andes. — Symposium Metallogeny and plate tectonics, St. John's, Newfoundland.

CLARK, S. P. (1966): Handbook of physical constants. — The Geological Society of America, 344 pp.

CONSTANTINIDES, D., PERTOLD, Z. (1974): Geological position of the Zlaté Hory — South deposit (in Czech). — Acta Univ. Carol. — Geologica, 145—154, Praha.

CORLISS, J. B. (1974): Sea water, sea-floor spreading, subduction and ore deposits. — Symposium Metallogeny and plate tectonics, St. John's, Newfoundland.

408

CRAWFORD, A. R. (1970): Continental drift and uncontinental thinking. — Econ. Geol. 65, 11—16, New Haven, Conn.

ČILLÍK, I., OGURČÁK, Á. (1965): Metodika prieskumu rudných ložísk. — Slovenské vydavateľstvo technickej literatúry, Bratislava, 392 pp.

DESPUJOLS, P., TERMIER, H. (1946): Introduction à l'étude de la métallogénie et à la prospection miniére. — Imprimerie officielle, Rabat, 199 pp.

DEWEY, J. F., BIRD, J. M. (1970): Mountain belts and the new global tecto ics. — J. Geophys. Res., 75, 2625—2647, Richmond, Va.

DLABAČ, M., ADAM, Z. (1959): Geologická interpretace reflexně seismického měření v Malé dunajské nížině. — VÚ ČND, Brno, Geofond. Praha.

EMMONS, W. H. (1940): The principles of economic geology. — McGraw-Hill, New York, London, pp. 529.

FREUND, H. (1954): Handbuch der Mikroskopie in der Technik. Bd. II. Mikroskopie der Boden-schätze, Teil 2, Mikroskopie der Erze, Aufbereitungsprodukte und Hüttenschlacken. — Umschau Verlag, Frankfurt am Main, 664 pp.

GARSON, M. S. (1974): Precambrian-Lower-Paleozoic plate tectonics and metalogenesis in the Red Sea region. — Symposium Metallogeny and plate tectonics, St. John's, Newfoundland.

GINZBURG, I. I. (1957): Theoretical principle of geochemical prospecting.
(Гинзбург, И. И.: Опыт разработки теоретических основ геохимических методов поисков. — Госгеолтехиздат, Москва, 298 стр.)

GLIKSON, A. Y. (1974): Early Precambrian shield elements: implications on the relevance of plate tectonics and the secular evolution of convection. — Symposium Metallogeny and plate tectonics, St. John's, Newfoundland.

GOVETT, G. S. J., GOVETT, M. H. (1972): Mineral resource supplies and the limits of economic growth. — Earth Science Reviews, 275—290, Amsterdam.

GRAY, A. (1958): The future of mineral exploration. — Bull. Inst. Min. Metall., 68, 23—36, London.

GRECHUKHIN, V. V. (1965): Geophysical methods for investigation of bore-holes in coal seams.
(Гречухин, В. В.: Геофизические методы исследования угольных скважин. — Недра, Москва, 486 стр.)

GREEN, H. (1972): Magmatic activity as the major process in the chemical evolution of the Earth's crust and mantle. — In A. R. Ritsema (editor): The Upper Mantle, Tectonophysics, 13, 47—71. Amsterdam.

GRIGORYAN, S. V. (1973): Primary geochemical aureoles in prospecting and exploration of hydro-thermal deposits.
(Григорян, С. В.: Первичные геохимические ореолы при поисках и разведке гидротермаль-ных месторождений. — Советская геология, 15—33, Москва)

GRIP, E. (1953): Tracing of glacial boulders in Sweden. — Econ. Geol., 48, 715—725, Lancaster, Pa.

GRUNTORÁD, J. (1967): Results of complex geophysical research in Zlaté Hory area. — Acta Universitatis Carolinae, Geologica No 2, 173—183. Praha.

GRUNTORÁD, J., JŮZEK, M., KNĚZ, J. (1967): Metoda vyzvané polarizace. — Učební text pro post-graduální studium. Přírodovědecká fakulta UK, Praha, 105 pp.

GRUNTORÁD, J. SKOPEC, J. (1963): Geofyzikální výzkum havlíčkobrodského rudního revíru. — Sborník geologických věd, Užitá geofyzika, řada UG, svazek 1, 33—67, Praha.

GUILD, P. W. (1972): Metallogeny and the new global tectonics. — Int. Geol. Congress, 24th session, Section 4, Mineral deposits, 17—24, Montreal.

ITSIKSON, M. I. (1953): Heavy minerals sampling during geological mapping and general prospecting.
(Ициксон, М. И.: Шлиховое опробование при геологической съемке и обзорных поисках. — Госгеолтехиздат, Москва, 58 стр.)

JANEČKA, J. (1964): Předběžné zhodnocení prací provedených na Sn-rudy a prognózní ocenění zásob ložisek Sn Českého masívu. — Geofond, Praha, 317 pp.

KHRUSHCHEV, N. A. (1961): Conclusions of conference on prospecting for concealed deposits in the countries of the Council of mutual economic aid.
(Хрущев, Н. А.: Итоги совещания по вопросу поисков скрытых полезных ископаемых стран-членов Совета экономической взаимопомощи (СЭВ). — Советская геология, 141—143, Москва)

KOMAROV, V. A. (1972): Electrical prospecting method of induced polarization.
(Комаров, В. А.: Электроразведка методом вызванной поляризации. — Недра, Ленинград, 214 стр.)

KOŘÁN, J. (1955): Přehledné dějiny československého hornictví I. — Nakl. ČSAV, Praha, 214 pp.

KOUTEK, J. (1964): Geologie československých rudních ložisek. I. Ložiska českého jádra. Učební texty vysokých škol. — Stát. pedagog. nakl., Praha, 116 pp.

KRAUSKOPF, K. B. (1967): Introduction to geochemistry. — McGraw-Hill Co., New York, Toronto and London, 721 pp.

KREITER, V. M. (1960): Prospecting and exploration of mineral deposits.
(Крейтер, В. И.: Поиски и разведка месторождений полезных ископаемых. — Госгеолтехиздат, Москва. 1-й том 332 стр., 2-й том, 1961, 390 стр.)

KREITER, V. M. (1964): Prospecting and exploration of mineral deposits.
(Крейтер, В. И.: Поиски и разведка месторождений полезных ископаемых. — Недра, Москва, 399 стр.)

KREITER, V. M., GORZHEVSKIJ, D. I., KOZERENKO, V. N. (1963): Favourable geological conditions for prospecting for industrial mineral deposits.
(Крейтер, В. И., Горжевский, Д. И., Козеренко, В. Н.: Группировка благоприятных геологических обстановок для поисков промышленных месторождений полезных ископаемых. — Геология рудных месторождений, 5, 76—87, Москва)

KRUTIKHOVSKAYA, Z. A., KUZHELOV, G. K. (1960): Application of geophysical methods for investigation of iron formation of the Ukrainian crystalline shield.
(Крутиховская, З. А., Кужелов, Г. К.: Применение геофизических методов для изучения железорудной формации украинского кристаллического щита. — Гос. науч.-тех. изд. по геом. и охр. недр, Москва, 129 стр.)

KRUTIKHOVSKAYA, SCHMIDT, N. G., KISELYOV M. I. (1967): Geophysical methods of prospecting and prognostic evaluation of iron deposits in the USSR. — Mining and Groundwater Geophysics, 363—370. Ottawa, Canada.

KUKAL, Z. (1973): Vznik pevnin a oceánů. — Academia, Praha, 254 pp.

KUREK, N. N. editor (1954): Altered rocks in the neighbourhood of deposits as prospecting guides.
(Курек, Н. Н.: Измененные околорудные породы и их поисковое значение. — Госгеолтехиздат, Москва, 270 стр.)

KUŽVART, M. (1967): Zkušenosti z prospekce nerostných surovin v rozvojových zemích. — Věstník ÚÚG, 42, 161—168, Praha.

KUŽVART, M., KONTA, J. (1968): Kaolin and laterite crusts of weathering in Europe. — Acta Univ. Carol. Geol. 1—2, 1—19, Praha.

LINDGREN, W. (1926): Magmas, dykes and veins. — Trans. Amer. Inst. Min. and Met. Eng., vol. 74.

MARKOV, K. K. (1960): Paleogeography.
(Марков, К. К.: Палеогеография. — Изд. Моск. унив., Москва, 268 стр.)

MAŠÍN, J., VÁLEK, R. (1963): Přehled užité geofyziky pro geology. — Stát. pedagog. nakl., Praha, 307 pp.

McKINSTRY, H. E. (1948): Mining geology. — Prentice-Hall, New York, 680 pp.

MILYAEV, A. S., FOKIN, A. N. (1963): Buried secondary aureoles of endogenous mineral deposits. — In: Prospecting for concealed deposits.
(Миляев, А. С., Фокин, А. Н.: Погребенные вторичные ореолы рассеяния эндогенных место-

рождений. — Глубинные поиски рудных месторождений, редакторы А. Н. Еремеева и
А. П. Соколова. — Госгеолтехиздат, Москва, 184 стр.)

MITCHELL, A. H., GARSON, M. S. (1972): Relationship of porphyry copper and circum-Pacific tin
deposits to paleo-Benioff zones. — Inst. Min. Metall., 81, B 10—B 25, London.

MRŇA, F. (1963): Geochemické metody při vyhledávání skrytých rudních ložisek. — Věstník ÚÚG,
28, 73—80, Praha.

NESTEROV, L. I., BERSUDSKII, D. D. et al. (1938): Short course of exploration geophysics for geologists.
(Нестеров, Л. И., Берсудский Д. Д. и др.: Краткий курс разведочной геофизики для геоло-
гов. — ГОНТИ, Москва, Ленинград)

ÖDMAN, O. H. (1942): Boliden, Sweden. On deposits as related to structural features. — Princeton
University Press, 166—169, New Jersey.

OZEROV, I. M. (1959): The plotting of heavy mineral concentrates and their analysis.
(Озеров, И. М.: Шлиховая съемка и анализ шлихов. — Гостоптехиздат, Ленинград, 377 стр.)

PAUTOT, G., AUZENDE, J. M., LE PICHON, X. (1970): Continuous deap sea salt layer along North
Atlantic margins related to early phase of rifting. — Nature, 227, 351—354, London.

PEARL, R. M. (1973): Handbook for prospectors. — McGraw-Hill Co., New York, Toronto and
London.

PEREIRA, J., DIXON, C. J. (1971): Mineralisation and plate tectonics. — Mineralium Deposita, 6,
404—405, Berlin, Heidelberg, New York.

PERTOLD, Z., CONSTANTINIDES, D. (1974): Mineralization of the Zlaté Hory — South deposit in
structural-metamorphic development of host rock (in Czech). — Acta Univ. Carol., Geologica
155—163, Praha.

PETRASCHEK, W. E. (1973): Some aspects of the relations between continental drift and metallogenic
provinces. — Implications of continental drift to the earth sciences (D. H. Tarling, S. K. Runcorn,
editors). Academic Press, London and New York, pp. 563—568.

PETRUSEVICH, M. N. (1954): Geological mapping and prospecting based on air reconnaissance.
(Петрисевич, М. Н.: Геолого-съемочные поисковые работы на основе аэрометодов. — Госгеолтехиздат, Москва, 106 стр.)

POMERANTSEV, V. V. (1961): Appraisal of deposits of ferrous and non-ferrous metals.
(Померанцев, В. В.: Оценка рудных месторождений цветных и черных металлов. — Госгортехиздат, Москва, 198 стр.)

POUBA, Z. (1955): Geneze ložisek okrů a malířských hlinek v plzeňské pánvi. — Universitas Carolina,
I. Scientific conference of the Geologic-geographic faculty, Charles University, p. 19.

— (1959): Geologické mapování. — Nakl. ČSAV, Praha, 523 pp.

RANKAMA, I. K., SAHAMA, G. (1950): Geochemistry. — Univ. of Chicago Press, Chicago, 912 p.p

ROUTHIER, P. (1963): Les gisements metallifères. — Tome I, II, Masson et Cie, Paris, 1282 pp.

ROUTHIER, P., BROUDER, P., FLEISCHER, R., MACQUER, J. C., PAVILLON, M. J. ROGER, G., ROUVIER, H
(1973): Some major concepts of metallogeny. — Mineral. Deposita, 8, 237—258, Berlin, Heidelberg, New York.

RUKHIN, L. B. (1962): Principles of general paleogeography.
(Рухин, Л. Б.: Основы общей палеогеографии. — Гостоптехиздат, Москва, 628 стр.)

RUSSEL, M. J. (1968): Structural control of base metal mineralization in Ireland in relation to continental drift. — Inst. Min. Metall. Transactions, 77, B 117—B 128, London.

SATTRAN, V., FIŠERA, M., KLOMÍNSKÝ, J. (1964): O genetickém vztahu endogenních ložisek cínu
a zlata k variskému magmatismu Českého masívu. — Věstník ÚÚG, 39, 435—439, Praha.

SAWKINS, F. J. (1974): Massive sulphide deposits in relation to geotectonics. — Symposium Metallogeny and plate tectonics, St. John's, Newfoundland.

SEIGEL, H. O. (1972): Ground investigation of airborne electromagnetic indications. — International
Geological Congress, Canada, Section 9, Exploration Geophysics, Montreal, p. 98—109.

SERPUKHOV, V. I. (1955): General principles of regional analysis.

(Серпухов, В. И.: Общие принципы регионального анализа. — Сов. геология, 43, 27—42, Москва)

SCHNEIDERHÖHN, H. (1952): Genetische Lagerstättengliederung auf geotektonischer Grundlage. — N. Jb. f. Min., Hf. 2, 3, 47—63, Stuttgart.

SILLITOE, R. H. (1972): A plate tectonic model for the origin of porphyry copper deposits. — Econ. Geol., 67, 184—197, New Haven, Conn.

— (1974): Mineralization in the Andes: a model for metallogeny of convergent plate margins. — Symposium Metallogeny and plate tectonics, St. John's, Newfoundland.

SHARP, W. E. (1974): A plate tectonic origin for diamond-bearing kimberlites. — Earth and Planetary Sci. Letters, 21, 351—354, Amsterdam.

SMIRNOV, S. S. (1951): Oxidation zone of mineral deposits.

(Смирнов, С. С.: Зона окисления месторождений. — Акад. наук СССР, Москва, 335 стр.)

SMIRNOV, V. I. (1955): Problems of prospecting for concealed deposits.

(Смирнов, В. И.: Проблемы поисков рудных месторождений не имеющих выхода на поверхности земли. — Советская геология, 49, 38—58, Москва)

— (1957): Geological principles of prospecting and exploration of mineral deposits.

(Смирнов, В. И.: Геологические основы поисков и разведок рудных месторождений. — Изд. Моск. унив., Москва, 587 стр.)

— (1962): Metallogeny of geosynclines.

(Смирнов, В. И.: Металлогения геосинклиналей. — Сборник Закономерности размещения полезных ископаемых, 5, 17—81, Изд. АН СССР, Москва)

SOKOLOV, K. P. (1966): Geophysical methods of exploration.

(Соколов, К. П.: Геофизические методы разведки. — Недра, Ленинград, 464 стр.)

STAROSTIN, V. I. (1965): The influence of physical and mechanical properties of rocks of Blyavinsk ore field on localization of sulphidic mineralization.

(Старостин, В. И.: Влияние физико-механических свойств горных пород Блявинского рудного поля на локализацию колчеданного оруденения. — Геол. рудных месторождений, 45—56, Москва)

STRAKHOV, N. M. (1962): Principles of the theory of lithogenesis.

(Страхов, Н. М.: Основы теории литогенеза. — Изд. АН СССР, Москва, 1-й том 212 стр., 2-й том 574 стр., 3-й том 550 стр.)

SÝKORA, L. (1959): Rostliny v geologickém výzkumu. — Nakl. ČSAV, Praha, 322 pp.

SZADECZKY—KARDOSS, E. (1941): Vorläufiges über den Kristallinitätsgrad der Eruptivsteine und seine Beziehung zur Erzverteilung. — Mitteilungen d. berg- und hüttenmänischen Abteilung Univ. für technische Wissenschaften, 251—272, Sopron.

ŠACOV, N. I. (1952): Hlubinné vrtání na naftu. — Průmyslové vydavatelství, Praha.

ŠUF, J. (1950): Geochemické naftové výzkumnictví v SSSR. — Přírodověd. sb. Ostravského kraje, 11, příloha, 96—123, Ostrava.

TARLING, D. H., RUNCORN, S. K., editors (1973): Implications of continental drift to the earth sciences. — Vol. 1, Academic Press, London, New York, 623 pp.

TATARINOV, P. M., GRUSHEVOI, V. G., LABAZINA, G. C., editors (1957): General principles of regional metallogenic analysis and methods of compiling metallogenic maps of folded regions.

(Татаринов, П. М., Грушевой, В. Г., Лабазина, Г. И. — редакторы: Общие принципы регионального металлогенического анализа и методика составления металлогенических карт для складчатых областей. — Материалы ВСЕГЕИ, новая серия, вып. 22, Госгеолтехиздат, Москва. 128 стр.)

THAYER, T. P. (1974): Metallogenic contrasts in the plutonic and volcanic rocks of the ophiolite assemblage. — Symposium Metallogeny and plate tectonics, St. John's, Newfoundland.

TOOMS, J. S. (1970): Review of knowledge of metalliferous brines and related deposits. — Inst. Min. Metal. Transactions, 79, B 116—B 126, London.

— (1974): Ethiopian mineral deposits and their relationship to opening of the Red Sea. — Symposium Metallogeny and plate tectonics, St. John's, Newfoundland.

Tooms, J. S., Webb, J. S. (1961): Geochemical prospecting investigations in the Northern Rhodesian Copperbelt. — Econ. Geol., 56, 815—846, Lancaster, Pa.

Torske, T. (1974): Metal provinces of Andean type continental margin in the Precambrian of South Norway. — Symposium Metallogeny and plate tectonics, St. John's, Newfoundland.

Veevers, J. J. (1958): Helicopter in the desert. — Chemical Engineering Mining Review, 50, 8, Melbourne, Sydney.

Verbeek, T., Dehenne, R., Bowdidge, C. (1972): Geophysical casehistory: The Thierry copper-nickel deposit in North-western Ontario, Canada. — International Geological Congress, Canada, Section 9, Exploration Geophysics, Montreal, p. 135—151.

Vokes, F. M., (1973): Metallogeny possibly related to continental break-up in Southwest Scandinavia. — Implications of continental drift to earth sciences. Vol. 1 (Tarling D. H., Runcorn S. K. ed.), 573—579, Academic Press, London, New York.

— (1974): Metallogeny relatable to continental drift in southern Scandinavia. — Symposium Metallogeny and plate tectonics, St. John's, Newfoundland.

Volosyuk, G. K., Safronov, N. J. (1971): Ore logging geophysics.
(Волосюк, Г. К., Сафронов, Н. И.: Скважинная рудная геофизика. — Недра, Ленинград, 534 стр.)

Voskuil, W. H. (1960): The search for mineral adequacy. — Virginia Minerals, 6, no 2, p. 1—7.

White, D. A. (1954): The stratigraphy and structure of the Mesabi Range in Minnesota. — Minnesota Geol. Surv. Bull. 38.

Williams, D. (1959—1960): Mineral exploration. — Proceedings of the Geologists Association, 70, 125—157, London.

Wright, G. M. (1959): Light helicopter reconnaissance in the barren grounds, Northwestern Territories. — Geol. Survey of Canada, Bull., 54, 7—14, Ottawa.

Wright, J. B., McCurra, P. (1973): Magmas, mineralisation and seafloor spreading. — Geol. Rundschau, 62, 116—125, Stuttgart.

Zeschke, G. (1961): Prospecting for ore deposits by panning heavy minerals from river sands. — Econ. Geol., 56, 1250—1257, Lancaster, Pa.

— (1962): Neftegazopromyslovaya geologiya. — Gostoptekhizdat, Moskva, 536 pp.

— (1964): Prospektion und feldmässige Beurteilung von Lagerstätten. — Springer Verlag, Wien, 307 pp.

Zhdanov, M. A. (1952): Methods of oil and gas reserves computation.
(Жданов, М. А.: Методы подсчета подземных запасов нефти и газа. — Госгеолиздат, Москва, 238 стр.)

REFERENCES TO EXPLORATION

AGTERBERG, F. P. (1974): Geomathematics. Mathematical background and geoscience applications. — Elsevier. Amsterdam, London, New York, 596 pp.

ALBOV, M. N. (1956): Mining geology.
(Альбов, М. Н.: Рудничная геология. — Гос. науч. тех. изд. лит. по цвет. и черной металлургии, Свердловск, 448 стр.)

— (1961): Sampling of ore deposits.
(Альбов, М. М.: Опробование рудных месторождений. — Гос. науч. тех. изд. литер. по горному делу, Москва, 255 стр.)

Anon. (1956): Modern mineral finding. — Mining Magazin, 95, no. 2, 85—86, London.

Anon. (1959): Symposium on saturation prospecting. — Annual General Meeting, Vancouver, April 1958, the Canadian Institute of Mining and Metallurgy, 1—30, Vancouver.

Anon. (1962): Směrnice Ústředního geologického úřadu. Směrnice a pokyny pro výzkum a průzkum ložisek nerostných surovin. — Ústř. geol. úřad, Praha, 170 pp.

AZHGIREI, G. D. (1964): Methods of prospecting and exploration of mineral deposits.
(Ажгирей, Г. Д.: Методы поисков и разведки полезных ископаемых. — Госгеолтехиздат, Москва, 462 стр.)

BARTALSKÝ, J. (1966): Vyšetrovanie variant predbežného a podrobného prieskumu ložísk nerastných surovín z hladiska ekonomickej efektívnosti. In: Seminár o metodike prieskumu železných a komplexných železných rúd, ČsVTS, Spiš, N. Ves, 70—80.

BIRYUKOV, V. I. (1962): Classification of exploration systems of solid mineral deposits.
(Бирюков, В. И.: Классификация систем разведки месторождений твердых полезных ископаемых. — Геология руд. месторождений, 1, 99—121, Москва)

— (1965): Regression of category of reserves.
(Бирюков, В. И.: Регрессия категории запасов. — Изв. высших учеб. заведений, Геология и разведка, 4, 86—93, Москва)

BOBRIEVICH, A. P. (1957): Siberian diamonds.
(Бобриевич, А. П.: Алмазы Сибири. — Госгеолтехиздат, Москва, 157 стр.)

BOGATSKII, V. V. (1962): Problems of sampling of ore deposits during the exploration and exploitation.
(Богатский, В. В.: Вопросы методики опробования рудных месторождений при разведке и эксплуатации. — Гос. науч. тех. изд. лит. по геологии и охране недр, Москва)

— (1963): Mathematical analysis of exploration grid.
(Богатский, В. В.: Математический анализ разведочной сети. — Гос. науч. тех. издат. лит. по геологии и охране недр, Москва. 211 стр.)

BOZHINSKII, A. P. (1962): Problems of preliminary exploration of placer deposits.
(Божинский, А. П.: Некоторые вопросы методики предварительной разведки россыпных месторождений. — Госгеолтехиздат, Москва, 85 стр.)

CHUMAKOV, I. D. (1965): Principles of determination of ratio of exploratory boreholes and underground workings.
(Чумаков, Ю. Д.: Принципы определения рационального соотношения бурового и горного способов разведки. — Разв. и охр. недр, 2, 6—12, Москва)

ČILLÍK, J. (1962): Technická projekcia usmernených vrtov v gemeridnom krystaliniku. — Geol. průzkum, 3, 72—74, Praha.

414

— (1963): Efektívnosť banských a vrtných prác v geologickom prieskume. — Geol. prúzkum 2, 40—42, Praha.

ČILLÍK, J., OGURČÁK, Š. (1964): Metodika prieskumu rudných ložísk. ÚGÚ, Bratislava, 392 pp.

DUBA, D. (1962): Dynamika podzemných vod. — SPN, Praha.

DUBA, D., MUCHA, I., JETEL, J. (1967): Hydraulika podzemných vôd hlbokých geologických štruktúr. — SPN, Praha.

FAIT, L. (1965): Disjunktívne poruchové systémy na sideritovom ložisku v Rudňanoch. — MP. 66 pp.

FEDORCHUK, V. P. (1964): Methods of prospecting and exploration of concealed mercury-antimony ores.

(Федорчук, В. П.: Методика поисков и разведки скрытого ртутно-сурьмяного оруденения. — Недра, Москва, 285 стр.)

FISHMAN, M. A. (1955): Technology of mineral raw materials.

(Фишман, М. А.: Технология полезных ископаемых. — Металлургиздат, Москва, 736 стр.)

FORRESTER, Y. D. (1949): Principles of field and mining geology. — J. Wiley and Sons, London, 647 pp.

GALKIN, B. I. (1962): Exploration of stockwork deposits of non-ferrous and rare metals.

(Галкин, Б. И.: Разведка штокверковых месторождений цветных и редких металлов. — Гос. науч. тех. изд. по геологии и охране недр, Москва, 233 стр.)

GERASIMENKO, G. I. (1958): Geological documentation and delineation during exploration and exploitation of coal deposits.

(Герасименко, Г. И.: Геологическая документация и геометризация при разведке и разработке угольных месторождений. — Углотехиздат, Москва, 122 стр.)

GRIFFITH, S. V. (1960): Alluvial prospecting and mining. — Pergamon Press, London.

HARRISON, H. L. H., (1954): Valuation of alluvial deposits. — Mining Publications, Salisbury House, London.

HAZEN, S. W. JR. (1968): Ore reserve calculations in ore reserve estimation and grade control. In: The Canadian Inst. of Mining and Metallurgy. — Spec. Vol. 9, 11—32.

HOZA, F. (1965): Kapesní příručka geologického průzkumu. SNTL. — Praha, 222 pp.

IVANOV, N. V. (1962): Mineralogical sampling according to types of ore and cross-sections.

(Иванов, Н. В.: Минералогическое опробование по типам руд и по типам разрезов. Сборник Вопросы методики опробования рудных месторождений при разведке и эксплуатации. — Госгеолтехиздат, Москва. стр. 57—65)

JANEČKA, J. (1966): Metodika výzkumu rozsypů. — Geol. průzkum, 11, 361—364, Praha.

KAMENSKI, G. N., KLIMENTOV, P. P., OVCHINIKOV, A. M. (1957): Hydrogeologie ložisek užitkových nerostů. — SNTL, Praha, 298 pp.

KASÍK, Š. (1966): Efektívnost těžby nerostných surovin. — SNTL, Praha, 142 pp.

KAUDEĽNYI, V. Y. (1957): Problems of methods of exploration of coal deposits.

(Каудельный, В. Ю.: К вопросу и методике разведки угольных месторождений. — Разв. и охрана недр, 5, 20—27, Москва)

KOWALCZYK, Z. (1965): Metodika fotogrametrické dokumentace ložisek, zpřístupněných hlubinnými díly. In: Mezinárodní seminář o metodách geometrie nerostných ložisek, ÚVR, Praha.

KAZAKOVSKII, D. A. (1945): Appraisal of accuracy of results in connection with delineation and reserves calculation of deposits.

(Казаковский, Д. А.: Оценка точности результатов в связи с геометризацией и подсчетом запасов месторождений. — Углетехиздат, Москва, 129 стр.)

KAZHDAN, A. B. (1974): Methodological principles of exploration of mineral deposits.

(Каждан, А. Б.: Методологические основы разведки полезных ископаемых. — Недра, Москва, 271 стр.)

KOBAKHIDZE, L. P. (1973): Economy of geological exploratory workings.

(Кобахидзе, Л. П.: Экономика геологоразведочных работ. — Недра, Москва, 303 стр.)

KOCH, G. S., LINK, R. F. (1970): Statistical analysis of geological data. — Wiley and Sons, New York, New York, 375 pp.

KODYMOVÁ, A. (1962): Říční rýžoviska a metodika jejich vyhledávání. — Geol. průzkum, 9, 275—277, Praha.

KOGAN, I. D. (1971): Calculation of reserves and geological appraisal of ore deposits.
(Коган, И. Д.: Подсчет запасов и геолого-промышленная оценка рудных месторождений. — Недра, Москва, 295 стр.)

KRAJČÍK, M. (1960): Priemyselná televízia v geológii. — Geol. průzkum 5, 135—138, Praha.

KREITER, V. M. (1961): Prospecting and exploration of mineral deposits.
(Крейтер, В. М.: Поиски и разведка месторождений полезных ископаемых. Том 2. — Гос. науч. тех. изд. лит. по геологии и охране недр, Москва, 390 стр.)

KRUKSHENK, M. D. (1967): Exploitation of sea placers.
(Крукшенк,М. Д.: Разработка морских россыпей.— Сборник Открытые горные выработки, 228—257, Москва)

KUKHARENKO, A. (1961): Mineralogy of placers.
(Кухаренко, А.: Минералогия россыпей. — Гос. науч. тех. изд. лит. по геологии и охране недр, Москва, 316 стр.)

McKINSTRY, H. E. (1957): Mining geology. — Prentice Hall, New York 680 pp.

MIRONOV, K. V. (1963): Geological and economical appraisal of coal deposits.
(Миронов, К.: Геолого-промышленная оценка угольных месторождений. — Гос. науч. тех. изд. по горному делу, Москва, 238 стр.)

NOVÁK, J. (1964): Metody geologického průzkumu. — SNTL, Praha, 218 pp.

PEKÁREK, O. (1960): Pokyny pro zpracování výpočtů zásob hnědého uhlí v oblasti SHR—SHD, 117 pp, Praha.

PETRÍK, F., ZEMAN, J. (1960): Návrh na stanovení výnosu jádra z uhelných slojí. — Geol. průzkum 5, p. 153.

POMERANTSEV, V. V. (1961): Appraisal of ore deposits of ferrous and non-ferrous metals.
(Померанцев, В. В.: Оценка рудных месторождений цветных и черных металлов. — Госгортехиздат, Москва, 199 стр.)

POPOV, I. V. (1959): Engineering geology.
(Попов, И. В.: Инженерная геология. — Изд. Моск. университета, Москва)

POUBA, Z. (1959): Geologické mapování. — NČSAV, Praha, 523 pp.

PRAŽSKÝ, J. (1964): Průzkumný vrt. — SNTL, Praha, 287 pp.

RAEBURN, C., MILNER, H. B. (1927): Alluvial prospecting. — T. Murby and Co. London.

ROST, R. (1956): Těžké minerály. — NČSAV, Praha, 238 pp.

RYZHOV, P. A., GUDKOV, V. M. (1966): Application of mathematical statistics in exploration of mineral deposits.
(Рыжов, П. А., Гудков, В. М.: Применение математической статистики при разведке недр — Недра, Москва 234 стр.)

SHARAPOV, I. P. (1965): Application of mathematical statistics in geology.
(Шарапов, И. П.: Применение математической статистики в геологии. — Недра, Москва, 259 стр.)

SHEKHTMAN, P. A. (1963): Application of principle of relative accuracy in exploration.
(Шехтман, П. А.: Применение в методике разведки принципа относительной точности. — Сборник Вопросы методики и экономики геологоразведочных работ. — САИГИМС, Ташкент, 3—91 стр.)

SHEVYAKOV, L. D., BREDIKHIN, A. N. (1957): Odvodňování dolů. — SNTL, Praha.

SKOK, V. I. (1959): Graphic methods of construction of geological sections.

(Скок, В. И.: Графические приемы построения геологических разрезов. — Разв. и охрана недр, 1, 15—18, Москва)

SMIRNOV, V. I. (1950): Calculation of mineral raw materials reserves.

(Смирнов, В. И.: Подсчет запасов минерального сырья. — Госгеолтехиздат, Москва, 341 стр.)

— (1957): Geological principles of prospecting and exploration of mineral deposits.

(Смирнов, В. И.: Геологические основы поисков и разведок рудных месторождений. — Изд. Моск. университета, Москва, 587 стр.)

— (1960): Calculation of reserves of mineral deposits.

(Смирнов, В. И.: Подсчет запасов месторождений полезных ископаемых. — Госгеолтехиздат, Москва, 671 стр.)

— (1965): The geology of mineral raw materials.

(Смирнов, В. И.: Геология полезных ископаемых. — Недра, Москва, 589 стр.)

STOČES, B. (1954): Důlní geologie. — NČSAV, Praha, 920 pp.

ŠTĚPÁNEK, M. (1960): První zkoušky s naší televizní sondou. — Geol. průzkum 5, 139—141. Praha.

VASILEV, S. P. (1955): Mining geology of coal deposits.

(Васильев, С. П.: Шахтная геология угольных месторождений. — Углетехиздат, Москва, 208 стр.)

VOLAROVICH, G. P. (1956): Handbook of methods of exploration and reserves calculation of gold deposits.

(Воларович, Г. П.: Руководство по методам разведки и подсчету запасов золоторудных месторождений. — Нигризолото, Москва, 255 стр.)

YAKZHIN, A. A. (1954): Sampling and calculation of reserves of solid mineral raw materials.

(Якжин, А. А.: Опробование и подсчет запасов твердых полезных ископаемых. — Гос. науч. тех. изд. лит. по геологии и охране недр, Москва, 295 стр.)

YANKOVICH, S. (1957): Oprobovanie i prorachun rezervi mineralnih sirovina. — Zavod za geol. i geofyz. istrazhivanya, Beograd, 379 pp.

— (1960): Ekonomska geologiya I. — Zavod za geol. i geofyz. istrazhivaniya, Beograd, 547 pp.

ZÁRUBA, Q., MENCL, V. (1957): Inženýrská geologie. — NČSAV, Praha, 425 pp.

ZENKOV, D. A. (1957): Methods of determination of density of the exploration grid.

(Зенков, Д. А.: Методы определения плотности разведочной сети, — Сов. геология, № 61, 130—143, Москва)

ZENKOV, D. A., SEMENOV, K. L. (1957): Vectorial method of delineation of mineral deposits.

(Зенков, Д. А., Семенов, К. Л.: Векторный метод оконтуривания тел полезных ископаемых. — Разв. и охрана недр, 7, 20—32, Москва)

ZESCHKE, G. (1964): Prospektion und feldmässige Beurteilung von Lagerstätten. — Springer-Verlag, Wien, 307 pp.

ŽEŽULKA, J. (1967): On reliability of reserves of solid mineral raw materials and coefficients of their calculation in conditional categories (on the example of ore deposits in Czechoslovakia).

(Жежулка, Ю.: О достоверности запасов твердых ископаемых и коэффициентах пересчета их в условную категорию, исходя из условий рудных месторождений Чехословакии. — Сов. геология, 4, 105—112, Москва)

INDEX

Compiled by B. Křίbek

Numbers of pages set up in bold face indicate pages of tables and figures.

428